Applied and Numerical Harmonic Analysis

More information about this series at http://www.springer.com/series/4968

W. David Joyner · Caroline Grant Melles

Adventures in Graph Theory

 Birkhäuser

W. David Joyner
Department of Mathematics
United States Naval Academy
Annapolis, MA
USA

Caroline Grant Melles
Department of Mathematics
United States Naval Academy
Annapolis, MA
USA

ISSN 2296-5009   ISSN 2296-5017   (electronic)
Applied and Numerical Harmonic Analysis
ISBN 978-3-319-88593-3   ISBN 978-3-319-68383-6   (eBook)
https://doi.org/10.1007/978-3-319-68383-6

Mathematics Subject Classification (2010): 05C10, 05C25, 57M15, 94C15

Printed on acid free paper

This book is published under the trade name Birkhäuser (www.birkhauser-science.com)
The registered company is Springer International Publishing AG
The registered company address is: Gewerbestrasse 11, 6330 Cham, Switzerland

To Garvin, Garvin, and John
　　　　　　　Caroline Grant Melles

To the memory of T.S. Michael
　　　　　　　W. David Joyner

ANHA Series Preface

The *Applied and Numerical Harmonic Analysis* (*ANHA*) book series aims to provide the engineering, mathematical, and scientific communities with significant developments in harmonic analysis, ranging from abstract harmonic analysis to basic applications. The title of the series reflects the importance of applications and numerical implementation, but richness and relevance of applications and implementation depend fundamentally on the structure and depth of theoretical underpinnings. Thus, from our point of view, the interleaving of theory and applications and their creative symbiotic evolution is axiomatic.

Harmonic analysis is a wellspring of ideas and applicability that has flourished, developed, and deepened over time within many disciplines and by means of creative cross-fertilization with diverse areas. The intricate and fundamental relationship between harmonic analysis and fields such as signal processing, partial differential equations (PDEs), and image processing is reflected in our state-of-the-art *ANHA* series.

Our vision of modern harmonic analysis includes mathematical areas such as wavelet theory, Banach algebras, classical Fourier analysis, time–frequency analysis, and fractal geometry, as well as the diverse topics that impinge on them.

For example, wavelet theory can be considered an appropriate tool to deal with some basic problems in digital signal processing, speech and image processing, geophysics, pattern recognition, biomedical engineering, and turbulence. These areas implement the latest technology from sampling methods on surfaces to fast algorithms and computer vision methods. The underlying mathematics of wavelet theory depends not only on classical Fourier analysis, but also on ideas from abstract harmonic analysis, including von Neumann algebras and the affine group. This leads to a study of the Heisenberg group and its relationship to Gabor systems, and of the metaplectic group for a meaningful interaction of signal decomposition methods. The unifying influence of wavelet theory in the aforementioned topics illustrates the justification for providing a means for centralizing and disseminating information from the broader, but still focused, area of harmonic analysis. This will be a key role of *ANHA*. We intend to publish with the scope and interaction that such a host of issues demands.

Along with our commitment to publish mathematically significant works at the frontiers of harmonic analysis, we have a comparably strong commitment to publish major advances in the following applicable topics in which harmonic analysis plays a substantial role:

Antenna theory	*Prediction theory*
Biomedical signal processing	*Radar applications*
Digital signal processing	*Sampling theory*
Fast algorithms	*Spectral estimation*
Gabor theory and applications	*Speech processing*
Image processing	*Time − frequency and*
Numerical partial differential equations	*time − scale analysis*
	Wavelet theory

The above point of view for the *ANHA* book series is inspired by the history of Fourier analysis itself, whose tentacles reach into so many fields.

In the last two centuries, Fourier analysis has had a major impact on the development of mathematics, on the understanding of many engineering and scientific phenomena, and on the solution of some of the most important problems in mathematics and the sciences. Historically, Fourier series were developed in the analysis of some of the classical PDEs of mathematical physics; these series were used to solve such equations. In order to understand Fourier series and the kinds of solutions they could represent, some of the most basic notions of analysis were defined, e.g., the concept of "function." Since the coefficients of Fourier series are integrals, it is no surprise that Riemann integrals were conceived to deal with uniqueness properties of trigonometric series. Cantor's set theory was also developed because of such uniqueness questions.

A basic problem in Fourier analysis is to show how complicated phenomena, such as sound waves, can be described in terms of elementary harmonics. There are two aspects of this problem: first, to find, or even define properly, the harmonics or spectrum of a given phenomenon, e.g., the spectroscopy problem in optics; second, to determine which phenomena can be constructed from given classes of harmonics, as done, for example, by the mechanical synthesizers in tidal analysis.

Fourier analysis is also the natural setting for many other problems in engineering, mathematics, and the sciences. For example, Wiener's Tauberian theorem in Fourier analysis not only characterizes the behavior of the prime numbers, but also provides the proper notion of spectrum for phenomena such as white light; this latter process leads to the Fourier analysis associated with correlation functions in filtering and prediction problems, and these problems, in turn, deal naturally with Hardy spaces in the theory of complex variables.

Nowadays, some of the theory of PDEs has given way to the study of Fourier integral operators. Problems in antenna theory are studied in terms of unimodular trigonometric polynomials. Applications of Fourier analysis abound in signal processing, whether with the fast Fourier transform (FFT), or filter design, or the adaptive modeling inherent in time-frequency-scale methods such as wavelet theory. The coherent states of mathematical physics are translated and modulated Fourier transforms, and these are used, in conjunction with the uncertainty principle, for dealing with signal reconstruction in communications theory. We are back to the raison d'être of the *ANHA* series!

College Park, USA   John J. Benedetto

Preface

For now, think of a graph $\Gamma = (V, E)$ as a pair of sets: V is a finite set of points, called vertices, and E is a finite set of pairs of vertices, called edges.

If the edges of Γ are labeled

$$E = \{e_1, e_2, \ldots, e_m\},$$

the vertices of Γ are labeled

$$V = \{v_1, v_2, \ldots, v_n\},$$

let

$$C^0(\Gamma, \mathbb{C}) = \{f \mid f : V \to \mathbb{C}\},$$

and let

$$C^1(\Gamma, \mathbb{C}) = \{f \mid f : E \to \mathbb{C}\},$$

where $\mathbb{C}$ denotes the field of complex numbers. We can identify $C^0(\Gamma, \mathbb{C})$ with the vector space $\mathbb{C}^n$ via the map $f \mapsto (f(v_1), f(v_2), \ldots, f(v_n))$. We can identify $C^1(\Gamma, \mathbb{C})$ with the vector space $\mathbb{C}^m$ via the map $f \mapsto (f(e_1), f(e_2), \ldots, f(e_m))$.

The goal this book is to illustrate and explore connections between graph theory and diverse fields of mathematics such as

- calculus on manifolds,
- group theory,
- algebraic curves,
- Fourier analysis,
- cryptography,

and other areas of combinatorics.

For an excellent (and free!) introduction to more basic discrete mathematics, also using SageMath, see the book by Doerr and Lavasseur [DL17].

Graph theory and calculus

How can these be similar? One deals with a finite set of vertices and edges, the
other deals with functions on the (infinite) real line.

You are all familiar with the usual definition of the derivative of a function
$f : \mathbb{R} \to \mathbb{R}$ (where $\mathbb{R}$ denotes the real numbers):

$$f'(a) = \lim_{\varepsilon \to 0} \frac{f(a+\varepsilon) - f(a)}{\varepsilon}, \quad a \in \mathbb{R}.$$

We want to replace this by either a function $f : V \to \mathbb{R}$, or a function
$g : E \to \mathbb{R}$, and find an analogous definition. Before proceeding, we need to
assume that the graph is "oriented". That is, we assume that each edge $e \in E$
has a "head" $h(e)$ and a "tail" $t(e)$. Roughly speaking, for any pair of neighboring vertices (i.e., two vertices which are connected by an edge), there is a
well-defined direction to travel from one to the other. If the vertices of a
graph is the set $V = \{0, 1, \ldots, n-1\}$ then the default orientation of an edge
$e = (u, v)$ is defined by $h(e) = \max\{u, v\}$, $t(e) = \min\{u, v\}$.

One way to proceed is to define, for $f : V \to \mathbb{R}$, the derivative by

$$f'(e) = f(h(e)) - f(t(e)).$$

In this case and for the usual definition from calculus, we look at a difference
of values of f at nearby points.

If $f \in C^0(\Gamma, \mathbb{C})$ is identified with the column vector

$$\vec{f} = \begin{pmatrix} f(v_1) \\ f(v_2) \\ \vdots \\ f(v_n) \end{pmatrix},$$

then the derivative map $f \to f'$ may be regarded as an $m \times n$ matrix, M.
Indeed, the entries of this matrix are

$$M_{i,j} = \begin{cases} 1, & \text{if } v_j = h(e_i), \\ -1, & \text{if } v_j = t(e_i), \\ 0, & \text{otherwise.} \end{cases}$$

(By the way, this matrix is the transpose of the incidence matrix, and will be
studied in more detail in later chapters.)

Another way to proceed is to define, for $g \in C^1(\Gamma, \mathbb{C})$, the derivative by

$$g'(v) = \sum_{e \in E, h(e)=v} g(e) - \sum_{e \in E, t(e)=v} g(e).$$

If we think of two edges as "nearby" if they share a vertex then this definition also is a difference of values of the function at nearby points.

If the edges and vertices are labeled as before, and if $g \in C^1(\Gamma, \mathbb{C})$ is identified with the column vector

$$\vec{g} = \begin{pmatrix} g(e_1) \\ g(e_2) \\ \vdots \\ g(e_m) \end{pmatrix},$$

then the derivative map $g \to g'$ may be regarded as an $n \times m$ matrix, B. Indeed, the entries of this matrix are

$$B_{i,j} = \begin{cases} 1, & \text{if } v_i = h(e_j), \\ -1, & \text{if } v_i = t(e_j), \\ 0, & \text{otherwise.} \end{cases}$$

Notice that $M_{i,j} = B_{j,i}$. In other words, one is the transpose matrix of the other, $B = M^t$. The two definitions which seemed so different are in fact in some sense dual to one another.

In §3.3 we introduce the notion of a graph morphism $\phi : \Gamma_2 \to \Gamma_1$ between connected graphs Γ_2 and Γ_1, whose distinguishing property is that it must behave well with respect to the corresponding incidence matrices B_2 and B_1. For example, in §3.3.3 we show that $B_2^t \Phi_V = \Phi_E B_1^t$, where Φ_V is a matrix describing the vertex map of ϕ, with rows indexed by vertices of Γ_2 and columns indexed by vertices of Γ_1, and Φ_E is a matrix describing the edge map of ϕ, with rows indexed by edges of Γ_2 and columns indexed by edges of Γ_1.

Graph theory and groups

There are several well-known constructions of graphs arising from a finite group G. These so-called Cayley graphs are named for Arthur Cayley[1]. It's not hard to show that the Cayley graph of a permutation group and a set of generators is connected.

---

[1] Cayley, 1821–1895, worked for 14 years as a lawyer before his election to the Sadleirian Professorship at Cambridge University.

Let X be a finite set, let $S = \{g_1, g_2, \ldots, g_n\}$ be a set of permutations of X, and let G be the permutation group generated by them:

$$G = \langle g_1, g_2, \ldots, g_n \rangle \subset S_X,$$

where S_X denotes the symmetric group on X. If S is, in addition, a *symmetric generating set* (meaning that it is closed under inverses) then we define the *Cayley graph* of G with respect to S to be the graph

$$\Gamma = Cay(G, S) = (V, E),$$

whose vertices V are the elements of G and whose edges are determined by the following condition: if x and y are vertices in $V = G$ then there is an edge $e = (x, y)$ from x to y in Γ if and only if $y = g_i x$, for some $i = 1, 2, \ldots, n$.

Cayley graphs encode group-theoretic information about G, as the following example shows.

Example. The Cayley graph of the Rubik's cube group. Let

$$G = \langle R, L, U, D, F, B \rangle \subset S_{54}$$

be the group of the $3 \times 3 \times 3$ Rubik's Cube generated by the quarter-turn moves, where S_{54} is the symmetric group on the 54 facets of the cube, and R denotes the move which rotates the Right face of the cube a quarter-turn counterclockwise (and similarly for Left, Up, Down, Front, Back).

Let

$$S = \{R, L, U, D, F, B, R^{-1}, L^{-1}, U^{-1}, D^{-1}, F^{-1}, B^{-1}\}.$$

The associated Cayley graph $\Gamma_{QT} = Cay(G, S)$ is called the Cayley graph of the cube in the quarter-turn metric. Each position of the cube corresponds to a unique element of the group G (i.e., the move you had to make to get to that position), hence to a unique vertex of the Cayley graph. Note each vertex of this graph has degree 12.

Let

$$S = \{R, L, U, D, F, B, R^2, L^2, U^2, D^2, F^2, B^2, R^{-1}, L^{-1}, U^{-1}, D^{-1}, F^{-1}, B^{-1}\}.$$

The associated Cayley graph $\Gamma_{FT} = Cay(G, S)$ is called the *Cayley graph of the cube in the faceturn metric*.

Moreover, a solution of the Rubik's cube is simply a path in the graph from the vertex associated to the present position of the cube to the vertex associated to the identity element. The number of moves in the shortest possible solution is simply the distance from the vertex associated to the present position of the cube to the vertex associated to the identity element. The diameter of the Cayley graph of G is the number of moves in the best possible solution in the worst possible case.

Thanks to years of hard work, led by computer scientist Tomas Rokicki, the diameter of Γ_{QT} is now known to be 26, and the diameter of Γ_{FT} is now known to be 20 [J15].

Graph theory and algebraic curves

There are a number of analogies between graphs and algebraic curves over a finite field. Each has a notion of a genus, there are harmonic mappings between graphs and between curves, each has a Picard group, each has a zeta function, and so on. Several sections in this book explore this connection, for example, §1.2.2, §1.2.3, and Chapter 3.

Let X be a smooth projective curve over an algebraically closed field F. Let $F(X)$ denote the function field of X (the field of rational functions on X) and $F(X)^\times = F(X) - \{0\}$ its group of units. If D is any divisor on X (i.e., a formal sum of points in X) then the *Riemann–Roch space* $L(D)$ is a finite dimensional F-vector space given by

$$L(D) = \{f \in F(X)^\times \mid div(f) + D \geq 0\} \cup \{0\},$$

where $div(f)$ denotes the (principal) divisor of the function f.

The Riemann–Roch space $L(D)$ may be regarded as a vector space of rational functions with prescribed zeroes and allowed poles on X. Let $\ell(D) = \dim(L(D))$ denote its dimension. We recall the *Riemann–Roch theorem*,

$$\ell(D) - \ell(K - D) = \deg(D) + 1 - g,$$

where K denotes a canonical divisor, $\deg(D)$ the degree of D, and g the genus of X.

There is an analogous Riemann–Roch theorem for graphs as well. The graph-theoretic analog is described in Chapter 3, "Graphs as Manifolds," below.

A finite graph has a Jacobian group, analogous to the Jacobian (abelian) variety of an algebraic curve (see §4.6). The Jacobian group, also called the critical group, has cardinality equal to the number of spanning trees of the graph (see Proposition 4.9.15 and Corollary 4.9.16).

In algebraic geometry, one studies nice maps between varieties that induce maps on certain structures on the varieties. What is the graph-theoretic analog of a nice map between algebraic curves? This is a harmonic morphism of graphs (see §3.3.2), which induces a surjective pushforward map and an injective pullback map of the corresponding Jacobian groups (see §4.8).

What is the graph-theoretic analog of the zeta function (see [La70]) of an algebraic curve over a finite field? One analog is the Duursma zeta function of a graph discussed in §1.2.3 below.

Graph theory and Fourier transforms

The theory of Fourier series is a part of a branch of mathematics called harmonic analysis. Given a manifold X with a Laplacian operator $\Delta = \Delta_X$, harmonic analysis, roughly speaking, seeks to express the functions on X as expansions in terms of the eigenfunctions of Δ, then to derive consequences from this "Fourier series." Can this general motif carry over to graph theory? In Chapter 2, we discuss some connections between the spectrum of the Laplacian $Q = Q_\Gamma$ of the graph and the graph Γ itself. In the simple case of circulant graphs, we even show (see §2.4) that functions on the graph can be expressed as expansions in terms of the eigenfunctions of the Laplacian.

Historically, Fourier series were discovered by J. Fourier, a French mathematical physicist[2].

To have a Fourier series, you must be given two things: (1) a period $P = 2L$, and (2) a suitably behaved function $f(x)$ defined on an interval of length $2L$, say $-L < x < L$. The Fourier series of $f(x)$ with period $2L$ is

$$FS(f)(x) = \sum_{n=-\infty}^{\infty} a_n \exp\left(\frac{n\pi x i}{L}\right),$$

where a_n is given by the integral formula[3],

$$a_n = \frac{1}{2L} \int_{-L}^{L} f(x) \exp\left(\frac{-n\pi x i}{L}\right) dx. \tag{1}$$

The Fourier operator $f \mapsto FS(f)$ has eigenvectors $\exp(\frac{n\pi x i}{L})$.

Graph-theoretic Fourier series

For the cycle graph on n vertices, Γ_n, the eigenvalues of the adjacency matrix are $\lambda_k = 2\cos(2\pi k/n)$, for $0 \le k \le n-1$, with eigenvectors $v_k = (\exp(\pi jki/n))_{j \in \{0,\ldots,n-1\}}$. These λ's are not all distinct but the multiplicities can be described as follows.

- (n even) The only eigenvalues of Γ_n which occur with multiplicity 1 are 2 and -2. The eigenvalues $2\cos(2\pi k/n)$, for $1 \le k \le \frac{n-2}{2}$, all occur with multiplicity 2.

---

[2] Physics was not Fourier's only profession. Indeed, Napoleon selected Fourier to be his scientific advisor during France's invasion of Egypt in the late 1700s, where he oversaw some large construction projects, such as highways and bridges. During this time, Fourier developed the theory of trigonometric series (now called Fourier series) to solve the heat equation.

[3] These formulas were not known to Fourier. To compute the Fourier coefficients he used sometimes ingenious roundabout methods involving large systems of equations.

• (*n* odd) The only eigenvalue of Γ_n which occurs with multiplicity 1 is 2. The eigenvalues $2\cos(2\pi k/n)$, for $1 \le k \le \frac{n-1}{2}$, all occur with multiplicity 2.

For example, the graph Γ_8 has eigenvalues (counted according to their multiplicity) 2, $\sqrt{2}$, $\sqrt{2}$, 0, 0, $-\sqrt{2}$, $-\sqrt{2}$, -2.

We identify the vertices V of a cycle graph Γ having n vertices with the abelian group of integers mod n, $\mathbb{Z}/n\mathbb{Z}$. If $\mathbb{C}$ denotes the field of complex numbers, let

$$C^0(\Gamma, \mathbb{C}) = \{f \mid f : \mathbb{Z}/n\mathbb{Z} \to \mathbb{C}\}.$$

This is a complex vector space which we can identify with the vector space $\mathbb{C}^n$ via the map $f \mapsto (f(0), f(1), \ldots, f(n-1))$.

Let $\zeta = \zeta_n = e^{2\pi i/n}$ denote an nth root of unity in $\mathbb{C}$. Recall, for $g \in C^0(\Gamma, \mathbb{C})$, the *discrete Fourier transform* of g is the function $\mathcal{F}_n g \in C^0(\Gamma, \mathbb{C})$ (also written $g^\wedge$) defined by

$$(\mathcal{F}_n g)(\lambda) = g^\wedge(\lambda) = \sum_{\ell \in \mathbb{Z}/n\mathbb{Z}} g(\ell) \zeta^{\ell\lambda}, \ \lambda \in \mathbb{Z}/n\mathbb{Z}.$$

If $G \in C^0(\Gamma, \mathbb{C})$, the *inverse discrete Fourier transform* of G is the function $(\mathcal{F}_n^{-1} G) \in C^0(\Gamma, \mathbb{C})$ (also written $G^\vee$) defined by

$$(\mathcal{F}_n^{-1} G)(\ell) = G^\vee(\ell) = \frac{1}{n} \sum_{\lambda \in \mathbb{Z}/n\mathbb{Z}} G(\lambda) \zeta^{-\ell\lambda}, \ \ell \in \mathbb{Z}/n\mathbb{Z}.$$

Using this, any function in $C^0(\Gamma, \mathbb{C})$ has an expansion in terms of the eigenvectors of the Laplacian.

Graph theory and cryptography

In Chapter 6, we explore a fascinating connection between graph theory, combinatorics, and cryptography.

Let $GF(p)$ be the prime field of characteristic p and let $d > 1$. A function $GF(2)^d \to GF(2)$ is a *Boolean function* on d variables. A function $GF(p)^d \to GF(p)$ is a *p-ary function* on d variables. Let

$$supp(f) = \{v \in V \mid f(v) \ne 0\}$$

denote the *support* of f, where $V = GF(p)^d$.

The set of invertible $n \times n$ matrices having coefficients in the field $GF(p)$ is denoted $GL(n, p)$. This set is a group under matrix multiplication. The group $GL(d, p)$ acts on the set of *p*-ary functions $f : GF(p)^d \to GF(p)$ by

$$g : f \mapsto f^g,$$

where

$$f^g(x) = f(gx), \quad x \in GF(p)^d.$$

It preserves the subspaces of (a) affine functions, (b) even functions, and (c) functions for which $f(0) = 0$.

Let p be a prime. A *general feedback shift register* is a map

$$C : GF(p)^d \to GF(p),$$

which, given initial values $x_0, \ldots, x_{d-1}$, gives rise to a sequence $x_0, \ldots, x_n, \ldots$, where $x_n = C(x_{n-d}, \ldots, x_{n-1})$ for $n \geq d$. When C is a linear function, iterating the map C generates a *linear feedback shift register* (LFSR). These are pseudorandom sequences in $GF(p)$ with long periods. Unfortunately, while LFSRs have some good pseudorandomness properties, they are not very secure. Here is one way to fix this. Consider a periodic sequence of period P,

$$x = x_0, \ldots, x_n, \ldots,$$

for example, a LFSR. Given $F : GF(p)^d \to GF(p)$, construct a filtered sequence by setting

$$y_i = F(x_{i-d}, \ldots, x_{i-1}),$$

for $i \geq d$, and $y_i = x_i$ for $0 \leq i \leq d - 1$. For suitably chosen F, these can be used to build secure stream ciphers.

What kind of function f makes a good filter function? A very nonlinear one. How to measure nonlinearity? One way is with a finite field analog of the Fourier transform. The *Walsh–Hadamard transform* of a $GF(p)$-valued function f is a complex-valued function on $V = GF(p)^d$ that can be defined as

$$W_f(u) = \sum_{x \in V} \zeta^{f(x) - \langle u, x \rangle},$$

where $\zeta = e^{2\pi i/p}$.

The Walsh transform of an affine function is a "spike" (supported at a single vector).

We call f *bent* if

$$|W_f(u)| = p^{n/2},$$

for all $u \in V$. In some sense, bent functions are maximally nonlinear. Bent functions are good filter functions (F above).

There is an interesting connection between bent functions and combinatorial structures called difference sets. The Dillon correspondence states that a function $f : GF(2)^d \to GF(2)$ is bent if and only if $f^{-1}(1)$ is an elementary Hadamard difference set of $GF(2)^n$.

There is also an interesting connection between bent functions and graph theory. The Bernasconi–Codenotti–VanderKam correspondence states that a function $f : GF(2)^d \to GF(2)$ is bent if and only if the Cayley graph of f is a strongly regular graph having parameters $(2^d, k, \lambda, \lambda)$ for some λ, where $k = |supp(f)|$. (In fact, the values of λ are known more explicitly.)

Is there a weighted analog, for $p > 2$, of the Dillon and Bernasconi–Codenotti–VanderKam correspondences? Questions of this sort are studied further in Chapter 6.

Graph theory and error-correcting codes

There are several well-known constructions of error-correcting codes that arise from a graph. Roughly speaking, an error-correcting code is a subset $C \subset GF(q)^n$ (where $GF(q)$ is a finite field having q elements), for which different elements are distinct enough that a small number of errors can be corrected by a suitable decoding algorithm.

For example, there is the binary code which is generated by the rows of the incidence matrix of a graph. There are also error-correcting codes over a finite field $GF(q)$ arising from cycles or cocycles of a graph. And there are expander codes constructed from specially constructed Cayley graphs.

Constructions in the other direction, graphs from codes, are less well known, but here's an example. Let $C \subset GF(2)^n$ denote an error-correcting code with (full rank) $k \times n$ generator matrix G. Let $f_i : GF(2)^n \to GF(2)$ denote the function supported on the i-th row of G, let

$$f = f_1 + \ldots + f_k,$$

and let Γ be the Cayley graph of f. In other words, the vertices of Γ are the vectors in $GF(2)^n$, and $e = (v, w) \in GF(2)^n \times GF(2)^n$ is an edge if and only if $f(v - w) \neq 0$. The code C corresponds in a natural way to the connected component of Γ containing the 0-vector. This is a k-regular, distance-regular graph.

There is a similar construction where we replace the rows of G by the columns of G, but we must assume that the columns don't contain the 0-vector. See §6.15 below for more details.

Note on using Sage: Some Sage commands used in this book use modules written for this book and uploaded to the github site for this book (https://github.com/springer-math/adventures-in-graph-theory).

Acknowledgments

We thank our friend and colleague T. S. Michael (who was diagnosed with pancreatic cancer and sadly passed away during the writing of this book), Amy Ksir, David Phillips, Charles Celerier, and Steven Walsh for their conversations and help.

Contents

Chapter 1
Introduction: Graphs—Basic Definitions

1.1 Graph theory

By a graph, we don't mean a chart or function plot or a diagram. We mean a combinatorial structure which consists of vertices, also called points or nodes, connected by edges, also called arcs. If you wish, think of vertices as destinations and edges as routes. A more precise statement is given below.

Graphs are used in many branches of science but the only application we consider here is to cryptography—the Biggs cryptosystem studied briefly in Chapter 4.

1.1.1 Basic definitions

We begin with a precise definition of a graph.

Definition 1.1.1. For any set S, let $S^{(2)} = \{\{s,t\} \mid s,t \in S, s \neq t\}$, i.e., $S^{(2)}$ is the set of unordered pairs of distinct elements of S. An *undirected simple graph* $\Gamma = (V, E)$ is an ordered pair of sets, where V is a set of *vertices* (possibly with weights attached) and $E \subseteq V^{(2)}$ is a set of *edges* (possibly with weights attached)[1]. We refer to $V = V_\Gamma$ as the *vertex set* of Γ, and $E = E_\Gamma$ as the *edge set*.

A *subgraph* of $\Gamma = (V, E)$ is a graph $\Gamma' = (V', E')$ formed from subsets $V' \subset V$ and $E' \subset E$ such that V' includes the endpoints of each $e \in E'$.

The *neighborhood* of a vertex v in a graph Γ is the set

$$N(v) = N_\Gamma(v) = \{v' \in V \mid (v, v') \text{ is an edge in } \Gamma\}.$$

---

[1]When there is no ambiguity, we will abuse notation and write $(u, v) \in V^{(2)}$ or $uv \in V^{(2)}$ if $\{u, v\} \in V^{(2)}$, when the meaning is clear.

© Springer International Publishing AG 2017
W.D. Joyner and C.G. Melles, *Adventures in Graph Theory*,
Applied and Numerical Harmonic Analysis,
https://doi.org/10.1007/978-3-319-68383-6_1

If $e \in E$ is an edge and $v \in V$ is a vertex on either "end" of e, then we say v is *incident* to e (or that e is *incident* to v). If u, v are vertices and $e = (u, v) \in E$ is an edge, then u and v are called *adjacent* or *neighboring* vertices. The vertices u and v are sometimes called the *endpoints* of e.

Graphs can be used to model electrical networks and other situations involving flows. Various matching problems can be stated in terms of graphs. If you think of vertices as gamblers and edges as bets between them, then the weight of an edge could be the cost of a bet between those gamblers, and the weight of a vertex could represent the numbers of chips that gambler has to play with. Edge weights in a graph could also represent costs or flow capacities.

There are situations in which we may wish to generalize the concept of a graph to allow more than one edge between a pair of vertices. Such graphs are called multigraphs.

Definition 1.1.2. An *undirected multigraph* Γ is an ordered pair of sets (V, E), where V is a set of *vertices* and E is a set of edges, together with a map $E \to V^{(2)}$, assigning to each edge an unordered pair of distinct vertices which are said to be incident to that edge.

Remark 1.1.3. A *loop* is an edge of the form (v, v), for some $v \in V$. Unless stated otherwise, none of the graphs in this book will have loops. A graph with no multiple edges or loops is called a *simple* graph.

Unless stated otherwise, we will assume that both the vertex set V and the edge set E are finite sets.

Let Γ be a graph. The cardinality $|V|$ of the vertex set V is called the *order* of Γ, and the cardinality $|E|$ of the edge set E is called the *size* of Γ.

Definition 1.1.4. An *orientation* of the edges is an injective function $\iota : E \to V \times V$. If $e = \{u, v\}$ and $\iota(e) = (u, v)$, then we call v the *head* of e and u the *tail* of e. We define the head and tail functions $h : E \to V$ and $t : E \to V$ by $h(e) = v$ and $t(e) = u$, i.e., $h(e)$ is the head of e and $t(e)$ is the tail of e. If the vertices of a graph are the set $V = \{0, 1, \ldots, n-1\}$, then the *default orientation* of an edge $e = (u, v)$ is defined by $h(e) = \max\{u, v\}$, $t(e) = \min\{u, v\}$. If the vertices of a graph are the set $V = \{0, 1, \ldots, n-1\}$ and if the edges of a graph are indexed $E = \{e_1, e_2, \ldots, e_m\}$, then we can associate to each orientation ι a vector,

$$\vec{o}_\iota = (o_1, o_2, \ldots, o_m) \in \{1, -1\}^m,$$

as follows: if $e_i = \{u, v\}$ and $h(e_i) = \max\{u, v\}$, then define $o_i = 1$, but if $h(e_i) = \min\{u, v\}$, then define $o_i = -1$. This $\vec{o}_\iota$ is called the *orientation vector* associated to ι.

If u and v are two vertices in a graph $\Gamma = (V, E)$, a u–v *walk*, W, is an alternating sequence of vertices and edges starting with u and ending at v,

$$W : e_0 = (v_0, v_1), e_1 = (v_1, v_2), \ldots, e_{k-1} = (v_{k-1}, v_k), \qquad (1.1)$$

where $v_0 = u$, $v_k = v$, and each $e_i \in E$. The number of edges in a walk is called its *length*, denoted $len(W) = k$. Notice that consecutive vertices in a walk are adjacent to each other. We are allowed to have repeated vertices and edges in a walk. A *cycle* is a walk with the same starting and ending vertex. A cycle is also called a *closed walk*. A graph $\Gamma = (V, E)$ is called *connected* provided it has the property that for any two vertices $u, v \in V$ there exists a u–v walk.

Definition 1.1.5. We define[2] the *genus* of a connected graph Γ to be

$$\mathrm{genus}(\Gamma) = |E| - |V| + 1.$$

Example 1.1.6. The graph with n vertices and an edge connecting each pair of vertices is called the *complete graph on n vertices* and denoted K_n. The graph K_4 is also called the *tetrahedron graph* and is shown in Figure 1.1. It has genus 3.

Figure 1.1: K_4 or the tetrahedron graph.

Example 1.1.7. The graph obtained from K_4 by removing one edge is called the *diamond graph* and is shown in Figure 1.2. It has genus 2.

Figure 1.2:   The diamond graph.

---

[2]This is not the usual definition of genus: the minimal integer n such that the graph can be embedded in a surface of genus n.

Example 1.1.8. A graph with n vertices $v_1, v_2, \ldots, v_n$ and edges (v_1, v_2), (v_2, v_3), $\ldots$, (v_{n-1}, v_n), (v_n, v_1) is called a *cycle graph* and denoted C_n. All cycle graphs have genus 1. The cycle graph C_5 is shown in Figure 1.3.

Figure 1.3: The cycle graph C_5.

Example 1.1.9. Suppose we start with a cycle graph on $n - 1$ vertices and add a vertex q and an edge from q to each vertex of the cycle graph. The resulting graph is called a *wheel graph* and denoted W_n. The vertices from the cycle graph are known as the *spoke vertices* and the corresponding edges from the spoke vertices to q are the *spokes*. The genus of W_n is $n - 1$. The wheel graph W_6 is shown in Figure 1.4.

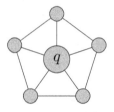

Figure 1.4: The wheel graph W_6.

Example 1.1.10. A graph consisting of two vertices and $n > 1$ edges connecting the vertices is called a *banana graph* or *dipole graph* of order n. It has genus $n - 1$. The case $n = 3$ is also called a *theta graph*. A banana graph with 4 edges is shown in Figure 1.5.

Figure 1.5: A banana graph.

There are various ways to describe a graph. For example, a simple graph Γ can be described using the vertices

$$V = V_\Gamma = \{0, 1, 2, \ldots, n - 1\},$$

and then Γ is uniquely determined by specifying an $n \times n$ matrix $A = (a_{ij})$, where $a_{ij} = 1$ if vertex i shares an edge with vertex j, and $a_{ij} = 0$ otherwise. This matrix A is called the (undirected, unweighted) *adjacency matrix* of the graph.

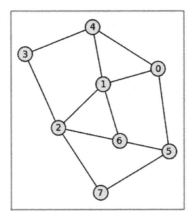

Figure 1.6: A graph created using Sage.

If no weights on the vertices or edges are specified, we usually assume all the weights are implicitly 1 and call the graph *unweighted*. In this case, the *degree* of a vertex $j \in V$ is the number of edges in Γ incident to j, denoted

$$\deg_\Gamma(j) = \sum_{e \in E,\ j \in e} 1.$$

The *degree sequence* of $\Gamma = (V, E)$ is the sequence of degrees, one for each vertex $v_i = i \in V$: $[d_0, d_1, \ldots, d_{n-1}]$, where $d_i = \deg_\Gamma(i)$.

A graph with edge weights, that is to say a function $wt : E \to \mathbb{R}_+$, is called a *weighted graph* or an *edge-weighted graph*. In the weighted case, the degree sometimes includes the weights:

$$\deg_{\Gamma, wt}(j) = \sum_{e \in E,\ j \in e} wt(e).$$

A graph $\Gamma = (V, E)$ having no cycles is called a *forest* (or a *tree* in the connected case). A *spanning tree* T of a connected graph is a tree that contains all the vertices of the graph. The smallest number of edges we can remove from Γ making it into a forest is called the *circuit rank*. For example, the circuit rank of the graph in Example 1.1.28 is 1. This is discussed further in §1.2.2.

Lemma 1.1.11. *Let $\Gamma = (V, E)$ denote a connected graph and let T be a spanning tree of Γ. For each edge $e \in E$ not in T, there is a unique cycle $cyc(T, e)$ of Γ containing only e and edges in T.*

This lemma defines $cyc(T, e)$, a so-called *fundamental cycle*. For a proof, see Biggs, Lemma 5.1 in [Bi93].

If Γ is a connected edge-weighted graph, where edge $e \in E$ has weight $w(e) \in \mathbb{R}_+$, then the *edge-weighted length* of a walk $W \subset E$ is

$$len(W) = \sum_{e \in W} w(e).$$

When the edge weight function is always 1, then we get the usual length. In that case, we define the *distance* between two neighboring vertices to be $dist(u, v) = 1$, if $(u, v) \in E$. In general, the *distance* between $u, v \in V$ is defined by

$$dist(u, v) = \begin{cases} \min_W \ len(W), & u \neq v, \\ 0, & u = v, \end{cases}$$

where the minimum runs over all u–v walks W.

The *girth* of Γ is the length of a shortest cycle (if it exists) contained in Γ. If no cycles exist in Γ the girth is, by convention, defined to be ∞.

If Γ is connected, the *diameter* is

$$diam(\Gamma) = \max_{u,v \in V} dist(u, v).$$

The *radius* of a connected graph is the minimum value

$$radius(\Gamma) = \min_{u \in V} \max_{v \in V} dist(u, v).$$

Example 1.1.12. For example, the girth of the Rubik's cube graph in the quarter turn metric Γ_{QT} (see the Example in the Preface) is 4, while the radius is 26.

A bijection $p : V \to V$ of the vertices which induces a bijection of the edges $P : E \to E$ is called an *automorphism* of $\Gamma = (V, E)$. The group of automorphisms of Γ is denoted as $Aut(\Gamma)$.

If $G \subset Aut(\Gamma)$ is a subgroup, then we say G *acts transitively* on Γ if for any two $v, w \in V$, there is a $g \in G$ such that $w = g(v)$.

We say $\Gamma = (V, E)$ is a *distance-transitive graph* provided, given any $v, w \in V$ at any distance i, and any other two vertices x and y at the same distance, there is an automorphism of Γ sending v to x and w to y.

We say Γ is a *distance-regular graph*, provided it is a regular graph such that, for any $v, w \in V$, the number of vertices at distance j from v, $|\{x \in V \mid dist(v, x) = j\}|$, and at distance k from w, $|\{x \in V \mid dist(w, x) = k\}|$, depends only upon j, k, and $i = dist(v, w)$.

Recall a connected graph is one for which, given any distinct vertices u, v, there is a walk in the graph connecting u to v. A *connected component* of

Γ is a subgraph which is connected and not the proper subgraph of another connected subgraph of Γ. The number of *connected components* of the graph Γ is denoted $c(\Gamma)$.

Definition 1.1.13. Let $\Gamma = (V, E)$ denote the *sensitivity graph* of a Boolean function $f : GF(2)^n \to GF(2)$, defined to be the (bipartite) graph containing only edges with different f-values:

$$V = GF(2)^n,$$

$$E = \{(u, v) \mid u, v \in GF(2)^n, f(u) \neq f(v)\}.$$

In general, the *sensitivity* of f, denoted $s(f)$, is the maximum vertex degree of the sensitivity graph, Γ.

When the sensitivity of f is small relative to the number of variables, it means that for every vertex $v \in GF(2)^n$, flipping one coordinate will change the value of f only for relatively few coordinates.

Example 1.1.14. Let $f(x)$ denote the Boolean function

$$f(x_0, x_1, x_2, x_3) = x_0 x_1 x_2 x_3 + x_0 x_1 x_2 + x_1 x_2 x_3 + x_0 x_1 + x_0 x_2.$$

The sensitivity graph Γ of this function on $GF(2)^4$ is depicted in Figure 1.7.

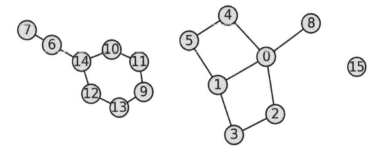

Figure 1.7: The sensitivity graph associated to f.

One of the major outstanding problems about Boolean functions is the *sensitivity conjecture*, which states that the minimum degree of a real polynomial that interpolates a Boolean function f is bounded above by some fixed power of the sensitivity of f.

Exercise 1.1. Compute the sensitivity of the graph in Figure 1.7.

A walk with no repeated vertices (except possibly the beginning and ending vertices, if they are the same) is called a *path*. A walk with no repeated edges is called a *trail*. Clearly a path is a trail, but the converse is not true in general. If there are $n = |V|$ vertices, then no path can have length more than n. A path of length n is called a *Hamiltonian path* and a graph which has a Hamiltonian path is called *traceable*. A path whose start and end vertices are the same is called a *simple cycle*. An *Eulerian cycle* is a trail whose start and end vertices are the same which visits every edge exactly once. A cycle of length n is called a *Hamiltonian cycle*. A graph which has an Eulerian cycle is called an *Eulerian graph* and a graph which has a Hamiltonian cycle is called a *Hamiltonian graph*.

Let $\Gamma = (V, E)$ be an undirected simple graph. If $m = |V|$, then the *adjacency matrix* $A = A_\Gamma$ is the $m \times m$ matrix with entries a_{ij} given by

$$a_{ij} = \begin{cases} 1, & \{i, j\} \in E, \\ 0, & \text{otherwise.} \end{cases}$$

Note that the diagonal entries a_{ii} are 0 (no loops). If Γ is an undirected edge-weighted graph,[3] the *weighted adjacency matrix* $A_{\Gamma, w}$ is the $m \times m$ matrix whose ij-th entry is the weight of the edge from i to j. The adjacency matrix of an undirected graph is symmetric.

We say f is a map from the graph $\Gamma_2 = (V_2, E_2)$ to the graph $\Gamma_1 = (V_1, E_1)$, denoted $f : \Gamma_2 \to \Gamma_1$, if f is associated to a map $f_V : V_2 \to V_1$ which satisfies the *edge-preservation condition*:

$$(i, j) \in E_2 \implies (f_V(i), f_V(j)) \in E_1.$$

Therefore, the map $f_E : E_2 \to E_1$ given by $f_E(i, j) = (f_V(i), f_V(j))$ is well-defined. We say f is an *isomorphism* if both f_V and f_E are bijections.

Lemma 1.1.15. *Two graphs Γ_1 and Γ_2 are isomorphic if and only if there is a permutation matrix P such that*

$$A_{\Gamma_2} = P^{-1} A_{\Gamma_1} P.$$

Proof. We may, without loss of generality, identify the vertex sets V_2 and V_1 with $\{0, 1, \ldots, n-1\}$. Suppose $f : \Gamma_2 \to \Gamma_1$ is an isomorphism, so $f_V : V_2 \to V_1$ may be regarded as a permutation.

If a matrix $P \in GL(n, \mathbb{Z})$ satisfies for each $n \times n$ matrix $M = (m_{ij})$, $P^{-1}MP = (m_{\rho(i), \rho(j)})$, for all i, j, for some permutation $\rho : \mathbb{Z}/n\mathbb{Z} \to \mathbb{Z}/n\mathbb{Z}$ then P is a permutation matrix. Every permutation matrix arises in this way.

Therefore, $A_{\Gamma_2} = P^{-1} A_{\Gamma_1} P$, for some P depending on f_V.

The converse is left to the reader.   □

---

[3]In the case of an unweighted multigraph Γ without loops, we associate to that graph an edge-weighted graph Γ' whose edge weights are given by the corresponding edge multiplicities of Γ.

The *characteristic polynomial* of Γ is

$$p_\Gamma(x) = \det(xI_n - A_\Gamma) = x^n + c_1 x^{n-1} + \cdots + c_{n-1}x + c_n,$$

where $n = |V|$, and I_n denotes the $n \times n$ identity matrix.

Since $A = A_\Gamma$ is symmetric, it is diagonalizable. Therefore, A has an orthonormal set of eigenvectors $\mathbf{v}_1, \ldots, \mathbf{v}_n$ corresponding to eigenvalues $\lambda_1, \ldots, \lambda_n$. If E_k is the $n \times n$ idempotent matrix

$$E_k = \mathbf{v}_k \cdot \mathbf{v}_k^t, \quad k = 1, \ldots, n,$$

then the *spectral decomposition* of A is

$$A = \sum_{i=1}^{n} \lambda_j E_i. \tag{1.2}$$

Example 1.1.16. Consider the cycle graph Γ having 4 vertices. This can be visualized as a square. The eigenvalues and eigenvectors of the adjacency matrix,

$$A = \begin{pmatrix} 0 & 1 & 0 & 1 \\ 1 & 0 & 1 & 0 \\ 0 & 1 & 0 & 1 \\ 1 & 0 & 1 & 0 \end{pmatrix},$$

are easy to compute and the spectral decomposition in (1.2) is easy to verify.

```
——————————————————————————— Sage ———————————————————————————
sage: Gamma = graphs.CycleGraph(4)
sage: A = Gamma.adjacency_matrix()
sage: v1 = vector(QQ,  [1/2,   1/2, 1/2,   1/2])
sage: v2 = vector(QQ,  [1/2,  -1/2, 1/2,  -1/2])
sage: v3 = vector(RR,  [0, 1/sqrt(2), 0, -1/sqrt(2)])
sage: v4 = vector(RR,  [1/sqrt(2), 0, -1/sqrt(2), 0])
sage: lambda1 = 2; A*v1 == lambda1*v1
True
sage: lambda2 = -2; A*v2 == lambda2*v2
True
sage: lambda3 = 0; A*v3 == lambda3*v3
True
sage: lambda4 = 0; A*v4 == lambda4*v4
True
sage: E1 = v1.tensor_product(v1)
sage: E2 = v2.tensor_product(v2)
sage: E3 = v3.tensor_product(v3)
sage: E4 = v4.tensor_product(v4)
sage: lambda1*E1 + lambda2*E2 + lambda3*E3 + lambda4*E4 == A.change_ring(RR)
True
```

Definition 1.1.17. The *incidence matrix* of an oriented graph Γ, with orientation ι, is an $n \times m$ matrix $B = B_{\Gamma,\iota} = (b_{ij})$, where m and n are the numbers of edges and vertices, respectively, such that $b_{ij} = 1$ if the vertex v_i is the head of edge e_j, $b_{ij} = -1$ if the vertex v_i is the tail of edge e_j, and $b_{ij} = 0$ otherwise. The *incidence matrix* of an undirected graph (or a graph without a prescribed orientation) Γ is an $n \times m$ matrix $B = B_\Gamma = (b_{ij})$ such that $b_{ij} = 1$ if the edge e_j is incident to vertex v_i, and $b_{ij} = 0$ otherwise.

When there is possible ambiguity, we call the matrix in the former case a signed incidence matrix and the latter an unsigned incidence matrix.

Exercise 1.2. For the graph Γ in Figure 1.6, solve the following problems.

(a) Find the adjacency matrix A of Γ.

(b) Fix an ordering of the edges and find the corresponding (unsigned) incidence matrix B of Γ.

(c) Is Γ Hamiltonian? If so, find a Hamiltonian cycle.

(d) Is Γ Eulerian? If so, find an Eulerian trail.

(e) Find the girth of Γ.

(f) Find the diameter of Γ.

Unfortunately, the literature is not consistent in the definition of some of the technical terms in graph theory. For example, the rank of a graph Γ is sometimes defined to be the rank of its adjacency matrix $A = A_\Gamma$ and sometimes defined to be the rank of its signed incidence matrix $B = B_{\Gamma,\iota}$ (as we'll see below, this rank does not depend on the orientation ι chosen). The lemma below concerns the latter definition.

We prove the following lemma for the oriented incidence matrix (see Godsil and Royle [GR01] for more details). In particular, the rank is independent of the orientation chosen.

Lemma 1.1.18. *Let Γ be a graph and B its incidence matrix. The rank of B is $n - c(\Gamma)$, where n is the number of vertices of Γ and $c(\Gamma)$ is its number of connected components.*

Proof. Suppose the matrix B is $n \times m$ and $z \in \mathbb{R}^n$ is in the left kernel of B: i.e., $zB = 0$. For each edge $(u, v) \in E$, $zB = 0$ implies $z_u = z_v$. Therefore, z is constant on connected components. This implies that the left kernel of B has dimension equal to the number of connected components of the graph. The result now follows from the rank plus nullity theorem from matrix theory. $\qquad\square$

Definition 1.1.19. If F is a field such as $\mathbb{R}$ or $GF(q)$ or a ring such as $\mathbb{Z}$, let

$$C^0(\Gamma, F) = \{f : V \to F\}, \quad C^1(\Gamma, F) = \{f : E \to F\},$$

be the sets of F-valued functions defined on V and E, respectively. These sets are sometimes called the *vertex space over F* and the *edge space over F*.

If F is a field, then these are F-inner product spaces with inner product

$$(f, g) = \sum_{x \in X} f(x)g(x), \qquad (X = V, \text{ respectively } X = E), \qquad (1.3)$$

and

$$\dim C^0(\Gamma, F) = |V|, \quad \dim C^1(\Gamma, F) = |E|.$$

Index the sets V and E in some arbitrary but fixed way and define, for $1 \le i \le |V|$ and $1 \le j \le |E|$,

$$f_i(v) = \begin{cases} 1, & v = v_i, \\ 0, & \text{otherwise}, \end{cases} \qquad g_j(e) = \begin{cases} 1, & e = e_j, \\ 0, & \text{otherwise}. \end{cases}$$

Lemma 1.1.20. *(a)* $\mathcal{F} = \{f_i\} \subset C^0(\Gamma, F)$ *is a basis.* *(b)* $\mathcal{G} = \{g_j\} \subset C^1(\Gamma, F)$ *is a basis.*

Proof. The proof is left as an exercise. $\qquad\qquad\qquad\qquad\qquad\qquad$ □

Define

$$B : C^1(\Gamma, F) \to C^0(\Gamma, F)$$

by

$$(Bf)(v) = \sum_{h(e)=v} f(e) - \sum_{t(e)=v} f(e). \qquad (1.4)$$

With respect to these bases $\mathcal{F}$ and $\mathcal{G}$, the matrix representing the linear transformation $B : C^1(\Gamma, F) \to C^0(\Gamma, F)$ is the incidence matrix (we use B to denote both the linear transformation and its matrix representation). Since both $C^1(\Gamma, F)$ and $C^0(\Gamma, F)$ are inner product spaces, we may define the *dual transformation* $B^* : C^0(\Gamma, F) \to C^1(\Gamma, F)$ defined by

$$(Bf, g)_{C^0(\Gamma, F)} = (f, B^*g)_{C^1(\Gamma, F)},$$

for all $f \in C^1(\Gamma, f)$ and $g \in C^0(\Gamma, F)$. The matrix representation of B^* is the transpose of the matrix representation of B.

Example 1.1.21. We use Sage to compute the incidence matrix of the house graph, depicted in Figure 1.8.

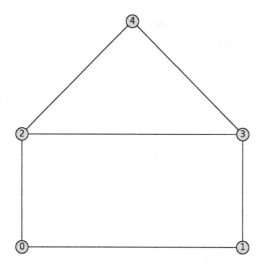

Figure 1.8:   The house graph.

```
──────────────────────────── Sage ────────────────────────────

sage: Gamma = graphs.HouseGraph()
sage: Gamma.edges(labels=None)
[(0, 1), (0, 2), (1, 3), (2, 3), (2, 4), (3, 4)]
sage: Gamma.vertices()
[0, 1, 2, 3, 4]
sage: B = Gamma.incidence_matrix(); B

[1 1 0 0 0 0]
[1 0 1 0 0 0]
[0 1 0 1 1 0]
[0 0 1 1 0 1]
[0 0 0 0 1 1]
sage: B.right_kernel()

Free module of degree 6 and rank 1 over Integer Ring
Echelon basis matrix:
[ 1 -1 -1  1  0  0]
sage: B.left_kernel()

Free module of degree 5 and rank 0 over Integer Ring
Echelon basis matrix:
[]
```

The unsigned incidence matrix is

$$
\begin{pmatrix}
1 & 1 & 0 & 0 & 0 & 0 \\
1 & 0 & 1 & 0 & 0 & 0 \\
0 & 1 & 0 & 1 & 1 & 0 \\
0 & 0 & 1 & 1 & 0 & 1 \\
0 & 0 & 0 & 0 & 1 & 1
\end{pmatrix},
$$

while the signed incidence matrix associated to the default orientation is

$$\begin{pmatrix} 1 & 1 & 0 & 0 & 0 & 0 \\ -1 & 0 & 1 & 0 & 0 & 0 \\ 0 & -1 & 0 & 1 & 1 & 0 \\ 0 & 0 & -1 & -1 & 0 & 1 \\ 0 & 0 & 0 & 0 & -1 & -1 \end{pmatrix}.$$

Exercise 1.3. Verify that the left kernel of a signed incidence matrix of an oriented connected simple graph is the space of constant functions.

If F is a field and Γ is a graph with orientation ι, then the kernel of the signed incidence map $B = B_{\Gamma,\iota} : C^1(\Gamma, F) \to C^0(\Gamma, F)$ is the *cycle space*[4],

$$\mathcal{Z} = \mathcal{Z}_{\Gamma,\iota} = \ker(B).$$

Via the natural basis, we may identify the cycle space with a subspace of F^m using the right kernel of the $n \times m$ matrix representation of the map B.

As the example below shows, this space does depend on the orientation selected.

Example 1.1.22. We use Sage to compute the cycle space of the house graph Γ, depicted in Figure 1.8, for two distinct orientations of Γ.

```
─────────────────────────── Sage ───────────────────────────
sage: Gamma = graphs.HouseGraph()
sage: Gamma.edges(labels=None)
[(0, 1), (0, 2), (1, 3), (2, 3), (2, 4), (3, 4)]
sage: Gamma.vertices()
[0, 1, 2, 3, 4]
sage: B1 = incidence_matrix(Gamma,6*[1])
sage: B2 = incidence_matrix(Gamma,[1,-1,1,-1,1,-1])
sage: B1.right_kernel()

Vector space of degree 6 and dimension 2 over Rational Field
Basis matrix:
[ 1 -1  1  0 -1  1]
[ 0  0  0  1 -1  1]
sage: B1.left_kernel()

Vector space of degree 5 and dimension 1 over Rational Field
Basis matrix:
[1 1 1 1 1]
sage: B2.right_kernel()

Vector space of degree 6 and dimension 2 over Rational Field
Basis matrix:
[ 1  1  1  0 -1 -1]
[ 0  0  0  1  1  1]
sage: B2.left_kernel()


Vector space of degree 5 and dimension 1 over Rational Field
Basis matrix:
[1 1 1 1 1]
sage: B1.right_kernel()==B2.right_kernel()
False
```

---

[4] Also called the *flow space* or the space of *circulation functions*.

The signed incidence matrix associated to the default orientation is

$$B_1 = \begin{pmatrix} 1 & 1 & 0 & 0 & 0 & 0 \\ -1 & 0 & 1 & 0 & 0 & 0 \\ 0 & -1 & 0 & 1 & 1 & 0 \\ 0 & 0 & -1 & -1 & 0 & 1 \\ 0 & 0 & 0 & 0 & -1 & -1 \end{pmatrix},$$

while the signed incidence matrix associated to the orientation $(1, -1, 1, -1, 1, -1)$ (see Definition 1.1.4) is

$$B_2 = \begin{pmatrix} 1 & -1 & 0 & 0 & 0 & 0 \\ -1 & 0 & 1 & 0 & 0 & 0 \\ 0 & 1 & 0 & -1 & 1 & 0 \\ 0 & 0 & -1 & 1 & 0 & -1 \\ 0 & 0 & 0 & 0 & -1 & 1 \end{pmatrix}.$$

Suppose $\Gamma = (V, E)$ is given an orientation ι and that $\vec{o}$ is the associated orientation vector. Order the edges

$$E = \{e_1, e_2, \ldots, e_m\},$$

in some arbitrary but fixed way. Consider a path from u to v in Γ $(u, v \in V)$, which we may regard as a subgraph $\Gamma' = (V', E')$ of Γ, $E' \subset E$, with a fixed orientation of the edges as in (1.1) so it is directed from $u = v_0$ to $v = v_k$. A *vector representation* of a path Γ' is an m-tuple

$$\vec{\Gamma'} = (a_1, a_2, \ldots, a_m) \in \{1, -1, 0\}^m, \tag{1.5}$$

where $a_i = 1$, if $e_i \in E'$ and the orientations of e_i in Γ and Γ' agree; $a_i = -1$, if $e_i \in E'$ and the orientations of e_i in Γ and Γ' are opposite; and $a_i = 0$, if $e_i \notin E'$. In particular, this defines a mapping

$$\{\text{cycles of } \Gamma = (V, E)\} \to \{1, -1, 0\}^m,$$

$$C \mapsto \vec{C}.$$

Example 1.1.23. Consider the house graph $\Gamma = (V, E)$ in Example 1.1.21. Label the edges as follows:

$$E = \{e_1 = (0, 1), e_2 = (0, 2), e_3 = (1, 3), e_4 = (2, 3), e_5 = (2, 4), e_6 = (3, 4)\},$$

oriented with the default orientation, so $\vec{o} = (1, 1, 1, 1, 1, 1)$. Let Γ' denote the cycle $2 \to 3 \to 4 \to 2$, oriented as indicated. The vector representation of this cycle is $\vec{\Gamma'} = (0, 0, 0, 1, -1, 1)$.

Lemma 1.1.24. *The cycle space of an oriented connected graph* $\Gamma = (V, E)$ *is the vector space spanned by the vector representations of the cycles of* Γ.

Proof. Let C be a cycle in Γ, let ι denote the given orientation, and let B denote the matrix representation of the oriented incidence map with respect to the standard basis. For each vertex v of C, there are exactly two edges of C incident to v. Regarding C as a subgraph of Γ, let $\vec{C} = \vec{C}(\iota)$ denote the associated vector representation as in (1.5). The i-th entry of $B\vec{C}$ is the dot product of $\vec{C}$ with the i-th row of B. The only nonzero entries of i-th row of B are those entries associated to the edges incident to the i-th vertex of Γ. Either exactly two of those incident edges are in C or none are. In the latter case, it's clear that the i-th entry of $B\vec{C}$ is 0. In the former case, there are two possibilities: either the two incident edges in C are oriented with the same sign or they aren't. If they are oriented with the same sign, then either both these edges go into the i-th vertex of Γ or both these edges go out of the i-th vertex of Γ. In each of these cases, it's clear that the i-th entry of $B\vec{C}$ is 0. Finally, suppose the two incident edges in C are oriented with opposite signs. This means one of these edges goes into the i-th vertex and the other goes out. This implies that both of the corresponding nonzero entries of the i-th row of B have the same sign, by definition of B. Again, the i-th entry of $B\vec{C}$ is 0.

Thus, the vector space spanned by the vector representations of the cycles is contained in the cycle space. $\qquad\qquad\qquad\qquad\qquad\qquad\qquad\square$

Indeed, let $T = (V, E_T)$ be a spanning tree of Γ. The set of fundamental cycles

$$cyc(T, g) \mid g \in E - E_T\},$$

defined via Lemma 1.1.11, is a set of linearly independent vectors in $\mathcal{Z}$. This is a basis of the cycle space $\mathcal{Z}$.

Definition 1.1.25. For any nontrivial partition

$$V = V_1 \cup V_2, \qquad V_i \neq \emptyset, \quad V_1 \cap V_2 = \emptyset,$$

the set of all edges $e = (v_1, v_2) \in E$, with $v_i \in V_i$ $(i = 1, 2)$, is called a *cocycle* (or *edge cut subgraph* or *disconnecting set* or *seg* or *edge cut set*) of Γ. A *cut* is a partition of the vertex set of $\Gamma = (V, E)$ into two subsets, $V = S \cup T$. The cocycle of such a cut is the set

$$\{(u, v) \in E \mid u \in S, \ v \in T\}$$

of edges that have one endpoint in S and the other endpoint in T. A cocycle with a minimal set[5] of edges is a *bond* of Γ.

Lemma 1.1.26. *Let* $\Gamma = (V, E)$ *denote a connected graph and let* T *be a spanning tree of* Γ. *For each edge* $h \in E$ *which is also an edge in* T, *there is a unique cocycle* $coc(T, h)$ *of* Γ *containing only* h *and edges not in* T.

This Lemma defines $coc(T, h)$.

Proof. The proof is left as an exercise.   □

Suppose $\Gamma = (V, E)$ is given an orientation ι and that $\vec{o}$ is the associated orientation vector. Order the edges

$$E = \{e_1, e_2, \ldots, e_m\},$$

in some arbitrary but fixed way. Consider a cut H of Γ (associated to a partition $V = V_1 \cup V_2$), which we may regard as a subgraph $H = (V', E')$ of Γ. We fix an orientation on H so that for any edge $(u, v) \in E'$, $u \in V_1$ and $v \in V_2$. A *vector representation* of H is an m-tuple

$$\vec{H} = (b_1, b_2, \ldots, b_m) \in \{1, -1, 0\}^m, \tag{1.6}$$

where

$$b_i = b_i(\Gamma', \iota) = \begin{cases} \vec{o}_i, & \text{if } e_i \in E', \\ 0, & \text{if } e_i \notin E'. \end{cases}$$

In particular, this defines a mapping

$$\{\text{cocycles of } \Gamma = (V, E)\} \to \{1, -1, 0\}^m,$$

$$H \mapsto \vec{H}.$$

The vector space spanned by the vector representations of the cocycles of Γ is the *cocycle space* (or *cut space* or *bond space*), $\mathcal{B}$.

We shall see later (see Lemma 3.2.2) that the orthogonal complement of the cycle space $\mathcal{Z}$ with respect to the inner product (1.3) is the cocycle space,

$$\mathcal{B} = \mathcal{Z}^\perp.$$

---

[5]In the sense of set theory. So, a bond is a cocycle that does not contain any other cocycle as a proper subset.

1.1.2 Simple examples

Example 1.1.27. The cube graph Γ is depicted Figure 1.9. It has incidence matrix

$$B = \begin{pmatrix}
0 & -1 & 0 & 0 & 0 & -1 & 0 & -1 & 0 & 0 & 0 & 0 \\
0 & 0 & 0 & -1 & 0 & 1 & -1 & 0 & 0 & 0 & 0 & 0 \\
-1 & 1 & -1 & 0 & 0 & 0 & 0 & 0 & 0 & 0 & 0 & 0 \\
1 & 0 & 0 & 1 & -1 & 0 & 0 & 0 & 0 & 0 & 0 & 0 \\
0 & 0 & 0 & 0 & 0 & 0 & 0 & 1 & 0 & 0 & -1 & -1 \\
0 & 0 & 0 & 0 & 0 & 0 & 1 & 0 & 0 & -1 & 0 & 1 \\
0 & 0 & 1 & 0 & 0 & 0 & 0 & 0 & -1 & 0 & 1 & 0 \\
0 & 0 & 0 & 0 & 1 & 0 & 0 & 0 & 1 & 1 & 0 & 0
\end{pmatrix}.$$

The cycle space $\mathcal{Z}$ has basis

$$(1,0,-1,0,1,0,0,0,0,-1,1,-1), (0,1,-1,0,0,0,1,0,0,-1,0,0),$$

$$(0,0,0,1,-1,0,0,1,0,0,-1,0), (0,0,0,0,0,1,-1,1,0,0,0,-1),$$

and $(0,0,0,0,0,0,0,0,1,-1,1,-1)$. The cocycle space $\mathcal{B}$ has basis

$$(1,0,0,0,0,-1,0,0,1,0,0,-1), (0,1,0,0,0,-1,0,0,0,-1,0,-1),$$

$$(0,0,1,0,0,0,0,0,-1,-1,0,0), (0,0,0,1,0,-1,0,0,0,0,-1,-1),$$

$$(0,0,0,0,1,0,0,0,-1,0,-1,0), (0,0,0,0,0,0,0,1,0,0,1,0,1),$$

$$(0,0,0,0,0,0,0,1,0,0,1,1).$$

Note that $\mathcal{B}^\perp = \mathcal{Z}$.

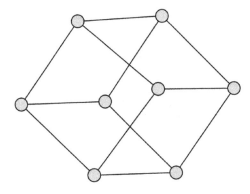

Figure 1.9: The cube graph in $\mathbb{R}^3$.

Exercise 1.4. Pick an orientation of the tetrahedron graph Γ depicted in Figure 1.10, for example the default orientation. Find a basis for the cycle space and a basis for the cocycle space of Γ.

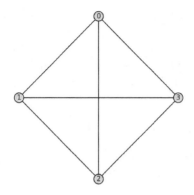

Figure 1.10:   The tetrahedron graph.

Example 1.1.28. Consider the graph in Figure 1.11.

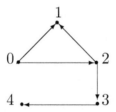

Figure 1.11:   An oriented graph with 5 vertices.

This graph has incidence matrix

$$
B = \begin{pmatrix}
-1 & -1 & 0 & 0 & 0 \\
0 & 1 & 1 & 0 & 0 \\
1 & 0 & -1 & -1 & 0 \\
0 & 0 & 0 & 1 & -1 \\
0 & 0 & 0 & 0 & 1
\end{pmatrix}.
$$

Example 1.1.29. Paley graphs are described in detail in §5.17. The Paley graph Γ having 9 vertices and 18 edges is depicted Figure 1.12. Its adjacency matrix is

$$A = \begin{pmatrix} 0 & 1 & 1 & 0 & 1 & 0 & 0 & 0 & 1 \\ 1 & 0 & 1 & 0 & 0 & 1 & 1 & 0 & 0 \\ 1 & 1 & 0 & 1 & 0 & 0 & 0 & 1 & 0 \\ 0 & 0 & 1 & 0 & 1 & 1 & 1 & 0 & 1 \\ 1 & 0 & 0 & 1 & 0 & 1 & 0 & 0 & 1 \\ 0 & 1 & 0 & 1 & 1 & 0 & 1 & 0 & 0 \\ 0 & 1 & 0 & 0 & 0 & 1 & 0 & 1 & 1 \\ 0 & 0 & 1 & 1 & 0 & 0 & 1 & 0 & 1 \\ 1 & 0 & 0 & 0 & 1 & 0 & 1 & 1 & 0 \end{pmatrix}.$$

The eigenvalues of A are $4, 1, 1, 1, 1, -2, -2, -2, -2$.

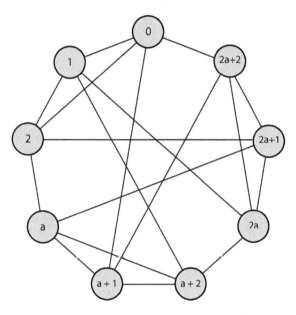

Figure 1.12: A Paley graph created using Sage.

Example 1.1.30. Consider the unoriented graph Γ below, with edges labeled as indicated, together with a spanning tree, depicted to its right, in Figure 1.13.

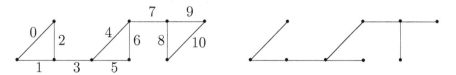

Figure 1.13: A graph and a spanning tree for it.

You can see from Figure 1.13 that

- by adding edge 2 to the tree, you get a cycle $0, 1, 2$ with vector representation
$$g_1 = (1, 1, 1, 0, 0, 0, 0, 0, 0, 0, 0),$$

- by adding edge 6 to the tree, you get a cycle $4, 5, 6$ with vector representation
$$g_2 = (0, 0, 0, 0, 1, 1, 1, 0, 0, 0, 0),$$

- by adding edge 10 to the tree, you get a cycle $8, 9, 10$ with vector representation
$$g_3 = (0, 0, 0, 0, 0, 0, 0, 0, 1, 1, 1).$$

The vectors $\{g_1, g_2, g_3\}$ form a basis of the cycle space of Γ.

Example 1.1.31. Consider the unoriented graph Γ below, with edges labeled as indicated, together with an example of a bond, depicted to its right, in Figure 1.14.

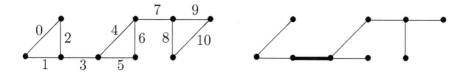

Figure 1.14: A graph and a cocycle of it.

You can see from Figure 1.14 that

- by removing edge 3 from the graph, you get a cocycle with vector representation
$$b_1 = (0, 0, 0, 1, 0, 0, 0, 0, 0, 0, 0),$$

- by removing edge 7 from the graph, you get a cocycle with vector representation
$$b_2 = (0, 0, 0, 0, 0, 0, 0, 1, 0, 0, 0),$$

- by removing edges $0, 1$ from the graph, you get a cocycle with vector representation
$$b_3 = (1, 1, 0, 0, 0, 0, 0, 0, 0, 0, 0),$$

- by removing edges $1, 2$ from the graph, you get a cocycle with vector representation
$$b_4 = (0, 1, 1, 0, 0, 0, 0, 0, 0, 0, 0),$$

- by removing edges $4, 5$ from the graph, you get a cocycle with vector representation
$$b_5 = (0,0,0,0,1,1,0,0,0,0,0),$$

- by removing edges $4, 6$ from the graph, you get a cocycle with vector representation
$$b_6 = (0,0,0,0,1,0,1,0,0,0,0),$$

- by removing edges $8, 9$ from the graph, you get a cocycle with vector representation
$$b_7 = (0,0,0,0,0,0,0,0,1,1,0),$$

- by removing edges $9, 10$ from the graph, you get a cocycle with vector representation
$$b_8 = (0,0,0,0,0,0,0,0,0,1,1).$$

The vectors $\{b_1, b_2, b_3, b_4, b_5, b_6, b_7, b_8\}$ form a basis of the cocycle space of Γ.

Note that these vectors are not orthogonal to the basis vectors of the cycle space in Example 1.1.30 unless we work over $GF(2)$.

Exercise 1.5. Let $V = \{0, 1, 2, \ldots, 10\}$, and, for all $u, v \in V$, let $(u, v) \in E$ if and only if $u < v$ and $v - u$ is a square (mod 11). Is the undirected graph $\Gamma = (V, E)$ connected? What is its order? What is its size?

Exercise 1.6. Find the cycle space of the graph in Figure 1.6.

Exercise 1.7. Find the cocycle space of the graph in Figure 1.6.

Definition 1.1.32. A graph Γ is said to be *planar* if it has an embedding in the plane (roughly speaking, if it can be drawn in the plane without edge crossings). A *dual graph* Γ^* of a planar graph Γ is a graph whose vertices correspond to the faces (or planar regions) of a such an embedding (including the infinite region). An edge of Γ^* is adjacent to two vertices of Γ^* if the corresponding regions of the embedding of Γ have a common edge.

Exercise 1.8. Show that the n-cycle graph C_n and the banana graph with n edges are dual graphs.

Good sources for further reading on basic graph theory are Biggs [Bi93], Bollobás [Bo98], Godsil and Royle [GR01], and Marcus [Mar08].

1.2 Some polynomial invariants for graphs

Polynomial invariants for graphs include the (vertex) chromatic polynomial, the (factorial, respectively, geometric) cover polynomial (for directed graphs), and one introduced above: the characteristic polynomial of the adjacency matrix.

Some others are introduced below.

1.2.1 The Tutte polynomial of a graph

Let $\Gamma = (V, E)$ be a graph. For any subset S of E, let $\Gamma_S = (V, S)$ be the spanning subgraph of Γ containing all vertices of Γ and the edges S.

Definition 1.2.1. We let

$$c(S) = \text{the number of connected components of } \Gamma_S.$$

The *rank* of S is defined to be

$$r(S) = |V| - c(S).$$

Thus the rank is the number of edges in a minimal spanning subgraph of Γ_S, i.e., in a spanning forest (where trees of 0 edges are allowed). The *nullity* of S is defined to be

$$n(S) = |S| - r(S) = |S| - |V| + c(S).$$

For example, if Γ is the complete graph on 4 vertices K_4 and if Γ_S is the subgraph consisting of a 3-cycle and a disjoint 4th vertex, then $c(\Gamma_S) = 2$ and $r(\Gamma_S) = 2$. If Γ_S is the subgraph with 4 vertices and no edges, we say that $c(\Gamma) = 4$ and $r(\Gamma) = 0$.

Definition 1.2.2. The *Tutte polynomial* of Γ is

$$T_\Gamma(x, y) = \sum_{S \subset E} (x-1)^{r(E)-r(S)} (y-1)^{|S|-r(S)},$$

where the sum is taken over all subsets S of E.

This formula can be rewritten in terms of the number of components and the nullity of spanning subgraphs as

$$T_\Gamma(x, y) = \sum_{S \subset E} (x-1)^{c(S)-c(E)} (y-1)^{n(S)}.$$

Definition 1.2.3. The *Whitney rank generating function* of Γ is defined as

$$R_\Gamma(x, y) = \sum_{S \subset E} x^{r(E)-r(S)} y^{|S|-r(S)}.$$

Thus
$$R_\Gamma(x-1, y-1) = T_\Gamma(x, y).$$

The Whitney rank generating polynomial may seem simpler to define than the Tutte polynomial, but the Tutte polynomial satisfies the nice duality formula
$$T_\Gamma(x, y) = T_{\Gamma^*}(y, x), \tag{1.7}$$

which relates the Tutte polynomial of any connected planar graph Γ with the Tutte polynomial of a dual graph Γ^*.

For example, recall from Exercise 1.8 that the n-cycle graph C_n and the banana graph with n edges are dual graphs. In Exercises 1.10 and 1.11, we will calculate the Tutte polynomials of a cycle graph and a banana graph and show that they satisfy Equation (1.7).

Example 1.2.4. For example, consider the complete graph on 4 vertices, K_4. There are 6 edges, so there are 2^6 subsets S of the edge set, with ranks given as follows:

- The empty set has $|S| = 0$ and $r(S) = 0$.

- There are 6 subsets with $|S| = 1$ and $r(S) = 1$.

- There are 15 subsets with $|S| = 2$ and $r(S) = 2$ (12 two-paths and 3 sets of two disconnected edges).

- There are 4 subsets with $|S| = 3$ and $r(S) = 2$ (the 4 three-cycles).

- There are 16 subsets with $|S| = 3$ and $r(S) = 3$ (12 three-paths and the 4 complements of the three-cycles).

- There are 15 subsets with $|S| = 4$ and $r(S) = 3$ (3 four-cycles and the 12 complements of the two-paths).

- There are 6 subsets with $|S| = 5$ and $r(S) = 3$.

- If S consists of all edges in K_4, $|S| = 6$ and $r(S) = 3$.

Thus
$$\begin{aligned}
T_{K_4}(x, y) &= (x-1)^3 + 6(x-1)^2 + 15(x-1) + 4(x-1)(y-1) \\
&\quad + 16 + 15(y-1) + 6(y-1)^2 + (y-1)^3 \\
&= x^3 + 3x^2 + 4xy + 2x + y^3 + 3y^2 + 2y.
\end{aligned}$$

Example 1.2.5. Computer calculations show that the Tutte polynomial of the cube graph on 8 vertices is

$$\begin{aligned}
T_{\text{cube}}(x, y) &= x^7 + 5x^6 + 15x^5 + 6x^4y + 29x^4 + 24x^3y + 40x^3 \\
&\quad + 12x^2y^2 + 52x^2y + 32x^2 + 8xy^3 + 39xy^2 + 46xy \\
&\quad + 11x + y^5 + 7y^4 + 20y^3 + 25y^2 + 11y.
\end{aligned}$$

We can see that calculating the Tutte polynomial by considering all subsets of $|E|$ would take a prohibitively long time for even moderately sized graphs. Fortunately, there are other methods. We will describe a method that uses only spanning subsets of E. This method uses a fixed ordering on the edges of E. It is not evident, a priori, that the polynomial generated by this method is independent of the ordering chosen, but it can be shown that it is equal to the Tutte polynomial. First, we give some definitions.

Definition 1.2.6. Let $\Gamma = (V, E)$ be a connected graph with a fixed ordering on E. Let T be a subset of E which forms a spanning tree of Γ. Let e be an edge in T. If we remove e from the tree T, the remaining graph has two components, with vertex sets V_1 and V_2, and with $V(\Gamma) = V_1 \cup V_2$. (It is possible that a component consists of a single vertex.) The *cut of* Γ *determined by* T *and* e consists of all edges of Γ from a vertex in V_1 to a vertex in V_2. The edge e is in the cut determined by T and e. We say that e is an *internally active edge* of T if it is the smallest edge in the cut, with respect to the given ordering on the edges of Γ. The number i of edges of T which are internally active in T is called the *internal activity* of T.

Definition 1.2.7. Let $\Gamma = (V, E)$ be a connected graph with a fixed ordering on E. Let T be a subset of E which forms a spanning tree of Γ. Let e be an edge which is not in T. If we add the edge e to T, we obtain a graph with exactly one cycle. We say that e is an *externally active edge* for T if it is the smallest edge in the cycle it determines, with respect to the given ordering on the edges of Γ. The number j of edges of $E \setminus T$ which are externally active for T is called the *external activity* of T.

The following theorem describes the Tutte polynomial in terms of spanning trees. The proof can be found in Bollobás [Bo98].

Theorem 1.2.8. *Let* $\Gamma = (V, E)$ *be a connected graph, with an ordering on the set of edges of* Γ. *Then the Tutte polynomial of* Γ *is given by*

$$T_\Gamma(x, y) = \sum_{i,j} t_{ij} x^i y^j,$$

where t_{ij} *denotes the number of spanning trees of internal activity* i *and external activity* j.

Corollary 1.2.9. *For a connected graph* Γ, $T_\Gamma(1, 1)$ *is the number of spanning trees of* Γ.

For example,

$$T_{K_4}(1, 1) = 16 \qquad \text{and} \qquad T_{\text{cube}}(1, 1) = 384.$$

Example 1.2.10. Consider the complete graph on 4 vertices K_4. Fix an ordering v_1, v_2, v_3, v_4 of the vertices and order the edges lexicographically: $e_1 = v_1v_2$, $e_2 = v_1v_3$, $e_3 = v_1v_4$, $e_4 = v_2v_3$, $e_5 = v_2v_4$, and $e_6 = v_3v_4$, as shown in Figure 1.15.

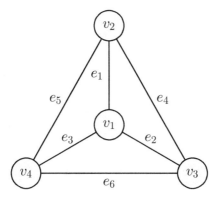

Figure 1.15: K_4 with an edge ordering.

A spanning tree must have exactly 3 edges. There are 20 ways to pick 3 edges from 6, but 4 of these, $\{e_1, e_2, e_4\}$, $\{e_1, e_3, e_5\}$, $\{e_2, e_3, e_6\}$, and $\{e_4, e_5, e_6\}$ give cycles, not spanning trees. This leaves 16 spanning trees.

As an example, consider the spanning tree $T = \{e_2, e_3, e_4\}$. First, we examine the edges in T to see which are internally active. Edge e_2 is not internally active, since e_1 is an edge in the cut determined by T and e_2. Edge e_3 is internally active, since e_3 is the smallest edge in the cut determined by T and e_3 (e_1 is not in this cut). Edge e_4 is not internally active, since e_1 is the smallest edge in the cut determined by T and e_2. Next, we examine the edges in $E \setminus T$ to see which are externally active. The edge e_1 must be externally active, since it is the smallest edge in any cycle of which it is a member. The edges e_5 and e_6 cannot be externally active in any cycle, since each cycle in K_4 has at least 3 edges, so must contain a smaller edge than e_5 or e_6.

We list the 16 spanning trees below, with the number i of internally active edges, and the number j of externally active edges for each.

- The tree $\{e_1, e_2, e_3\}$ has $i = 3$ and $j = 0$, so $t_{30} = 1$.

- The trees $\{e_1, e_2, e_5\}$, $\{e_1, e_2, e_6\}$, and $\{e_1, e_3, e_4\}$ have $i = 2$ and $j = 0$, so $t_{20} = 3$.

- The trees $\{e_1, e_4, e_5\}$ and $\{e_1, e_4, e_6\}$ have $i = 1$ and $j = 0$ so $t_{10} = 2$.

- The trees $\{e_1, e_3, e_6\}$, $\{e_1, e_5, e_6\}$, $\{e_2, e_3, e_4\}$, and $\{e_2, e_3, e_5\}$ have $i = 1$ and $j = 1$, so $t_{11} = 4$.

- The trees $\{e_2, e_4, e_5\}$ and $\{e_2, e_4, e_6\}$ have $i = 0$ and $j = 1$, so $t_{01} = 2$.

- The trees $\{e_2, e_5, e_6\}$, $\{e_3, e_4, e_5\}$, and $\{e_3, e_4, e_6\}$ have $i = 0$ and $j = 2$, so $t_{02} = 3$.

- The tree $\{e_3, e_5, e_6\}$ has $i = 0$ and $j = 3$, so $t_{03} = 3$.

Thus

$$T_{K_4}(x, y) = \sum_{i,j} t_{ij} x^i y^j$$
$$= x^3 + 3x^2 + 2x + 4xy + 2y + 3y^2 + y^3$$

as above.

Another approach to calculating the Tutte polynomial is to use reduction formulas under deletion and contraction of edges (see, e.g., Bollobás [Bo98]].

Several other evaluations of the Tutte polynomial are known (see D. Welsh [We99], for example).

Lemma 1.2.11. *The Tutte polynomial has the following additional properties.*

(a) $T_\Gamma(1, 2)$ *is the number of connected spanning subgraphs of* Γ,

(b) $T_\Gamma(2, 1)$ *is the number of spanning forests of* Γ,

(c) $T_\Gamma(2, 2)$ *is the number of spanning subgraphs of* Γ,

(d) $T_\Gamma(1 - k, 0)$ *is the number of proper k-vertex colorings of* Γ,

(e) $T_\Gamma(2, 0)$ *is the number of acyclic orientations of* Γ*, and*

(f) $T_\Gamma(0, -2)$ *is the number of Eulerian orientations of* Γ.

From Merino [Me04], it is known in general that $T_\Gamma(1, 0)$ represents the number of critical configurations of level 0 (discussed further in Chapter 4). Using Sage, it is easy to compute this quantity.

Example 1.2.12. Using Sage, we compute the Tutte polynomial of the cube graph and the tetrahedron graph.

```
                        ──── Sage ────

sage: Gamma = graphs.CubeGraph(3)
sage: Gamma.tutte_polynomial()
x^7 + 5*x^6 + 15*x^5 + 6*x^4*y + y^5 + 29*x^4 + 24*x^3*y + 12*x^2*y^2
 + 8*x*y^3 + 7*y^4 + 40*x^3 + 52*x^2*y + 39*x*y^2 + 20*y^3 + 32*x^2
 + 46*x*y + 25*y^2 + 11*x + 11*y
sage: Gamma.tutte_polynomial()(x=1,y=0)
133
sage: Gamma = graphs.TetrahedralGraph()
sage: Gamma.tutte_polynomial()
x^3 + y^3 + 3*x^2 + 4*x*y + 3*y^2 + 2*x + 2*y
sage: Gamma.tutte_polynomial()(x=1,y=0)
6
```

Exercise 1.9. Show that the Tutte polynomial of a tree with m edges is x^m.

Exercise 1.10. Show that the Tutte polynomial of the cycle graph C_n is

$$T_{C_n}(x, y) = x^{n-1} + x^{n-2} + \cdots x + y.$$

Exercise 1.11. Show that the Tutte polynomial of a banana graph Γ_n with n edges is

$$T_{\Gamma_n}(x, y) = x + y + y^2 + \cdots + y^{n-1},$$

i.e., $T_{\Gamma_n}(x, y) = T_{C_n}(y, x)$.

Exercise 1.12. Show that the Tutte polynomial of the diamond graph (K_4 with one edge removed) is $x^3 + 2x^2 + 2xy + x + y^2 + y$.

Exercise 1.13. Suppose that Γ is a connected graph with at least one edge and Γ_1 and Γ_2 are subgraphs of Γ with a single vertex (and no edges) in common, such that Γ is the union of Γ_1 and Γ_2. Show that

$$T_\Gamma(x, y) = T_{\Gamma_1}(x, y) T_{\Gamma_2}(x, y).$$

1.2.2 The Ihara zeta function

Recall the *circuit rank* $\chi(\Gamma)$ of an undirected graph Γ is the minimum number edges to remove from Γ, making it into a forest.

Lemma 1.2.13. The circuit rank is given by the simple formula

$$m - n + c,$$

where m is the size of Γ (i.e., number of edges), n is the order of Γ (i.e., the number of vertices), and c is the number of connected components. The circuit rank is equal to the dimension of the cycle space of Γ.

This is well known and the proof is omitted.
Let

$$p = (u_0, \ldots, u_{len(p)-1}, u_0)$$

be a cycle of the graph $\Gamma = (V, E)$ of length $len(p)$. We say p is a *prime cycle* if it is a closed cycle of Γ having no backtracks and is not a multiple of a cycle with a smaller number of edges.

The *Ihara zeta function* can be defined by the product formula:

$$\frac{1}{\zeta_\Gamma(u)} = \prod_p (1 - u^{len(p)}).$$

This product is taken over all prime cycles p of the graph[6].

It is known that for regular graphs the Ihara zeta function is a rational function.

Lemma 1.2.14. *(Ihara, Sunada) If Γ is k-regular with adjacency matrix A, then*

$$\zeta_\Gamma(u) = \frac{1}{(1 - u^2)^{\chi(\Gamma)-1} \det(I - Au + (k-1)u^2 I)},$$

where $\chi(\Gamma)$ is the circuit rank of Γ.

By Lemma 1.2.13, the *circuit rank* of Γ is given by the formula

$$\chi(\Gamma) = |E| - |V| + c,$$

where c denotes the number of connected components of Γ.

Example 1.2.15. For the cycle graph having 4 vertices, we have

$$\chi(\Gamma) = 4 - 4 + 1 = 1,$$

and

$$A = \begin{pmatrix} 0 & 1 & 0 & 1 \\ 1 & 0 & 1 & 0 \\ 0 & 1 & 0 & 1 \\ 1 & 0 & 1 & 0 \end{pmatrix},$$

so

$$\zeta_\Gamma(u)^{-1} = \det(I - Au + (2-1)u^2 I)$$

$$= \det \begin{pmatrix} u^2 & -u & 0 & -u \\ -u & u^2 & -u & 0 \\ 0 & -u & u^2 & -u \\ -u & 0 & -u & u^2 \end{pmatrix}$$

$$= (u^2 + 1)^2 (u + 1)^2 (u - 1)^2.$$

---

[6]This formulation in a graph-theoretic setting is actually due to Sunada [Su86], rather than Ihara.

Here is some Sage code verifying this.

```
                                   ─── Sage ───
sage: t = var('t')
sage: Gamma = graphs.CycleGraph(4)
sage: A = Gamma.adjacency_matrix()
sage: I4 = identity_matrix(4)
sage: factor(det(I4 - t*A + (2-1)*t^2*I4))
(t^2 + 1)^2*(t + 1)^2*(t - 1)^2
sage: Gamma.ihara_zeta_function_inverse()
t^8 - 2*t^4 + 1
```

Note these last two outputs agree.

Indeed, there is this generalization (see Terras [Te10]).

Proposition 1.2.16. *(Hashimoto, Bass) Assume $\Gamma = (V, E)$ is a connected graph having no vertices of degree 1 and $|E| \geq |V|$. Let A denote the adjacency matrix of Γ and let Δ_0 denote the $|V| \times |V|$ diagonal matrix whose entries are the values of $\deg(v) - 1$, $v \in V$. Then*

$$\zeta_\Gamma(u) = \frac{1}{(1 - u^2)^{\chi(\Gamma)-1} \det(I - Au + \Delta_0 u^2 I)}.$$

Exercise 1.14. Show that for a graph Γ that is a square with one diagonal, we have

$$\zeta_\Gamma(u)^{-1} = (1 - u^2)(1 - u)(1 + u^2)(1 + u + 2u^2)(1 - u^2 - 2u^3).$$

1.2.3 The Duursma zeta function

In a fascinating series of papers starting in the 1990s, Iwan Duursma introduced a new class of zeta functions. They are intended to be for linear codes, but they were implicitly introduced for other objects as well.

The Duursma zeta function of a code

Before we define the Duursma zeta function of a graph, we introduce the Duursma zeta function of a linear block code, following Duursma [Duu01], [Duu03a], [Duu03b], and [Duu04]. For more on codes, see Biggs [Bi08].

Let q denote a prime power. A *linear (error-correcting) code* C over $GF(q)$ is a subspace of $GF(q)^n$, where $GF(q)^n$ is provided with the standard basis. The parameter n is called the *length* of the code C. The *Hamming weight* of $v \in GF(q)^n$,

$$wt(v) = |\{i \mid v_i \neq 0\}|,$$

is the number of nonzero coordinates of $v = (v_1, \ldots, v_n)$. The *weight enumerator* of C is

$$A_C(x, y) = \sum_{i=0}^{n} A_i x^i y^{n-i},$$

where A_i is the number of codewords in C of weight i. The minimum value on $C - \{0\}$,

$$d_C = \min_{c \in C, c \neq 0} wt(c),$$

is called the *minimum distance* of the code C.

Let C be an $[n, k, d]_q$ code, i.e., a linear code over $GF(q)$ of length n, dimension k (as a vector space over $GF(q)$), and minimum distance $d = d_C$. The *Singleton bound* states that $k + d \leq n + 1$ (see for example [HP10] for a proof). If equality is satisfied in this bound, then we call C a *minimum distance separable* or *MDS* code.

Motivated by analogies with local class field theory, in [Duu04] Iwan Duursma introduced the *zeta function* $Z = Z_C$ associated to a linear code C over a finite field,

$$Z(T) = \frac{P(T)}{(1 - T)(1 - qT)}, \tag{1.8}$$

where $P(T)$ is a polynomial of degree $n + 2 - d - d^{\perp}$ (the Duursma zeta polynomial defined below)[7].

If $C^{\perp}$ denotes the dual code of C, with parameters $[n, n - k, d^{\perp}]$, then the *MacWilliams identity* relates the weight enumerator of $C^{\perp}$ to that of C:

$$A_{C^{\perp}}(x, y) = |C|^{-1} A_C(x + (q - 1)y, x - y).$$

Definition 1.2.17. A polynomial $P(T)$ for which

$$\frac{(xT + (1 - T)y)^n}{(1 - T)(1 - qT)} P(T) = \cdots + \frac{A_C(x, y) - x^n}{q - 1} T^{n-d} + \cdots .$$

is called a *Duursma zeta polynomial of C*.

The *Duursma zeta function* is defined in terms of the zeta polynomial by means of (1.8) above.

Lemma 1.2.18. *If we expand* $\frac{(xT + y(1 - T))^n}{(1 - T)(1 - qT)}$ *in powers of T, we find it is equal to*

---

[7]In general, if C is an $[n, k, d]$-code, then we use $[n, k^{\perp}, d^{\perp}]$ for the parameters of the dual code, $C^{\perp}$. It is a consequence of Singleton's bound that $n + 2 - d - d^{\perp} \geq 0$, with equality when C is an MDS code.

$$b_{0,0}y^n T^0 + (b_{1,1}xy^{n-1} + b_{1,0}y^n)T^1$$
$$+ (b_{2,2}x^2 y^{n-2} + b_{2,1}xy^{n-1} + b_{2,0}y^n)T^2 + \cdots$$
$$+ (b_{n-d,n-d}x^{n-d}y^d + b_{n-d,n-d-1}x^{n-d-1}y^{d+1} + \cdots + b_{n-d,0}y^n)T^{n-d} + \cdots ,$$

where $b_{i,j}$ are coefficients given by

$$b_{k,\ell} = \sum_{i=\ell}^{k} \frac{q^{k-i+1-1} - 1}{q-1} \binom{n}{i}\binom{i}{\ell}.$$

Proof. Define c_j by

$$\frac{(xT + (1-T)y)^n}{(1-T)(1-qT)} = \sum_{k=0}^{\infty} c_k(x,y)T^k.$$

It is not hard to see that

$$\frac{1}{(1-T)(1-qT)} = \sum_{j=0}^{\infty} \frac{q^{j+1} - 1}{q-1}T^j,$$

and of course

$$(xT + (1-T)y)^n = \sum_{i=0}^{n} \binom{n}{i} y^{n-i}(x-y)^i T^i.$$

Therefore,

$$c_k(x,y) = \sum_{i+j-k} \frac{q^{j+1} - 1}{q-1} \binom{n}{i} y^{n-i}(x-y)^i. \tag{1.9}$$

To finish the proof, use Equation (1.9) above and compare coefficients. $\square$

Proposition 1.2.19. *The Duursma zeta polynomial $P = P_C$ exists and is unique, provided $d^\perp \geq 2$.*

For the proof, see Proposition 96 in Joyner and Kim [JK11].

The Duursma zeta function of a graph

This section will explore some of the properties of the analogous zeta function for graphs and give examples using the software package Sage.

Let $\Gamma = (V, E)$ be a graph. One way to define the zeta function of Γ is via the binary code defined by the Duursma zeta function of the cycle space of Γ over $GF(2)$. Indeed, the matroid attached to a graph is typically that vector matroid attached to the (oriented) incidence matrix of the graph (see Duursma [Duu04]).

As is well known (see for example, Dankelmann, Key, and Rodrigues [DKR13]), the binary code $B = B_\Gamma$ generated by the incidence matrix of Γ is the cocycle space of Γ over $GF(2)$, and the dual code $B^\perp$ is the cycle space $Z = Z_\Gamma$ of Γ:

$$B_\Gamma^\perp = Z_\Gamma.$$

Proposition 1.2.20. Let $F = GF(2)$. The cycle code of $\Gamma = (V, E)$ is a linear binary block code of length $|E|$, dimension $g(\Gamma)$ (namely, the genus of Γ), and minimum distance equal to the girth of Γ. If $C \subset GF(2)^{|E|}$ is the cycle code associated to Γ and C^* is the cocycle code associated to Γ, then C^* is the dual code of C. In particular, the cocycle code of Γ is a linear binary block code of length $|E|$, and dimension $r(\Gamma) = |E| - n(\Gamma)$.

This follows from Hakimi and Bredeson [HB68] (see also Jungnickel and Vanstone [JV10]) in the binary case[8].

Proof. Let d denote the minimum distance of the code C. Let g denote the girth of Γ, i.e., the smallest cardinality of a cycle in Γ. If K is a cycle in Γ, then the vector $\mathrm{vec}(K) \in GF(2)^{|E|}$ is an element of the cycle code $C \subset GF(2)^{|E|}$. This implies $d \leq g$.

In the other direction, suppose K_1 and K_2 are cycles in Γ with associated support vectors $v_1 = \mathrm{vec}(K_1)$, $v_2 = \mathrm{vec}(K_2)$. Assume that at least one of these cycles is a cycle of minimum length, say K_1, so the weight of its corresponding support vector is equal to the girth g. The only way that $wt(v_1 + v_2) < \min\{wt(v_1), wt(v_2)\}$ can occur is if K_1 and K_2 have some edges in common. In this case, the vector $v_1 + v_2$ represents a subgraph which is either a cycle or a union of disjoint cycles. In either case, by minimality of K_1, these new cycles must be at least as long. Therefore, $d \geq g$, as desired.
□

Definition 1.2.21. Define p_Γ to be the *Duursma zeta polynomial* $P = P_\mathcal{B}$ of Γ, in the notation of Proposition 1.2.19, where $\mathcal{B} = \mathcal{B}_\Gamma$ is the cocycle space of Γ. Define the *Duursma zeta function* of Γ by

$$\zeta_\Gamma(t) = \frac{p_\Gamma(t)}{(1-t)(1-2t)}. \tag{1.10}$$

Exercise 1.15. Check that the Duursma zeta polynomial of the cycle graph Γ_4 with 4 vertices is

$$p_{\Gamma_4}(t) = \frac{2}{5}t^2 + \frac{2}{5}t + \frac{1}{5}.$$

Example 1.2.22. The binary code given by the cycle space of the wheel graph W_5 with 5 vertices has zeta polynomial

---
[8]It is likely true in the nonbinary case as well, but no proof seems to be in the literature.

$$\frac{2}{7}t^4 + \frac{2}{7}t^3 + \frac{3}{14}t^2 + \frac{1}{7}t + \frac{2}{14}.$$

The roots lie on the circle of radius $1/\sqrt{2}$. The binary code given by the cycle space of the wheel graph W_6 with 6 vertices has zeta polynomial

$$\frac{1}{3}t^6 + \frac{11}{42}t^5 + \frac{1}{7}t^4 + \frac{1}{12}t^3 + \frac{1}{14}t^2 + \frac{11}{168}t + \frac{1}{24}.$$

Remark 1.2.23. It is well known that the cycle space of a planar graph Γ is the cocycle space of its dual graph, $\Gamma^\perp$, and vice versa.

The (Duursma) *genus* of the cocycle space is the quantity

$$\gamma = \gamma_\Gamma = m + 1 - \mathrm{rank}(B) - d_\Gamma,$$

where $m = |E|$ denotes the number of edges in Γ (the number of columns of the incidence matrix of Γ), B is the incidence matrix over $GF(2)$, and d_Γ denotes the minimum distance of the cocycle space $\mathcal{B}_{GF(2)}$ (as a binary linear code over $GF(2)$ in the standard basis). It's well known that for connected Γ,

$$\mathrm{rank}(B) = |V| - 1,$$

and it's known[9], that

$$d_\Gamma = \lambda(\Gamma),$$

where $\lambda(\Gamma)$ denotes the edge connectivity of Γ, i.e., the minimum cardinality of an edge cut. Therefore,

$$\gamma = |E| + 2 - |V| - \lambda(\Gamma).$$

In any case, define the *normalized Duursma zeta function* by

$$z_\Gamma(t) = t^{1-\gamma}\zeta_\Gamma(t).$$

If Γ is a planar graph, and if F denotes the set of faces of Γ, then the formula for the Euler characteristic gives us

$$|E| + 2 - |V| = |F|,$$

so that in this case, $\gamma = |F| - \lambda(\Gamma)$. If Γ is a planar graph, then we also have the functional equation

$$z_{\Gamma^\perp}(t) = z_\Gamma(1/2t), \qquad (1.11)$$

---

[9]See Theorem 1 in Dankelmann, Key, and Rodrigues [DKR13].

where $\Gamma^{\perp}$ denotes the planar dual graph of Γ.

The following question asks when an analog of the Riemann hypothesis holds.

Open Question 1. *For which planar graphs Γ satisfying $\Gamma^{\perp} \cong \Gamma$, are all the roots of the Duursma zeta polynomial on the circle $|t| = 2^{-1/2}$?*

1.2.4 Graph theory and mathematical blackjack

In this section, we illustrate a connection between graph theory, (set-theoretic) combinatorics, and a combinatorial game called mathematical blackjack.

An *m-(sub)set* is a (sub)set with m elements. For integers $k < m < n$, a *Steiner system* $S(k, m, n)$ is an n-set X and a set S of m-subsets having the property that any k-subset of X is contained in exactly one m-set in S.

Example 1.2.24. Let

$$\mathbf{P}^1(GF(11)) = \{\infty, 0, 1, 2, \ldots, 9, 10\}$$

denote the projective line over the finite field $GF(11)$ with 11 elements. Let

$$Q = \{0, 1, 3, 4, 5, 9\}$$

denote the quadratic residues and 0 and let

$$L = \langle \alpha, \beta \rangle \cong PSL(2, GF(11)),$$

where $\alpha(y) = y + 1$ and $\beta(y) = -1/y$. Let

$$S = \{\lambda(Q) \mid \lambda \in L\}.$$

In other words, each element of S is a subset of $\mathbf{P}^1(GF(11))$ of size 6.

Lemma 1.2.25. *S is a Steiner system of type $(5, 6, 12)$.*

If $X = \{1, 2, \ldots, 12\}$, a Steiner system $S(5, 6, 12)$ is a set of 6-sets, called *hexads*, with the property that any set of 5 elements of X is contained in ("can be completed to") exactly one hexad.

A *decomposition* of a graph Γ is a set of subgraphs $H_1, \ldots, H_k$ that partition the edges of Γ. If all the H_i are isomorphic to a given group H, then we say the decomposition is an H decomposition. In chapter 6, we shall run into graph decompositions in our discussion of the Cayley graph of a p-ary function.

From the perspective of Steiner systems, if H is any connected subgraph of K_n (the complete graph on n vertices), an H-decomposition of K_n gives rise to a $S(k, m, n)$, where m is the number of vertices of H and k is the smallest number such that any k vertices of H must contain at least 2 neighbors.

Question 1.1. *Is the Steiner system $S(5, 6, 12)$ described above associated to a graph decomposition of K_{12}?*

The MINIMOG description

This section is devoted to Conway's [Co84] construction of $S(5, 6, 12)$ using a 3×4 array called the MINIMOG.

The *tetracode words* are

0	0	0	0	0	+	+	+	0	-	-	-
+	0	+	-	+	+	-	0	+	-	0	+
-	0	-	+	-	+	0	-	-	-	+	0

With "0" $= 0$, "+" $= 1$, "-" $= 2$, these vectors form a linear code over $GF(3)$. (This notation is Conway's. One must remember here that "+" + "+" = "-" and "-" + "-" = "+"!) They may also be described as the set of all 4-tuples in $GF(3)$ of the form

$$(0, a, a, a), \quad (1, a, b, c), \quad (2, c, b, a),$$

where abc is any cyclic permutation of 012.

The *MINIMOG in the shuffle numbering* is the 3×4 array

$$
\begin{array}{cccc}
6 & 3 & 0 & 9 \\
5 & 2 & 7 & 10 \\
4 & 1 & 8 & 11
\end{array}
$$

We label the rows as follows:

- the first row has label 1,
- the second row has label $+$, and
- the third row has label $-$.

0	6	3	0	9
+	5	2	7	10
-	4	1	8	11

A *col* (or column) is a placement of three + signs in a column of the array:

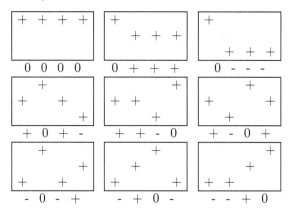

A *tet* (or tetrad) is a placement of 4 + signs having entries corresponding (as explained below) to a tetracode.

Each line in $GF(3)^2$ with finite slope occurs once in the 3×3 part of some tet. The *odd man out* for a column is the label of the row corresponding to the nonzero digit in that column; if the column has no nonzero digit, then the odd man out is a "?". Thus the tetracode words associated in this way to these patterns are the odd men out for the tets.

The *signed hexads* are the combinations 6-sets obtained from the MINI-MOG from patterns of the form

$$\text{col-col, col+tet, tet-tet, col+col-tet.}$$

Lemma 1.2.26. *(Conway [CS99], Chapter 11, p. 321) If we ignore signs, then from these signed hexads we get the 132 hexads of a Steiner system $S(5,6,12)$. These are all possible 6-sets in the shuffle labeling for which the odd men out form a part (in the sense that an odd man out "?" is ignored or regarded as a "wild-card") of a tetracode word and the column distribution is not $0, 1, 2, 3$ in any order[10].*

Example 1.2.27. Associated to the col-col pattern

+		+		+	-
+	-	+	=	+	-
+		+		+	-

---

[10] That is to say, the following cannot occur: some column has 0 entries, some column has exactly 1 entry, some column has exactly 2 entries, and some column has exactly 3 entries.

is the tetracode 0 0 ? ? and the signed hexad $\{-1, -2, -3, 4, 5, 6\}$ and the hexad $\{1, 2, 3, 4, 5, 6\}$.

Associated to the col+tet pattern

$$
\begin{array}{|c|}
\hline
+ \\
+ \\
+ \\
\hline
\end{array}
\;-\;
\begin{array}{|c|}
\hline
+ \\
+\ +\ + \\
\\
\hline
\end{array}
\;=\;
\begin{array}{|c|}
\hline
+\ + \\
-\ +\ + \\
+ \\
\hline
\end{array}
$$

is the tetracode $0\ +\ +\ +$ and the signed hexad $\{1, -2, 3, 6, 7, 10\}$ and the hexad $\{1, 2, 3, 6, 7, 10\}$.

Furthermore, it is known (see Conway [Co84]) that the Steiner system $S(5, 6, 12)$ in the shuffle labeling has the following properties.

- There are 11 hexads with total 21 and none with lower total.

- The complement of any of these 11 hexads in $\{0, 1, \ldots, 11\}$ is another hexad.

- There are 11 hexads with total 45 and none with higher total.

Mathematical blackjack

Mathematical blackjack is a 2-person combinatorial game whose rules will be described below. What is remarkable about it is that a winning strategy, discovered by Conway and Ryba (see Conway and Sloane [CS86] and Kahane and Ryba [KR01]), depends on knowing how to determine hexads in the Steiner system $S(5, 6, 12)$.

Mathematical blackjack is played with 12 cards, labeled $0, \ldots, 11$ (for example, *king, ace*, 2, 3, $\ldots$, 10, *jack*, where the *king* is 0 and the *jack* is 11). Divide the 12 cards into two piles of 6 (to be fair, this should be done randomly). Each of the 6 cards of one of these piles is to be placed face up on the table. The remaining cards are in a stack which is shared and visible to both players. If the sum of the cards face up on the table is less than or equal to 21, then no legal move is possible[11] so you must shuffle the cards and deal a new game.

- Players alternate moves.

- A move consists of exchanging a card on the table with a lower card from the other pile.

- The player whose move makes the sum of the cards on the table under 21 loses.

The winning strategy (given below) for this game is due to Conway and Ryba (see [CS86] and [KR01]). There is a Steiner system $S(5, 6, 12)$ of hexads

---

[11] Conway [Co76] calls such a game $0 = \{\,|\,\}$; in this game, the first player automatically loses and so a good player will courteously offer you the first move!

in the set $\{0, 1, \ldots, 11\}$. This Steiner system is associated to the MINIMOG in the shuffle numbering.

Proposition 1.2.28. *(Ryba) For this Steiner system, the winning strategy is to choose a move which is a hexad from this system.*

This result is proven in Kahane and Ryba [KR01].

If you are unfortunate enough to be the first player starting with a hexad from $S(5, 6, 12)$ then, according to this strategy and properties of Steiner systems, there is no winning move. In a randomly dealt game, there is a probability of

$$\frac{132}{\binom{12}{6}} = 1/7$$

that the first player will be dealt such a hexad, hence a losing position. In other words, we have the following result.

Lemma 1.2.29. *The probability that the first player has a win in mathematical blackjack (with a random initial deal) is $6/7$.*

Example 1.2.30. We play a game.

- Initial deal: $0, 2, 4, 6, 7, 11$. The total is 30. The pattern for this deal is

  where $\bullet$ is a $\pm$. No combinations of $\pm$ choices will yield a tetracode odd men out, so this deal is not a hexad.

- First player replaces 7 by 5: $0, 2, 4, 5, 6, 11$. The total is now 28. (Note this is a square in the picture at 1.) This corresponds to the col+tet

+		+
+	+	
-		+

  with tetracode odd men out $-\ +\ 0\ -$.

- Second player replaces 11 by 7: $0, 2, 4, 5, 6, 7$. The total is now 24. Interestingly, this 6-set corresponds to the pattern

  (hence possible with tetracode odd men out $0\ +\ +\ ?$, for example). However, it has column distribution $3, 1, 2, 0$, so it cannot be a hexad.

- First player replaces 6 by 3: $0, 2, 3, 4, 5, 7$. (Note this is a cross in the picture at 0.) This corresponds to the tet-tet pattern

	+	-
+	-	+
-		

with tetracode odd men out $- - + 0$. Cards total 21. First player wins.

In addition to the references mentioned above, see also Curtis [Cu76] and [Cu84] for further reading related to mathematical blackjack.

Chapter 2
Graphs and Laplacians

2.1 Motivation

In this chapter, we are interested in exploring questions such as the following.

Question 2.1. *If a group G acts on a graph Γ, what is the relationship between the spectrum of Γ and the spectrum of the quotient Γ/G?*

If G is a group acting on graphs Γ_1 and Γ_2, then a G-equivariant map $\Gamma_1 \to \Gamma_2$ is a map that respects the action of G on these graphs.

Question 2.2. *If G is a group acting on graphs Γ_1 and Γ_2, and if there is a G-equivariant map $\Gamma_1 \to \Gamma_2$, how are the Laplacians of Γ_1 and Γ_2 related?*

2.2 Basic results

Let's start with some motivation for the definition of the Laplacian.

Recall that if $\Gamma = (V, E)$ is a graph and F is a ring, then $C^0(\Gamma, F)$ is the set of all F-valued functions on the vertex set V of Γ, and $C^1(\Gamma, F)$ is the set of all F-valued functions on the edge set E of Γ. If the graph Γ is a large square lattice grid, and if $f \in C^0(\Gamma, F)$, then the usual definition of the Laplacian,

$$\frac{\partial^2 f}{\partial x^2} + \frac{\partial^2 f}{\partial y^2},$$

corresponds to the discrete *Laplacian Q on f*

$$(Qf)(v) = \sum_{w:d(w,v)=1} [f(w) - f(v)] \tag{2.1}$$

© Springer International Publishing AG 2017
W.D. Joyner and C.G. Melles, *Adventures in Graph Theory*,
Applied and Numerical Harmonic Analysis,
https://doi.org/10.1007/978-3-319-68383-6_2

where $d(w, v)$ is the graph distance function on $V \times V$. Indeed,

$$\frac{\partial^2 f}{\partial x^2} = \lim_{\epsilon \to 0} \frac{[f(x + \epsilon, y) - f(x, y)] + [f(x - \epsilon, y) - f(x, y)]}{\epsilon^2}$$

and

$$\frac{\partial^2 f}{\partial y^2} = \lim_{\epsilon \to 0} \frac{[f(x, y + \epsilon) - f(x, y)] + [f(x, y - \epsilon) - f(x, y)]}{\epsilon^2}$$

so taking $\epsilon = 1$ gives the desired discrete analog of $f_{xx} + f_{yy}$ on the grid graph. This operator only depends on "local" properties. That is, $(Qf)(v)$ depends only on the neighbors of v in the graph Γ. It may come as a surprise to find out that Q governs a number of "global properties" of Γ as well, such as connectivity. We shall see these and other fascinating properties of Q below.

Recall from §1.1 that, given an orientation on Γ, there is a linear transformation

$$B : C^1(\Gamma, F) \to C^0(\Gamma, F)$$

given by

$$(Bf)(v) = \sum_{h(e)=v} f(e) - \sum_{t(e)=v} f(e) \tag{2.2}$$

and a dual linear transformation

$$B^* : C^0(\Gamma, F) \to C^1(\Gamma, F).$$

Recall also (see Lemma 1.1.20) that there are natural bases for the spaces $C^0(\Gamma, F)$ and $C^1(\Gamma, F)$. If we choose orderings of the vertices and edges of Γ, then the linear transformation B is given with respect to these bases by the incidence matrix, which we also denote by B. The dual B^* of this linear transformation is given by the transpose B^t of the incidence matrix.

The *vertex Laplacian* (or simply "the Laplacian") is the linear transformation $Q = Q_\Gamma : C^0(\Gamma, F) \to C^0(\Gamma, F)$ defined by

$$Q = BB^*, \tag{2.3}$$

where B is the linear transformation of Equation (2.2) above, and B^* is its dual.

The matrix representation of the Laplacian will also be denoted Q. We leave it as an exercise to show that the linear transformation $Q = BB^*$ is independent of the orientation chosen on Γ.

Exercise 2.1. Suppose that Γ is given an orientation, and denote by B the $n \times m$ incidence matrix of Γ with respect to this orientation (and some order-ings of the vertex and edge sets of Γ). Denote by Q the matrix representation of the Laplacian. Show that the matrix BB^t is independent of the orientation chosen and that $BB^t = Q$.

Recall that if the graph Γ has vertex set

$$V = V_\Gamma = \{0, 1, 2, \ldots, n-1\},$$

then the (undirected, unweighted) adjacency matrix of Γ is the $n \times n$ matrix $A = (a_{ij})$, where $a_{ij} = 1$ if vertex i shares an edge with vertex j, and $a_{ij} = 0$ otherwise. There is a simple connection between the Laplacian and the adjacency matrix.

Lemma 2.2.1. *For an oriented graph Γ with unsigned adjacency matrix A, there is a natural basis of $C^0(\Gamma, F)$ for which the matrix representation of the Laplacian is given by*

$$Q = \Delta - A,$$

where Δ denotes the diagonal matrix of the degrees of the vertices of V:

$$\Delta = \begin{pmatrix} d_0 & 0 & 0 & \cdots & 0 \\ 0 & d_1 & 0 & \cdots & 0 \\ 0 & 0 & d_2 & \cdots & 0 \\ \vdots & \vdots & \vdots & \ddots & \vdots \\ 0 & 0 & 0 & \cdots & d_{n-1} \end{pmatrix},$$

where $d_i = \deg_\Gamma(i)$, for $i \in V$.

Proof. Let $Q = (Q_{ij})$. Since $Q = BB^t$, where $B = (b_{ij})$ is the incidence matrix, Q_{ii} is the inner product of the i-th row of the incidence matrix with itself. This simply counts the number of edges incident to vertex i, so $Q_{ii} = \deg_\Gamma(i)$. If $i \neq j$ then Q_{ij} is the inner product of the i-th row of the incidence matrix with the j-th row. This is nonzero if i and j are both incident to the same edge, and zero if they are not. Indeed, if i and j are both incident to edge k then either

$$b_{ik} = 1, \quad b_{jk} = -1, \quad b_{\ell,k} = 0 \text{ for all } \ell \neq i, j,$$

or

$$b_{ik} = -1, \quad b_{jk} = 1, \quad b_{\ell,k} = 0 \text{ for all } \ell \neq i, j.$$

In either case, the k-th entry of the i-th row times the k-th entry of the j-th row equals -1. Summing over all edges k gives $Q_{ij} = -a_{ij}$. $\square$

Example 2.2.2. Consider the graph in Figure 2.1.

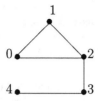

Figure 2.1: A graph with 5 vertices.

This graph has Laplacian matrix

$$Q = BB^t = \begin{pmatrix} 2 & -1 & -1 & 0 & 0 \\ -1 & 2 & -1 & 0 & 0 \\ -1 & -1 & 3 & -1 & 0 \\ 0 & 0 & -1 & 2 & -1 \\ 0 & 0 & 0 & -1 & 1 \end{pmatrix}.$$

Example 2.2.3. Consider the 3×3 grid graph in Figure 2.2.
Sage can be used to calculate the Laplacian matrix of this graph.

```
─────────────────────────── Sage ───────────────────────────
sage: Gamma = graphs.GridGraph([3,3])
    ## this is the 3x3 grid graph with 9 vertices
sage: B = incidence_matrix(Gamma, 12*[-1])
sage: B
[-1 -1  0  0  0  0  0  0  0  0  0  0]
[ 0  1 -1 -1  0  0  0  0  0  0  0  0]
[ 0  0  0  1 -1  0  0  0  0  0  0  0]
[ 1  0  0  0  0 -1 -1  0  0  0  0  0]
[ 0  0  1  0  0  0  1 -1 -1  0  0  0]
[ 0  0  0  0  1  0  0  0  1 -1  0  0]
[ 0  0  0  0  0  1  0  0  0  0 -1  0]
[ 0  0  0  0  0  0  0  1  0  0  1 -1]
[ 0  0  0  0  0  0  0  0  0  1  0  1]
```

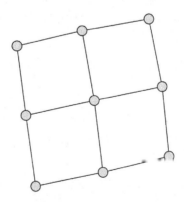

Figure 2.2: The 3×3 grid graph.

```
sage: B*transpose(B)
[ 2 -1  0 -1  0  0  0  0  0]
[-1  3 -1  0 -1  0  0  0  0]
[ 0 -1  2  0  0 -1  0  0  0]
[-1  0  0  3 -1  0 -1  0  0]
[ 0 -1  0 -1  4 -1  0 -1  0]
[ 0  0 -1  0 -1  3  0  0 -1]
[ 0  0  0 -1  0  0  2 -1  0]
[ 0  0  0  0 -1  0 -1  3 -1]
[ 0  0  0  0  0 -1  0 -1  2]
sage: Gamma.laplacian_matrix()
[ 2 -1  0 -1  0  0  0  0  0]
[-1  3 -1  0 -1  0  0  0  0]
[ 0 -1  2  0  0 -1  0  0  0]
[-1  0  0  3 -1  0 -1  0  0]
[ 0 -1  0 -1  4 -1  0 -1  0]
[ 0  0 -1  0 -1  3  0  0 -1]
[ 0  0  0 -1  0  0  2 -1  0]
[ 0  0  0  0 -1  0 -1  3 -1]
[ 0  0  0  0  0 -1  0 -1  2]
```

The "4" in the center of the Laplacian matrix illustrates the fact that there are four edges emanating from the central vertex of the 3×3 grid graph in Figure 2.2.

Lemma 2.2.4. *For any vector* $\mathbf{x} = (x_0, \ldots, x_{n-1})$, *we have*
$$\mathbf{x}^t Q \mathbf{x} = \sum_{(i,j) \in E, j > i} (x_i - x_j)^2.$$

In other words, the quantity $\mathbf{x}^t Q \mathbf{x}$ measures how far away the vector $\mathbf{x}$ is from being constant.

Proof. Indeed,
$$\mathbf{x}^t Q \mathbf{x} = \mathbf{x}^t B B^t \mathbf{x} = (B^t \mathbf{x}) \cdot (B^t \mathbf{x}) = \sum_{(i,j) \in E, j > i} (x_i - x_j)^2.$$

$\square$

Example 2.2.5. Consider the tetrahedral graph, Γ. Using Sage, it is easy to verify that the identity in Lemma 2.2.4 holds in this example.

```
                              Sage
sage: Gamma = graphs.TetrahedralGraph()
sage: Q = Gamma.laplacian_matrix(); Q
[ 3 -1 -1 -1]
[-1  3 -1 -1]
[-1 -1  3 -1]
[-1 -1 -1  3]
sage: x0,x1,x2,x3 = var("x0,x1,x2,x3")
sage: x = vector(SR, [x0,x1,x2,x3])
sage: expand(x.dot_product(Q*x))
3*x0^2 - 2*x0*x1 + 3*x1^2 - 2*x0*x2 - 2*x1*x2 + 3*x2^2 - 2*x0*x3
  - 2*x1*x3 - 2*x2*x3 + 3*x3^2
sage: expand(x.dot_product(Q*x))-expand((x1-x0)^2+(x2-x1)^2+(x3-x2)^2
  +(x2-x0)^2+(x3-x0)^2+(x3-x1)^2)
0
```

For a given orientation of Γ, the *edge Laplacian* is the linear transformation $Q_e = Q_{e,\Gamma} : C^1(\Gamma, F) \to C^1(\Gamma, F)$ defined by

$$Q_e = B^* B, \tag{2.4}$$

where B is the linear transformation of Equation (2.2) and B^* is its dual. Unlike the vertex Laplacian, Q_e depends on the orientation.

The following proposition describes the kernel of the Laplacian matrix of a connected graph.

Proposition 2.2.6. *If Γ is a connected graph, the kernel of the Laplacian matrix Q consists of all multiples of the all 1's vector $\mathbf{1} = (1, 1, \ldots, 1)$, i.e., $\mathbf{1}$ is an eigenvector of Q corresponding to the eigenvalue 0, and the eigenspace of 0 is 1-dimensional.*

Proof. Each row of B^t contains 1 once and -1 once and all other entries of the row are 0. Thus $B^t \mathbf{1} = \mathbf{0}$, the zero vector. Furthermore, if x is a vector in the kernel of BB^t, then $x^t BB^t x = \mathbf{0}$, so $B^t x = \mathbf{0}$. But if x is in the kernel of B^t, then x takes the same value on the head and tail vertices of each edge. Since Γ is assumed to be connected, x must take the same value on all vertices of Γ. $\qquad\square$

Corollary 2.2.7. *If Γ is a connected graph, the rank of the Laplacian matrix Q is $n - 1$, where n is the number of vertices of Γ.*

Definition 2.2.8. The *spectrum* of a graph Γ is the multi-set of eigenvalues of the (unsigned) adjacency matrix $A = A_\Gamma$. We sometimes denote the spectrum of Γ by $\sigma(\Gamma)$. The *Laplacian spectrum* of a graph Γ is the set of eigenvalues of the Laplacian matrix $Q = Q_\Gamma$. The *characteristic polynomial* p_Γ of a graph Γ is the characteristic polynomial of A.

Let Γ be a simple connected graph. If Γ has n vertices, let

$$\lambda_0 = 0 \leq \lambda_1 \leq \cdots \leq \lambda_{n-1} \tag{2.5}$$

denote the eigenvalues of Q. By Lemma 2.2.4, Q is positive semidefinite. This implies $\lambda_i(Q) \geq 0$ for all i. Consequently, $\lambda_0 = 0$, because the vector $\mathbf{v}_0 = (1, 1, \ldots, 1)$ satisfies $Q\mathbf{v}_0 = \mathbf{0}$.

Lemma 2.2.9. *If Γ is a k-regular graph, and the Laplacian Q of Γ has eigenvalues $\lambda_0, \lambda_1, \ldots, \lambda_{n-1}$, then the adjacency matrix A of Γ has eigenvalues $k - \lambda_0, k - \lambda_1, \ldots, k - \lambda_{n-1}$.*

Proof. This is an immediate corollary of Lemma 2.2.1. $\qquad\square$

Recall, the diameter of the graph Γ is the maximum distance between any two vertices of Γ.

Lemma 2.2.10. *Let Γ be a simple connected graph with n vertices and diameter d. Then*

$$\lambda_1 \geq (dn)^{-1},$$

where λ_1 is as in Equation (2.5).

Proof. Since Q is symmetric, $\mathbb{R}^n$ has an orthonormal basis of eigenvectors of Q.

It is a consequence of well-known facts from linear algebra[1] that we have

$$\lambda_1 = \min_{\mathbf{v},\ \mathbf{v}\cdot\mathbf{1}=0} \frac{\mathbf{v}\cdot Q\mathbf{v}}{\mathbf{v}\cdot\mathbf{v}}.$$

Let $\mathbf{f} = (f_0, f_1, \ldots, f_{n-1})$ denote an eigenvector of Q satisfying

$$\lambda_1 = \frac{\sum_{(i,j)\in E}(f_j - f_i)^2}{\sum_{i\in V} f_i^2}.$$

Let j_0 denote a vertex such that $|f_{j_0}| = \max_{i\in V}|f_i|$. We know that $\mathbf{f}$ is orthogonal to the all 1's vector (which is an eigenvector for $\lambda_0 = 0$), so there exists $j_1 \in V$ such that $f_{j_0}f_{j_1} < 0$.

Let P denote a shortest path from j_0 to j_1. We denote the adjacent vertices in P by $i_0 = j_0, i_1, \ldots, i_{r-1}, i_r = j_1$. The number of edges in P is at most d, so $r \leq d$. Note

$$f_{i_r} - f_{i_0} = f_{i_r} - f_{i_{r-1}} + f_{i_{r-1}} - f_{i_{r-2}} + \cdots + f_{i_1} - f_{i_0}$$

$$= (f_{i_r} - f_{i_{r-1}}, f_{i_{r-1}} - f_{i_{r-2}}, \ldots, f_{i_1} - f_{i_0}) \cdot (1, 1, \ldots, 1)$$

$$\leq ((f_{i_r} - f_{i_{r-1}})^2 + (f_{i_{r-1}} - f_{i_{r-2}})^2 + \cdots + (f_{i_1} - f_{i_0})^2)^{1/2} \cdot \sqrt{d},$$

so

$$(f_{i_r} - f_{i_{r-1}})^2 + (f_{i_{r-1}} - f_{i_{r-2}})^2 + \cdots + (f_{i_1} - f_{i_0})^2 \geq \frac{(f_{i_r} - f_{i_0})^2}{d}.$$

We therefore have

---

[1]See Biggs, §8c, [Bi93] for further details.

$$\lambda_1 = \frac{\sum_{(i,j)\in E}(f_j - f_i)^2}{\sum_{i\in V} f_i^2}$$

$$\geq \frac{\sum_{(i,j)\in P}(f_j - f_i)^2}{n|f_{j_0}|^2}$$

$$\geq \frac{(f_{j_1} - f_{j_0})^2}{dn|f_{j_0}|^2}$$

$$\geq \frac{1}{dn},$$

$$(2.6)$$

as desired. $\qquad\qquad\qquad\qquad\qquad\qquad\qquad\qquad\qquad\square$

Remark 2.2.11. (1) A similar result, for the "normalized Laplacian", can be found in Chung [Ch92], Lemma 1.9.

(2) The above argument, with a more detailed analysis, can be pushed to prove a stronger result (due to B. McKay):

$$\lambda_1 \geq \frac{4}{dn}.$$

See Mohar [Mo91b] for details. See also Spielman [Sp10].

Example 2.2.12. A barbell graph (see Figure 2.3) illustrates how good McKay's lower bound really is.

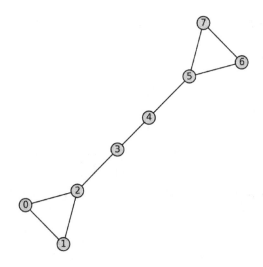

Figure 2.3: A barbell graph created using Sage.

```
────────────────────────── Sage ──────────────────────────
sage: Gamma = graphs.BarbellGraph(3,2)
sage: d = Gamma.diameter()
sage: n = len(Gamma.vertices())
sage: Q = Gamma.laplacian_matrix()
sage: spec = Q.eigenvalues()
sage: spec.sort()
sage: spec
[0, 0.1863934973516692?, 1, 2.4706834198711161?, 3, 3, 4, 4.3429230827777170?]
sage: 4.0/(d*n)
0.100000000000000
```

Let Γ be a graph, and let ρ be an element of the automorphism group $Aut(\Gamma)$ of Γ. If f is an eigenfunction of Q (regarded as a vector indexed by the vertices V of Γ), we define ρf to be the eigenfunction whose entries are permuted according to the action of ρ^{-1} on V, i.e., $(\rho f)(v) = f(\rho^{-1}(v))$ for any vertex v of Γ.

Definition 2.2.13. A *representation* of G is a homomorphism $\pi : G \rightarrow GL(n, \mathbb{C})$. A vector subspace $W \subset \mathbb{C}^n$ is called *G-invariant* if $\pi(g)w \in W$ for all $g \in G$ and all $w \in W$. The restriction of π to a G-invariant subspace is known as a *subrepresentation*. A representation is said to be *irreducible* if it has only trivial subrepresentations. A *character* χ of G is a trace of a representation π, denoted

$$\chi(g) = \mathrm{tr}(\pi(g)), \quad g \in G,$$

i.e., $\chi(g)$ is the trace of the matrix $\pi(g) \in GL(n, \mathbb{C})$.

Lemma 2.2.14. *The eigenspaces of the Laplacian of a graph Γ are representations of the automorphism group $Aut(\Gamma)$. In other words, if f is an eigenfunction of $Q = Q_\Gamma$ corresponding to an eigenvalue λ, then ρf is also an eigenfunction of Q with eigenvalue λ.*

Proof. We have

$$Q(\rho f)(v) = \sum_{(v,w) \in E} (\rho f(v) - \rho f(w))$$

$$= \sum_{(v,w) \in E} (f(\rho^{-1}(v)) - f(\rho^{-1}(w)))$$

$$= \sum_{(\rho^{-1}v, w) \in E} (f(\rho^{-1}(v)) - f(w))$$

$$= Qf(\rho^{-1}(v))$$

$$= \lambda f(\rho^{-1}(v))$$

$$= \lambda(\rho f)(v).$$

□

Lemma 2.2.15. *Every row sum and column sum of Q is zero.*

Proof. Since Q is symmetric, each column sum agrees with the corresponding row sum. Indeed, in the row sum corresponding to vertex v, the degree of v is summed with a "-1" for each neighbor of v. These cancel, giving us a sum of 0, as desired. ∎

Example 2.2.16. The *star graph $Star_n$* is a graph on $n+1$ vertices $v_0, \ldots, v_n$ with edges (v_0, v_i), for $i = 1, \ldots, n$. For example, $Star_5$ is depicted in Figure 2.4.

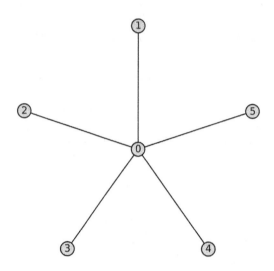

Figure 2.4: A star graph created using Sage.

Exercise 2.2. (a) Show that the eigenvalues of the adjacency matrix of $Star_n$ are $-\sqrt{n}$, 0 (with multiplicity $n-1$), $\sqrt{n}$.

(b) Show that the eigenvalues of the Laplacian matrix of $Star_n$ are 0, 1 (with multiplicity $n-1$), $n+1$.

Example 2.2.17. Consider the Paley graph on 9 vertices, Γ, depicted in Figure 1.12.

This graph has incidence matrix

$$B = \begin{pmatrix}
-1 & -1 & -1 & -1 & 0 & 0 & 0 & 0 & 0 & 0 & 0 & 0 & 0 & 0 & 0 & 0 & 0 & 0 \\
0 & 0 & 0 & 1 & -1 & -1 & -1 & 0 & 0 & 0 & 0 & 0 & 0 & 0 & 0 & 0 & 0 & 0 \\
0 & 0 & 1 & 0 & 0 & 1 & -1 & -1 & 0 & 0 & 0 & 0 & 0 & 0 & 0 & 0 & 0 & 0 \\
0 & 0 & 0 & 0 & 0 & 0 & 0 & 1 & 1 & 1 & -1 & 0 & 0 & 0 & 0 & 0 & 0 & 0 \\
0 & 1 & 0 & 0 & 0 & 0 & 0 & 0 & 0 & 0 & 1 & -1 & -1 & 0 & 0 & 0 & 0 & 0 \\
0 & 0 & 0 & 0 & 0 & 1 & 0 & 0 & 0 & 0 & 0 & 0 & 1 & -1 & 0 & 0 & 0 & 0 \\
0 & 0 & 0 & 0 & 1 & 0 & 0 & 0 & 0 & 0 & 0 & 0 & 0 & 1 & -1 & -1 & 0 \\
0 & 0 & 0 & 0 & 0 & 0 & 0 & 1 & 0 & 1 & 0 & 0 & 0 & 0 & 0 & 1 & -1 \\
1 & 0 & 0 & 0 & 0 & 0 & 0 & 0 & 0 & 0 & 0 & 1 & 0 & 0 & 1 & 0 & 1
\end{pmatrix}$$

and Laplacian matrix

$$Q = BB^t = \begin{pmatrix} 4 & -1 & -1 & 0 & -1 & 0 & 0 & 0 & -1 \\ -1 & 4 & -1 & 0 & 0 & -1 & -1 & 0 & 0 \\ -1 & -1 & 4 & -1 & 0 & 0 & 0 & -1 & 0 \\ 0 & 0 & -1 & 4 & -1 & -1 & 0 & -1 & 0 \\ -1 & 0 & 0 & -1 & 4 & -1 & 0 & 0 & -1 \\ 0 & -1 & 0 & -1 & -1 & 4 & -1 & 0 & 0 \\ 0 & -1 & 0 & 0 & 0 & -1 & 4 & -1 & -1 \\ 0 & 0 & -1 & -1 & 0 & 0 & -1 & 4 & -1 \\ -1 & 0 & 0 & 0 & -1 & 0 & -1 & -1 & 4 \end{pmatrix}.$$

The eigenvalues of Q are

$$0, 3, 3, 3, 3, 6, 6, 6, 6.$$

This graph is just one case of a very interesting family of graphs named after Raymond Paley. More information on them is given, for example, in §5.17 below. We shall see more of this remarkable graph and its cousins, later.

Example 2.2.18. Let Cyc_n denote the cycle graph with n vertices. The characteristic polynomial of $\Gamma = Cyc_n$ is

$$p_\Gamma(x) = 2T_n(x/2) - 2,$$

where $T_n(x)$ is the n-th Chebyshev polynomial of the first kind (see Stevanovic [St.14]).

For an (undirected) graph Γ, denote the eigenvalues of the Laplacian by

$$\lambda_0(Q) \le \lambda_1(Q) \le \cdots \le \lambda_{n-1}(Q).$$

Recall the incidence matrix B and its transpose B^t can be regarded as homomorphisms

$$B : C^1(\Gamma, \mathbb{Z}) \to C^0(\Gamma, \mathbb{Z}) \quad \text{and} \quad B^t : C^0(\Gamma, \mathbb{Z}) \to C^1(\Gamma, \mathbb{Z}).$$

Therefore, we can regard the Laplacian $Q = BB^t$ as a homomorphism $C^0(\Gamma, \mathbb{Z}) \to C^0(\Gamma, \mathbb{Z})$.

Lemma 2.2.19. *The adjacency matrix of Γ has an eigenvector of all 1's if and only if Γ is regular.*

Proof. By definition, the sum of the entries in the i-th row of A is equal to the number of vertices which share an edge with vertex i. This is, of course, the degree of i. Let $\mathbf{1} = \mathbf{1}_n \in \mathbb{Z}^n$ denote the vector of all 1's. If $\deg(v) = \deg_\Gamma(v)$ denotes the degree of a vertex v then we have shown

$$A\mathbf{1} = \begin{pmatrix} \deg(0) \\ \deg(1) \\ \vdots \\ \deg(n-1) \end{pmatrix}.$$

The vector on the right-hand side of the above equation is a scalar vector (i.e., all the components are the same) if and only if Γ is regular. $\square$

The *index* of Γ is the largest eigenvalue of Γ. The index has an eigenvector which consists of nonnegative components.

Lemma 2.2.20. *Suppose Γ has connected components $\Gamma_1, \ldots, \Gamma_r$.*

(a) *Possibly after reordering the vertices of Γ, Q is a block diagonal matrix, where each block is the respective Laplacian matrix for a corresponding component. In other words, Q is permutation conjugate to a block diagonal matrix.*

(b) *If $n = |V|$, define the vector $\mathbf{v}_j \in \mathbb{R}^n$ to be the vector whose component associated to a vertex in Γ_j is $= 1$ and all other components $= 0$. Then $\mathbf{v}_j$ is an eigenvector of Q having eigenvalue 0.*

Proof. This proof is left as an exercise in every other textbook on graph theory, so we should not be any different. Exercise! $\square$

Lemma 2.2.21. *The graph Γ is connected if and only if the index of Γ occurs with multiplicity 1 and it has an eigenvector which consists of strictly positive components.*

Proof. This follows from the Perron–Frobenius Theorem: If an $n \times n$ matrix has nonnegative entries then it has a nonnegative real eigenvalue λ which has maximum absolute value among all eigenvalues. This eigenvalue λ has a nonnegative real eigenvector. If, in addition, the matrix has no block triangular decomposition (i.e., it does not contain a $k \times (n-k)$ block of 0s disjoint from the diagonal), then λ has multiplicity 1 and the corresponding eigenvector is positive. $\square$

Exercise 2.3. Show that the multiplicity of $\lambda = 0$ as an eigenvalue of the Laplacian Q is the number of connected components in the graph.

Remark 2.2.22. Let Q^* denote a reduced Laplacian matrix (obtained by removing any row and the corresponding column of the Laplacian matrix Q) of a connected graph Γ. Then the critical group of Γ is isomorphic to

$\mathbb{Z}^{n-1}/\mathrm{Col}(Q^*)$, where $\mathrm{Col}(Q^*)$ denotes the $\mathbb{Z}$-span of the columns of Q^* and n is the number of vertices of Γ. For more details, see Proposition 4.6.2 in Chapter 4 on chip-firing.

For further reading on the topics of this section, see also Biyikogu, Leydold, and Stadler [BLS07] and Mohar [Mo91a].

2.3 The Moore–Penrose pseudoinverse

Throughout this section, we assume that the underlying graph Γ is connected.

The Moore–Penrose pseudoinverse of the Laplacian matrix Q is a type of generalized inverse of Q. There are other generalized inverses. Generalized inverses are sometimes classified by the additional properties they have, beyond that of the definition below.

After some preliminaries, we will give a construction of the Moore–Penrose pseudoinverse of the Laplacian matrix. An alternative construction will also be given (see Lemma 2.3.10).

Definition 2.3.1. If M and L are matrices such that $MLM = M$, then L is said to be a *generalized inverse* of M.

Let J be the $n \times n$ matrix, all of whose entries are 1. Let $\mathbf{1}$ be the n-vector, all of whose entries are 1.

Remark 2.3.2. Note that

- $QJ = JQ = 0$ (the all 0's matrix).

- $J^2 = nJ$.

- If x is any n-vector and $\deg(x) = \sum_{i=1}^{n} x_i$, then $Jx = \deg(x)\mathbf{1}$.

- In particular, $J\mathbf{1} = n\mathbf{1}$.

- If s has degree 0, then $Js = \mathbf{0}$ (the all 0's vector) since $\deg(s) = 0$.

Lemma 2.3.3. $Q + \frac{1}{n}J$ *is nonsingular.*

We will prove this lemma at the end of §2.3, after giving an alternative construction of the Moore–Penrose pseudoinverse.

Remark 2.3.4. Note that $(Q + \frac{1}{n}J)J = J = J(Q + \frac{1}{n}J)$.

Definition 2.3.5. The *Moore–Penrose pseudoinverse* of Q is defined to be

$$Q^+ = \left(Q + \frac{1}{n}J\right)^{-1} - \frac{1}{n}J.$$

Example 2.3.6. The Paley graph on 9 vertices Γ has Laplacian matrix

$$Q = \begin{pmatrix}
4 & -1 & -1 & 0 & -1 & 0 & 0 & 0 & -1 \\
-1 & 4 & -1 & 0 & 0 & -1 & -1 & 0 & 0 \\
-1 & -1 & 4 & -1 & 0 & 0 & 0 & -1 & 0 \\
0 & 0 & -1 & 4 & -1 & -1 & 0 & -1 & 0 \\
-1 & 0 & 0 & -1 & 4 & -1 & 0 & 0 & -1 \\
0 & -1 & 0 & -1 & -1 & 4 & -1 & 0 & 0 \\
0 & -1 & 0 & 0 & 0 & -1 & 4 & -1 & -1 \\
0 & 0 & -1 & -1 & 0 & 0 & -1 & 4 & -1 \\
-1 & 0 & 0 & 0 & -1 & 0 & -1 & -1 & 4
\end{pmatrix},$$

whose Moore–Penrose pseudoinverse is

$$Q^+ = \begin{pmatrix}
\frac{2}{9} & 0 & 0 & -\frac{1}{18} & 0 & -\frac{1}{18} & -\frac{1}{18} & -\frac{1}{18} & 0 \\
0 & \frac{2}{9} & 0 & -\frac{1}{18} & -\frac{1}{18} & 0 & 0 & -\frac{1}{18} & -\frac{1}{18} \\
0 & 0 & \frac{2}{9} & 0 & -\frac{1}{18} & -\frac{1}{18} & -\frac{1}{18} & 0 & -\frac{1}{18} \\
-\frac{1}{18} & -\frac{1}{18} & 0 & \frac{2}{9} & 0 & 0 & -\frac{1}{18} & 0 & -\frac{1}{18} \\
0 & -\frac{1}{18} & -\frac{1}{18} & 0 & \frac{2}{9} & 0 & -\frac{1}{18} & -\frac{1}{18} & 0 \\
-\frac{1}{18} & 0 & -\frac{1}{18} & 0 & 0 & \frac{2}{9} & 0 & -\frac{1}{18} & -\frac{1}{18} \\
-\frac{1}{18} & 0 & -\frac{1}{18} & -\frac{1}{18} & -\frac{1}{18} & 0 & \frac{2}{9} & 0 & 0 \\
-\frac{1}{18} & -\frac{1}{18} & 0 & 0 & -\frac{1}{18} & -\frac{1}{18} & 0 & \frac{2}{9} & 0 \\
0 & -\frac{1}{18} & -\frac{1}{18} & -\frac{1}{18} & 0 & -\frac{1}{18} & 0 & 0 & \frac{2}{9}
\end{pmatrix}.$$

Proposition 2.3.7. *The Moore–Penrose pseudoinverse Q^+ has the following properties:*

   i. Q^+ is symmetric.

  ii. $Q^{++} = Q$.

 iii. $\left(Q + \frac{1}{n}J\right)^{-1} = Q^+ + \frac{1}{n}J$.

 iv. $JQ^+ = Q^+J = 0$ (the all 0's matrix).

  v. $QQ^+ = Q^+Q = I - \frac{1}{n}J$.

 vi. $QQ^+Q = Q$ and $Q^+QQ^+ = Q^+$.

vii. $Q^+ = B^+(B^+)^t$, where $B^+ = Q^+B$ and B is the incidence matrix.

Proof. The first three properties are immediate, from the definition. From property (iii), we have

$$J\left(Q + \frac{1}{n}J\right)\left(Q^+ + \frac{1}{n}J\right) = JI$$

which expands to

$$JQQ^+ + \frac{1}{n}JQJ + \frac{1}{n}J^2Q^+ + \frac{1}{n^2}J^3 = J.$$

Noting that $JQ = 0$ and $J^2 = nJ$, we obtain

$$0 + 0 + JQ^+ + J = J,$$

so that $JQ^+ = 0$. Taking transposes, and noting that both J and Q^+ are symmetric, gives $Q^+J = 0$ also.

Using property (iii) again, we have

$$\left(Q + \frac{1}{n}J\right)\left(Q^+ + \frac{1}{n}J\right) = I.$$

Expanding gives

$$QQ^+ + \frac{1}{n}QJ + \frac{1}{n}JQ^+ + \frac{1}{n^2}J^2 = I.$$

Simplifying and using property (iv) gives

$$QQ^+ + 0 + 0 + \frac{1}{n}J = I$$

so that $QQ^+ = I - \frac{1}{n}J$. Taking transposes and noting that Q and Q^+ are symmetric gives $Q^+Q = I - \frac{1}{n}J$.

Using property (v), we have

$$QQ^+Q = Q\left(I - \frac{1}{n}J\right) = Q - \frac{1}{n}QJ = Q.$$

The proof that $Q^+QQ^+ = Q^+$ is similar.

Finally, we note that $B^+(B^+)^t = Q^+BB^t(Q^+)^t = Q^+QQ^+$ (since Q^+ is symmetric) which equals Q^+ by property (vi). □

Corollary 2.3.8. *If s is a vector whose entries sum to 0, then*

$$QQ^+s = Q^+Qs = s \qquad and \qquad s^tQQ^+ = s^tQ^+Q = s^t.$$

Proof. By part (v) of Proposition 2.3.7, $QQ^+s = Q^+Qs = (I - \frac{1}{n}J)s$. When the entries of s sum to 0, $Js = 0$, so $QQ^+s = Q^+Qs = Is = s$. The proof of the second statement is similar, and it also follows from the fact that Q and Q^+ are symmetric. □

Example 2.3.9. The cycle graph on 3 vertices $\Gamma = C_3$ has Laplacian matrix

$$Q = \begin{pmatrix} 2 & -1 & -1 \\ -1 & 2 & -1 \\ -1 & -1 & 2 \end{pmatrix}$$

whose Moore–Penrose pseudoinverse is

$$Q^+ = \begin{pmatrix} \frac{2}{9} & -\frac{1}{9} & -\frac{1}{9} \\ -\frac{1}{9} & \frac{2}{9} & -\frac{1}{9} \\ -\frac{1}{9} & -\frac{1}{9} & \frac{2}{9} \end{pmatrix}.$$

The signed incidence matrix attached to the edges

$$E = \{e_0 = (1,0), e_1 = (2,0), e_2 = (2,1)\}$$

having orientation $[1, 1, 1]$ is

$$B = \begin{pmatrix} 1 & 1 & 0 \\ -1 & 0 & 1 \\ 0 & -1 & -1 \end{pmatrix},$$

so that

$$B^+ = Q^+ B = \begin{pmatrix} \frac{1}{3} & \frac{1}{3} & 0 \\ -\frac{1}{3} & 0 & \frac{1}{3} \\ 0 & -\frac{1}{3} & -\frac{1}{3} \end{pmatrix}.$$

Exercise 2.4. Show that the matrices of Example 2.3.9 satisfy properties (i)-(vii) of Proposition 2.3.7. Also, show that B^+ is not a generalized inverse of B, i.e., show that $BB^+B \neq B$.

We will now give another construction of the Moore–Penrose pseudoinverse.

Since the Laplacian matrix Q is a real $n \times n$ symmetric matrix, we can choose a basis of $\mathbb{R}^n$ consisting of n orthonormal eigenvectors $w_1, w_2, \ldots, w_n$ corresponding to eigenvalues $\lambda_1, \lambda_2, \ldots, \lambda_n$ of Q. For Γ a connected graph, the rank of Q is $n - 1$, so we may assume that $\lambda_1 = 0$ and $w_1 = \frac{1}{\sqrt{n}}\mathbf{1}$, i.e., every entry of w_1 is $\frac{1}{\sqrt{n}}$. The nonzero eigenvalues are all positive, since $Q = BB^t$. Let U be the orthogonal matrix whose columns are the eigenvectors $w_1, w_2, \ldots, w_n$ and let Σ be the diagonal matrix whose diagonal entries are $\lambda_1 = 0, \lambda_2, \lambda_3, \ldots, \lambda_n$. Then $UU^t = U^tU = I$ and

$$Q = U\Sigma U^t. \tag{2.7}$$

We define the pseudoinverse of the diagonal matrix Σ to be the diagonal matrix Σ^+ whose i-th diagonal entry is the reciprocal of the i-th diagonal entry of Σ, if this entry is nonzero, and zero otherwise, i.e., the diagonal entries of Σ^+ are $0, \frac{1}{\lambda_2}, \frac{1}{\lambda_3}, \ldots, \frac{1}{\lambda_n}$.

We define Q^+ to be the matrix given by

$$Q^+ = U\Sigma^+ U^t. \tag{2.8}$$

The matrix Q^+ is called a pseudoinverse of Q. We will show that Q^+ is equal to the Moore–Penrose pseudoinverse defined above, and is thus independent of the choice of matrix U. Note that from Equation (2.8), it is clear that Q^+ is symmetric and has the same rank $n-1$ as Q. Furthermore, $QQ^+ = Q^+Q$ and the matrices Q and Q^+ have the same eigenvectors. Thus the all 1's vector $\mathbf{1}$ is a basis for the kernel of Q^+ and $Q^+J = JQ^+ = 0$, the all 0's matrix.

Lemma 2.3.10. *The matrix Q^+ given by Equation (2.8) is the Moore–Penrose pseudoinverse, i.e.,*

$$U\Sigma^+ U^t = \left(Q + \frac{1}{n}J\right)^{-1} - \frac{1}{n}J. \tag{2.9}$$

Proof. By Equation (2.7), proving Equation (2.9) is equivalent to proving

$$\left(U\Sigma^+ U^t + \frac{1}{n}J\right)\left(U\Sigma U^t + \frac{1}{n}J\right) = I.$$

The left side of the previous equation expands to give

$$U\Sigma^+ U^t U\Sigma U^t + \frac{1}{n}U\Sigma^+ U^t J + \frac{1}{n}JU\Sigma U^t + \frac{1}{n^2}J^2$$

$$= U\Sigma^+ \Sigma U^t + \frac{1}{n}UU^t Q^+ J + \frac{1}{n}JQUU^t + \frac{1}{n}J$$

$$= U\Sigma^+ \Sigma U^t + \frac{1}{n}J$$

since $Q^+J = JQ = 0$. Let $I_{(1,1)}$ denote the matrix, all of whose entries are 0, except for the $(1,1)$-entry, which is 1. Note that $\Sigma^+\Sigma = I - I_{(1,1)}$. Then

$$U\Sigma^+ \Sigma U^t + \frac{1}{n}J = U(I - I_{(1,1)})U^t + \frac{1}{n}J$$

$$= I - UI_{(1,1)}U^t + \frac{1}{n}J.$$

It is not hard to check that $UI_{(1,1)}U^t = \frac{1}{n}J$, from which the required identity follows. □

Exercise 2.5. Prove that the formula

$$\left(Q - \frac{1}{n}J\right)^{-1} + \frac{1}{n}J$$

also gives the Moore–Penrose pseudoinverse of Q.

We will now prove Lemma 2.3.3.

Proof. Let $Q = U\Sigma U^t$ be a decomposition of Q as above. The diagonal elements of Σ are $0, \lambda_2, \lambda_3, \ldots, \lambda_n$ and the first column of U is the eigenvector $\frac{1}{\sqrt{n}}\mathbf{1}$ corresponding to $\lambda_1 = 0$. We can decompose J as $J = U\Sigma_J U^t$ where $\Sigma_J = I_{(1,1)}$ is the diagonal matrix with 1 in the $(1,1)$-entry and all other entries zero. Then $Q + \frac{1}{n}J = U\Sigma' U^t$, where Σ' is the diagonal matrix with diagonal entries $1, \lambda_2, \lambda_3, \ldots, \lambda_n$. Thus, $Q + \frac{1}{n}J$ is nonsingular. $\square$

2.4 Circulant graphs

Recall that a *circulant matrix* is a square matrix where each row vector is a cyclic shift one element to the right relative to the preceding row vector, such as

$$C = \begin{pmatrix} c_0 & c_{n-1} & \cdots & c_1 \\ c_1 & c_0 & \cdots & c_2 \\ \vdots & \vdots & & \vdots \\ c_{n-1} & c_{n-2} & \cdots & c_0 \end{pmatrix}.$$

Circulant matrices have the property that $\mathbf{v}_k = (\zeta^{jk}/\sqrt{n} \mid j = 0, \ldots, n-1)$ is an eigenvector with eigenvalue

$$\lambda_k(C) = \sum_{j=0}^{n-1} \zeta^{-jk} c_j = \sum_{j=1}^{n} \zeta^{jk} c_{n-j},$$

for each $k = 0, \ldots, n-1$.

A graph Γ is called *circulant* if its vertices can be reindexed in such a way that its adjacency matrix is a circulant matrix. For example, a cycle graph is a circulant graph.

Example 2.4.1. Consider the Möbius ladder graph on 8 vertices, Γ, depicted in Figure 2.5.

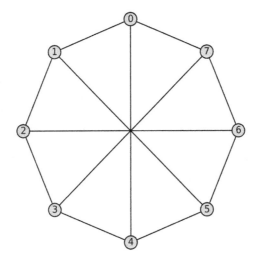

Figure 2.5: A Möbius ladder graph created using Sage.

This graph has adjacency matrix

$$A_\Gamma = \begin{pmatrix} 0 & 1 & 0 & 0 & 1 & 0 & 0 & 1 \\ 1 & 0 & 1 & 0 & 0 & 1 & 0 & 0 \\ 0 & 1 & 0 & 1 & 0 & 0 & 1 & 0 \\ 0 & 0 & 1 & 0 & 1 & 0 & 0 & 1 \\ 1 & 0 & 0 & 1 & 0 & 1 & 0 & 0 \\ 0 & 1 & 0 & 0 & 1 & 0 & 1 & 0 \\ 0 & 0 & 1 & 0 & 0 & 1 & 0 & 1 \\ 1 & 0 & 0 & 1 & 0 & 0 & 1 & 0 \end{pmatrix},$$

incidence matrix

$$B = \begin{pmatrix} -1 & -1 & -1 & 0 & 0 & 0 & 0 & 0 & 0 & 0 & 0 & 0 \\ 0 & 0 & 1 & -1 & -1 & 0 & 0 & 0 & 0 & 0 & 0 & 0 \\ 0 & 0 & 0 & 0 & 1 & -1 & -1 & 0 & 0 & 0 & 0 & 0 \\ 0 & 0 & 0 & 0 & 0 & 0 & 1 & -1 & -1 & 0 & 0 & 0 \\ 0 & 1 & 0 & 0 & 0 & 0 & 0 & 0 & 1 & -1 & 0 & 0 \\ 0 & 0 & 0 & 1 & 0 & 0 & 0 & 0 & 0 & 1 & -1 & 0 \\ 0 & 0 & 0 & 0 & 0 & 1 & 0 & 0 & 0 & 0 & 1 & -1 \\ 1 & 0 & 0 & 0 & 0 & 0 & 0 & 1 & 0 & 0 & 0 & 1 \end{pmatrix},$$

and Laplacian matrix

$$Q = BB^t = \begin{pmatrix} 3 & -1 & 0 & 0 & -1 & 0 & 0 & -1 \\ -1 & 3 & -1 & 0 & 0 & -1 & 0 & 0 \\ 0 & -1 & 3 & -1 & 0 & 0 & -1 & 0 \\ 0 & 0 & -1 & 3 & -1 & 0 & 0 & -1 \\ -1 & 0 & 0 & -1 & 3 & -1 & 0 & 0 \\ 0 & -1 & 0 & 0 & -1 & 3 & -1 & 0 \\ 0 & 0 & -1 & 0 & 0 & -1 & 3 & -1 \\ -1 & 0 & 0 & -1 & 0 & 0 & -1 & 3 \end{pmatrix}.$$

It is a circulant graph.

2.4.1 Cycle graphs

For the cycle graph on n vertices, Γ_n, the eigenvalues are $2\cos(2\pi k/n)$, for $0 \le k \le n-1$. Since these are not distinct, some can occur with multiplicities. n even: The only eigenvalues of Γ_n which occur with multiplicity 1 are 2 and -2. The eigenvalues $2\cos(2\pi k/n)$, for $1 \le k \le \frac{n-2}{2}$, all occur with multiplicity 2.
n odd: The only eigenvalue of Γ_n which occurs with multiplicity 1 is 2. The eigenvalues $2\cos(2\pi k/n)$, for $1 \le k \le \frac{n-1}{2}$, all occur with multiplicity 2.
    For example, the graph Γ_8 is depicted in Figure 2.6

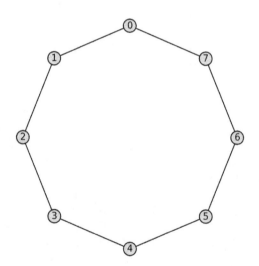

Figure 2.6:   A cycle graph created using Sage.

The adjacency matrix is circulant:

$$A_{\Gamma_8} = \begin{pmatrix} 0 & 1 & 0 & 0 & 0 & 0 & 0 & 1 \\ 1 & 0 & 1 & 0 & 0 & 0 & 0 & 0 \\ 0 & 1 & 0 & 1 & 0 & 0 & 0 & 0 \\ 0 & 0 & 1 & 0 & 1 & 0 & 0 & 0 \\ 0 & 0 & 0 & 1 & 0 & 1 & 0 & 0 \\ 0 & 0 & 0 & 0 & 1 & 0 & 1 & 0 \\ 0 & 0 & 0 & 0 & 0 & 1 & 0 & 1 \\ 1 & 0 & 0 & 0 & 0 & 0 & 1 & 0 \end{pmatrix}.$$

The eigenvalues (counted according to their multiplicity are 2, $\sqrt{2}$, $\sqrt{2}$, 0, 0, $-\sqrt{2}$, $-\sqrt{2}$, -2.

Let Γ_1 be the cycle graph with n vertices, let Γ_2 be the cycle digraph (directed graph) with n vertices (with edges oriented counterclockwise around the cycle), let G_i denote the automorphism group of Γ_i, $i = 1, 2$.

Lemma 2.4.2. *The automorphism group G_1 of the cycle graph Γ_1 is the dihedral group of order $2n$, D_n.*

Proof. We may assume that the vertices are labeled

$$V = \{0, 1, \ldots, n-1\},$$

and that the edges are

$$E = \{(0,1), (1,2), \ldots, (n-1, 0)\}.$$

Clearly, the "rotation" (written in disjoint cycle notation) belongs to the automorphism group, i.e., $(0, 1, \ldots, n-1) \in G_1$. Clearly, the "reflection" (written in disjoint cycle notation) belongs to the automorphism group, i.e., $(0, n-1)(1, n-2) \cdots \in G_1$. The rotation and reflection generate D_n. The remainder of the proof is left as Exercise 2.6. □

Exercise 2.6. Complete the proof of Lemma 2.4.2 above by showing that, for $n > 2$, there are no other automorphisms of the cycle graph Γ_1 beyond the elements in the dihedral group, D_n. (Hint: Let $g \in G_1$. Suppose $g : 0 \mapsto i$. Then it must send 1 and $n-1$ to a neighbor of i.)

Exercise 2.7. Show that the automorphism group G_2 of the cycle digraph (directed graph) Γ_2 is the cyclic group of order n, C_n.

2.4.2 Relationship to convolution operators

We identify the vertices V of a circulant graph Γ having n vertices with the abelian group of integers mod n, $\mathbb{Z}/n\mathbb{Z}$. If $\mathbb{C}$ denotes the field of complex numbers, let

$$C^0(\Gamma, \mathbb{C}) = \{ f \mid f : \mathbb{Z}/n\mathbb{Z} \to \mathbb{C} \}.$$

This is a complex vector space which we can identify with the vector space $\mathbb{C}^n$ via the map $f \longmapsto (f(0), f(1), \dots, f(n-1))$.

Define *convolution* by

$$
\begin{array}{ccc}
C^0(\Gamma, \mathbb{C}) \times C^0(\Gamma, \mathbb{C}) & \to & C^0(\Gamma, \mathbb{C}) \\
(f, g) & \longmapsto & f * g,
\end{array}
\tag{2.10}
$$

where

$$(f * g)(k) = \sum_{\ell \in \mathbb{Z}/n\mathbb{Z}} f(\ell) g(k - \ell).$$

This is commutative: $f * g = g * f$.

Let $\zeta = \zeta_n$ denote a primitive n^{th} root of unity in $\mathbb{C}$. Recall, for $g \in C^0(\Gamma, \mathbb{C})$, the *discrete Fourier transform* $\mathcal{F}_n$ of g is defined by

$$(\mathcal{F}_n g)(\lambda) = g^\wedge(\lambda) = \sum_{\ell \in \mathbb{Z}/n\mathbb{Z}} g(\ell) \zeta^{\ell\lambda}, \qquad \lambda \in \mathbb{Z}/n\mathbb{Z}.$$

Define the *inverse discrete Fourier transform* of G by

$$(\mathcal{F}_n^{-1} G)(\ell) = G^\vee(\ell) = \frac{1}{n} \sum_{\lambda \in \mathbb{Z}/n\mathbb{Z}} G(\lambda) \zeta^{-\ell\lambda}, \qquad \ell \in \mathbb{Z}/n\mathbb{Z}.$$

The following lemma states the basic and very useful fact that the Fourier transform of a convolution is the product of the Fourier transforms.

Lemma 2.4.3. *For any $f, g \in C^0(\Gamma, \mathbb{C})$, we have*

$$(f * g)^\wedge(\lambda) = f^\wedge(\lambda) g^\wedge(\lambda).$$

Proof.

$$
\begin{aligned}
(f * g)^\wedge(\lambda) &= \sum_{\ell \in \mathbb{Z}/n\mathbb{Z}} \sum_{k \in \mathbb{Z}/n\mathbb{Z}} f(k) g(\ell - k) \zeta^{\ell\lambda} \\
&= \sum_{k \in \mathbb{Z}/n\mathbb{Z}} f(k) \sum_{\ell \in \mathbb{Z}/n\mathbb{Z}} g(\ell - k) \zeta^{\ell\lambda} \\
&= \sum_{k \in \mathbb{Z}/n\mathbb{Z}} f(k) \zeta^{k\lambda} \sum_{\ell' \in \mathbb{Z}/n\mathbb{Z}} g(\ell') \zeta^{\ell'\lambda} \\
&= f^\wedge(\lambda) g^\wedge(\lambda).
\end{aligned}
$$

$\square$

Definition 2.4.4. For $h \in C^0(\Gamma, \mathbb{C})$, define $T_h : C^0(\Gamma, \mathbb{C}) \to C^0(\Gamma, \mathbb{C})$ by

$$T_h(f) = (h \otimes f^\wedge)^\vee,$$

where $\otimes$ denotes the componentwise product of two vectors:

$$(a_0, a_1, \ldots, a_{n-1}) \otimes (b_0, b_1, \ldots, b_{n-1}) = (a_0 b_0, a_1 b_1, \ldots, a_{n-1} b_{n-1}).$$

A linear transformation $M : C^0(\Gamma, \mathbb{C}) \to C^0(\Gamma, \mathbb{C})$ of the form $M = T_h$, for some $h \in C^0(\Gamma, \mathbb{C})$, is called a *Fourier multiplier operator*.

Let $\tau : C^0(\Gamma, \mathbb{C}) \to C^0(\Gamma, \mathbb{C})$ denote the *translation map*: $(\tau f)(x) = f(x + 1)$ (addition in $\mathbb{Z}/n\mathbb{Z}$). Note τ sends $(x_0, x_1, \ldots, x_{n-1})$ to $(x_1, x_2, \ldots, x_{n-1}, x_0)$. A transformation $T : C^0(\Gamma, \mathbb{C}) \to C^0(\Gamma, \mathbb{C})$ which commutes with τ is called *translation invariant* (or translation equivariant).

Define the *convolution operator associated to g*,

$$\mathcal{T}_g : C^0(\Gamma, \mathbb{C}) \to C^0(\Gamma, \mathbb{C}),$$

by $\mathcal{T}_g(f) = f * g$.

Exercise 2.8. Show that a Fourier multiplier operator $M = T_h$ is a linear transformation of the form $\mathcal{F}_n^{-1} D \mathcal{F}_n$, where D is an $n \times n$ diagonal matrix. (Hint: The diagonal elements of D may be taken to be values of h on the elements of $\mathbb{Z}/n\mathbb{Z}$.)

Recall a matrix A is *circulant* if and only if there is an n such that $A_{k,\ell} = A_{k+1 \,(\text{mod } n),\, \ell+1 \,(\text{mod } n)}$, for all $0 \le k \le n - 1$, $0 \le \ell \le n - 1$.

The following result appears as Theorem 2.19 of Frazier [Fr99]. It charac terizes the Fourier multiplier operators on $C^0(\Gamma, \mathbb{C})$.

Theorem 2.4.5. *Let $T : C^0(\Gamma, \mathbb{C}) \to C^0(\Gamma, \mathbb{C})$ denote a linear operator. The following statements are equivalent:*

1. *T is translation invariant.*

2. *The matrix $[T]$ representing T in the standard basis is circulant.*

3. *T is a convolution operator.*

4. *T is a Fourier multiplier operator.*

5. *The matrix B representing T in the Fourier basis is diagonal.*

We will prove the equivalence of the first three items in the following three lemmas. We leave the proof of the equivalence of the remaining items as an exercise.

Lemma 2.4.6. *The convolution operator $\mathcal{T}_g$ is translation invariant. In other words, the diagram*

$$
\begin{array}{ccc}
C^0(\Gamma,\mathbb{C}) & \xrightarrow{\ \tau\ } & C^0(\Gamma,\mathbb{C}) \\
{\scriptstyle \mathcal{T}_g}\big\downarrow & & {\scriptstyle \mathcal{T}_g}\big\downarrow \\
C^0(\Gamma,\mathbb{C}) & \xrightarrow{\ \tau\ } & C^0(\Gamma,\mathbb{C})
\end{array}
$$

commutes, for all $g \in C^0(\Gamma,\mathbb{C})$.

Proof. For $k \in \mathbb{Z}/n\mathbb{Z}$, we have

$$
\begin{aligned}
\mathcal{T}_g(\tau(f))(k) &= (\tau(f) * g)(k) \\
&= \sum_{\ell \in \mathbb{Z}/n\mathbb{Z}} (\tau(f))(\ell) g(k-\ell) \\
&= \sum_{\ell \in \mathbb{Z}/n\mathbb{Z}} f(\ell+1) g(k-\ell) \\
&= \sum_{\ell' \in \mathbb{Z}/n\mathbb{Z}} f(\ell') g(k-\ell'+1) \\
&= \tau(\mathcal{T}_g(f))(k).
\end{aligned}
$$

$\square$

Lemma 2.4.7. *A linear transformation $T : C^0(\Gamma,\mathbb{C}) \to C^0(\Gamma,\mathbb{C})$ is translation invariant if and only if the matrix representing it in the standard basis is circulant.*

Proof. Since $T : C^0(\Gamma,\mathbb{C}) \to C^0(\Gamma,\mathbb{C})$ is linear, it is represented by an $n \times n$ matrix

$$
Tf = A\vec{f},
$$

where $\vec{f} = (f(0), f(1), \ldots, f(n-1))^t$. In other words,

$$
T(f)(k) = \sum_{\ell \in \mathbb{Z}/n\mathbb{Z}} f(\ell) A_{k,\ell},
$$

for $k \in \mathbb{Z}/n\mathbb{Z}$. For $k \in \mathbb{Z}/n\mathbb{Z}$, we have

$$
\begin{aligned}
T(\tau(f))(k) &= \sum_{\ell \in \mathbb{Z}/n\mathbb{Z}} A_{k,\ell}(\tau(f))(\ell) \\
&= \sum_{\ell \in \mathbb{Z}/n\mathbb{Z}} A_{k,\ell} f(\ell+1) \\
&= \sum_{\ell' \in \mathbb{Z}/n\mathbb{Z}} A_{k,\ell'-1} f(\ell'),
\end{aligned}
$$

where the 2nd subscript of $A_{i,j}$ is taken mod n. On the other hand,

$$\tau(T(f))(k) = \sum_{\ell \in \mathbb{Z}/n\mathbb{Z}} A_{k+1,\ell} f(\ell).$$

It follows that the linear transformation T is translation invariant if and only if its matrix A is circulant: $A_{k,\ell} = A_{k+1\,(\mathrm{mod}\ n),\ \ell+1\,(\mathrm{mod}\ n)}$, for all $0 \le k \le n-1$, $0 \le \ell \le n-1$. $\qquad\square$

Lemma 2.4.8. *A linear transformation $T : C^0(\Gamma,\mathbb{C}) \to C^0(\Gamma,\mathbb{C})$ is a convolution map if and only if the matrix representing it in the standard basis is circulant.*

Proof. It is not hard to see the connection between maps given by circulant matrices and convolution operators. Suppose that the matrix representing T in the standard basis is circulant. By Lemma 2.4.7, T is also translation invariant. Define $g \in C^0(\Gamma,\mathbb{C})$ by

$$g(-k) = A_{0,k\,(\mathrm{mod}\ n)}, \qquad \text{for } k \in \mathbb{Z}/n\mathbb{Z}.$$

Then

$$T(f)(0) = \sum_{\ell \in \mathbb{Z}/n\mathbb{Z}} g(-\ell) f(\ell),$$

for all $f \in C^0(\Gamma,\mathbb{C})$. Replacing f by a translation $\pmod n$ (since T is translation invariant and g is periodic with period n) gives

$$T(f)(k) = \sum_{\ell \in \mathbb{Z}/N\mathbb{Z}} g(k-\ell) f(\ell),$$

for all $f \in C^0(\Gamma,\mathbb{C})$. In other words, T is a convolution map. This construction can be reversed: a convolution map corresponds to a circulant matrix transformation. $\qquad\square$

Exercise 2.9. For the circulant matrices T and maps f given below, verify that $\tau T(f) = T(\tau(f))$.

(a)

$$T = \begin{pmatrix} 0 & 1 & 0 & 0 & 0 \\ 0 & 0 & 1 & 0 & 0 \\ 0 & 0 & 0 & 1 & 0 \\ 0 & 0 & 0 & 0 & 1 \\ 1 & 0 & 0 & 0 & 0 \end{pmatrix} \quad \text{and } f = (1,2,3,4,5).$$

(b)

$$T = \begin{pmatrix} 0 & 1 & 2 & 0 & 0 \\ 0 & 0 & 1 & 2 & 0 \\ 0 & 0 & 0 & 1 & 2 \\ 2 & 0 & 0 & 0 & 1 \\ 1 & 2 & 0 & 0 & 0 \end{pmatrix} \quad \text{and } f = (1,2,3,4,5).$$

Exercise 2.10. Prove that the last two items of Theorem 2.4.5 are equivalent to the first three items.

In graph-theoretic terms, a Fourier multiplier operator is a linear transformation of the form $B_\Gamma^{-1} D B_\Gamma$, where D is an $n \times n$ diagonal matrix and B_Γ is a matrix of eigenvectors of the Laplacian of Γ.

Question 2.3. *To what extent can the result above about Fourier multiplier operators be generalized to arbitrary graphs?*

2.5 Expander graphs

Let $\Gamma = (V, E)$ be a graph, let S be a subset of V, and let

$$\partial S = \{(u, v) \in E(G) \mid u \in S, v \in V \setminus S\}$$

denote the *edge boundary* of S. This is the cocycle associated to the partition $S \cup (V \setminus S)$.

The *edge expansion* $h(\Gamma)$ is defined as

$$h(\Gamma) = \min_{0 < |S| \leq \frac{|V|}{2}} \frac{|\partial S|}{|S|}.$$

We say Γ has the *expander property* when each subset $S \subset V$ has a "relatively large" edge expansion, as specified in the definition below.

The second smallest eigenvalue of the Laplacian matrix Q of Γ, $\lambda_1(Q)$, is called the *spectral gap*. For example, if $\Gamma = (V, E)$ is a k-regular graph, then it is known that all eigenvalues λ of the adjacency matrix A of Γ satisfy $-k \leq \lambda \leq k$ (so the eigenvalues of Q satisfy $0 \leq \lambda \leq 2k$, by Lemma 2.2.9). Moreover, $\lambda = k$ is an eigenvalue of A (with the all 1's vector as an eigenvector). Therefore, in the regular case, the spectral gap measures the gap from the so-called trivial eigenvalue k of A to the next one (i.e., to $k - \lambda_1(Q)$, by Lemma 2.2.9).

Definition 2.5.1. Let $\Gamma = (V, E)$ be a k-regular graph.
We call

$$\gamma_\Gamma = \frac{\lambda_1(Q)}{k},$$

the *relative spectral gap* of Γ. We say Γ is a (k, r) expander if, for each $S \subset V$,

$$\frac{|\partial S|}{|S|} \geq kr \left(1 - \frac{|S|}{|V|}\right).$$

The following result can be found in Roth [Ro06], §13.3.

Theorem 2.5.2. *If Γ is a k-regular graph then Γ is a (k,r)-expander for each r with $0 \leq r \leq 1 - \gamma_\Gamma$.*

Example 2.5.3. Consider the 4-regular Paley graph on 9 vertices $\Gamma = (V, E)$, depicted in Figure 1.12.
If $S = \{0, 1, 2\}$ then

$$\partial S = \{(0, a + 1), (0, 2a + 2), (1, a + 2), (1, 2a), (2, a), (2, 2a + 1)\}.$$

The edge expansion is 2. Recall from Example 2.2.17 that the Laplacian spectrum is $\{0, 3, 3, 3, 3, 6, 6, 6, 6\}$, so $\lambda_1(Q) = 3$ and $\gamma_\Gamma = 3/4$. By the above theorem, the inequality

$$h(\Gamma) \geq kr(1 - |S|/|V|),$$

for all nonempty subsets $S \subset V$, holds for $0 \leq r \leq 1 - \gamma_\Gamma$. In this case, $k = 4$, and so $kr(1 - |S|/|V|) \leq 32r/9$.

Example 2.5.4. Consider the 6-regular graph on 16 vertices $\Gamma = (V, E)$, depicted in Figure 2.7 (Example 2.6.3 below). The eigenvalues of the adjacency matrix are

$$6, 2, 2, 2, 2, 2, 2, -2, -2, -2, -2, -2, -2, -2, -2, -2.$$

According to Sage, the edge expansion is $h(\Gamma) = 7/2$. The Laplacian spectrum is $\{0, 4, 4, 4, 4, 4, 4, 8, 8, 8, 8, 8, 8, 8, 8, 8\}$, so $\lambda_1(Q) = 4$ and $\gamma_\Gamma = 4/6 = 2/3$. By the above theorem, the inequality

$$h(\Gamma) \geq kr(1 - |S|/|V|),$$

for all nonempty subsets $S \subset V$, holds for $0 \leq r \leq 1 - \gamma_\Gamma$. In this case, $k = 6$, and so $kr(1 - |S|/|V|) \leq 15/8$.

Let $\Gamma = (V, E)$ be a connected k-regular graph with $n = |V|$ vertices, and let

$$k = \lambda_0(A) \geq \lambda_1(A) \geq \cdots \geq \lambda_{n-1}(A)$$

be the eigenvalues of its adjacency matrix $A = A_\Gamma$. If there exists $\lambda_i = \lambda_i(A)$ with $|\lambda_i| < k$, define

$$\lambda(\Gamma) = \max_{|\lambda_i| < k} |\lambda_i|.$$

We call Γ a *Ramanujan graph* if

$$\lambda(\Gamma) \leq 2\sqrt{k-1}.$$

Example 2.5.5. The above Example 2.5.4 is a 6-regular graph Γ satisfying

$$\lambda(\Gamma) = 2 \leq 2\sqrt{k-1} = 2\sqrt{5}.$$

Therefore, it is an example of a Ramanujan graph.

The excellent book by Davidoff, Sarnak, and Valette [DSV03] contains a detailed construction of an infinite family of Ramanujan graphs. For instance, if p, q are odd primes with $q > 2\sqrt{p}$ and p is not a quadratic residue (mod q), they construct a symmetric generating set $S_{p,q}$ of $PGL(2, q)$, such that the Cayley graph $Cay(PGL(2, q), S_{p,q})$ (this notation is defined at the beginning of §2.6 below) is a $(p + 1)$-regular Ramanujan graph. The details of the construction require more number-theoretic background than we have introduced here. We refer to [DSV03] for details.

2.6 Cayley graphs

Let G be a finite multiplicative group. Let $S \subset G$ be a subset which satisfies $S = S^{-1}$ and $1 \notin S$. The *Cayley graph* of (G, S) is the graph $\Gamma = Cay(G, S)$ whose vertices are $V = G$ and whose edges E are defined by those pairs (g_1, g_2) such that $g_2 g_1^{-1} \in S$.

This is a k-regular graph having degree $k = |S|$. By Lemma 2.2.1, the eigenvalues $\lambda_i(Q)$ of the Laplacian Q are related to the eigenvalues $\lambda_i(A)$ of the (unweighted) adjacency matrix by

$$\lambda_i(Q) = k - \lambda_i(A).$$

Exercise 2.11. Show that a Cayley graph $\Gamma = Cay(G, S)$ is regular with degree $|S|$.

Exercise 2.12. Show that a Cayley graph $\Gamma = Cay(G, S)$ is connected if and only if S generates G.

Example 2.6.1. The Sage commands to produce the types of Cayley graphs we are interested in are a bit tricky. You will want to be sure to select the "simple" option and to select symmetric generating sets.

```
──────────────────── Sage ────────────────────

sage: G = AdditiveAbelianGroup([6])
sage: Gp = G.permutation_group()
sage: g1, g2 = Gp.gens()
sage: S = [(g1*g2)^2,(g1*g2)^(-2),(g1*g2)^3]; S
[(3,5,4), (3,4,5), (1,2)]
sage: A = Gp.cayley_graph(generators=S,simple=True).adjacency_matrix()
sage: Gamma1 = Graph(A)
sage: AG1 = Gamma1.automorphism_group()
sage: AG1.cardinality()
12
```

```
──────────────────── Sage ────────────────────

sage: G = SymmetricGroup(3)
sage: S = G.gens()+[G.gens()[0]^(-1)]; S
[(1,2,3), (1,2), (1,3,2)]
sage: A = G.cayley_graph(generators=S,simple=True).adjacency_matrix()
sage: Gamma2 = Graph(A)
sage: AG2 = Gamma2.automorphism_group()
sage: AG2.cardinality()
12
```

These two Cayley graphs are isomorphic. With these options selected, we see that the graph has the desired symmetry. More of this example will be given later (e.g., Example 4.7.8 below).

There is also an edge-weighted analog of this definition.

For $g \in G$, the *conjugacy class* of g is the subset

$$Cl_G(g) = \{x^{-1}gx \mid x \in G\}.$$

The set of conjugacy classes will be denoted G_*. A function $f : G \to \mathbb{C}$ is called a *class function* if it is constant on conjugacy classes. In other words, f is a class function if and only if the restriction $f|_\gamma$ is a constant (possibly depending on γ), for each $\gamma \in G_*$.

Let $\alpha : G \to \mathbb{Z}$ be a given class function. The *edge-weighted Cayley graph* associated to (G, S, α), is the graph $\Gamma = Cay(G, S)$, where edge (g_1, g_2) has weight $\alpha(g_2 g_1^{-1})$. We denote this graph by $Cay(G, S, \alpha)$. By convention, if $\alpha(g_2 g_1^{-1}) = 0$ then we say that the 0-weighted edge (g_1, g_2) does not exist.

Let $n = |G|$. A subgroup H of S_n *acts* on Γ if and only if it is an automorphism group of the unweighted graph Γ^* and each graph automorphism $h \in H$ also preserves the edge weights. In particular, the H-orbit of any edge of Γ consists of edges which have the same edge weight. The set of such actions on Γ forms a subgroup of the automorphism group of Γ^*.

2.6.1 Cayley graphs on abelian groups

The eigenvalues of the Cayley graph in the abelian case are easy to determine.

Proposition 2.6.2. *Let G be a finite abelian group written multiplicatively, let $\chi : G \to \mathbb{C}^\times$ be a homomorphism of G, and let $S \subset G$ be a symmetric set. Let A be the adjacency matrix of the Cayley graph $\Gamma = Cay(G, S)$. Consider the vector $x \in \mathbb{C}^G$ such that $x_a = \chi(a)$, where $\mathbb{C}^G$ denotes the vector space of complex-valued functions on G. Then x is an eigenvector of A, with eigenvalue $\chi(S) = \sum_{s \in S} \chi(s)$.*

Proof. This follows from the proof in the non-abelian case, given in the next section. $\qquad\square$

Example 2.6.3. Consider the Cayley graph Γ of $\mathbb{Z}/4\mathbb{Z} \times \mathbb{Z}/4\mathbb{Z}$, with generator set $S = \{\pm(0,1), \pm(1,0), \pm(1,1)\}$. This has Laplacian matrix

$$
\begin{pmatrix}
6 & -1 & 0 & -1 & -1 & -1 & 0 & 0 & 0 & 0 & 0 & 0 & -1 & 0 & 0 & -1 \\
-1 & 6 & -1 & 0 & 0 & -1 & -1 & 0 & 0 & 0 & 0 & 0 & -1 & -1 & 0 & 0 \\
0 & -1 & 6 & -1 & 0 & 0 & -1 & -1 & 0 & 0 & 0 & 0 & 0 & -1 & -1 & 0 \\
-1 & 0 & -1 & 6 & -1 & 0 & 0 & -1 & 0 & 0 & 0 & 0 & 0 & 0 & -1 & -1 \\
-1 & 0 & 0 & -1 & 6 & -1 & 0 & -1 & -1 & -1 & 0 & 0 & 0 & 0 & 0 & 0 \\
-1 & -1 & 0 & 0 & -1 & 6 & -1 & 0 & 0 & -1 & -1 & 0 & 0 & 0 & 0 & 0 \\
0 & -1 & -1 & 0 & 0 & -1 & 6 & -1 & 0 & 0 & -1 & -1 & 0 & 0 & 0 & 0 \\
0 & 0 & -1 & -1 & -1 & 0 & -1 & 6 & -1 & 0 & 0 & -1 & 0 & 0 & 0 & 0 \\
0 & 0 & 0 & 0 & -1 & 0 & 0 & -1 & 6 & -1 & 0 & -1 & -1 & -1 & 0 & 0 \\
0 & 0 & 0 & 0 & -1 & -1 & 0 & 0 & -1 & 6 & -1 & 0 & 0 & -1 & -1 & 0 \\
0 & 0 & 0 & 0 & 0 & -1 & -1 & 0 & 0 & -1 & 6 & -1 & 0 & 0 & -1 & -1 \\
0 & 0 & 0 & 0 & 0 & 0 & -1 & -1 & -1 & 0 & -1 & 6 & -1 & 0 & 0 & -1 \\
-1 & -1 & 0 & 0 & 0 & 0 & 0 & 0 & -1 & 0 & 0 & -1 & 6 & -1 & 0 & -1 \\
0 & -1 & -1 & 0 & 0 & 0 & 0 & 0 & -1 & -1 & 0 & 0 & -1 & 6 & -1 & 0 \\
0 & 0 & -1 & -1 & 0 & 0 & 0 & 0 & 0 & -1 & -1 & 0 & 0 & -1 & 6 & -1 \\
-1 & 0 & 0 & -1 & 0 & 0 & 0 & 0 & 0 & 0 & -1 & -1 & -1 & 0 & -1 & 6
\end{pmatrix}
$$

and characteristic polynomial

$$(x - 6) \cdot (x - 2)^6 \cdot (x + 2)^9.$$

The graph Γ is depicted in Figure 2.7.

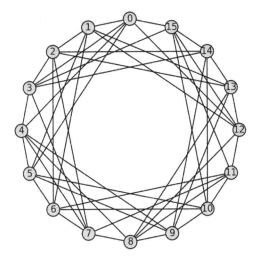

Figure 2.7: The Cayley graph of $\mathbb{Z}/4\mathbb{Z} \times \mathbb{Z}/4\mathbb{Z}$, with generator set $S = \{\pm(0,1), \pm(1,0), \pm(1,1)\}$, created using Sage.

```
———————————————————————————————————— Sage ————————————

sage: G = AdditiveAbelianGroup([4,4])
sage: GP = G.permutation_group()
sage: g0 = GP.gens()[0]
sage: g1 = GP.gens()[1]
sage: S = [g0, g1, g0^(-1), g1^(-1), (g0*g1)^(-1), g0*g1]
sage: Gamma = GP.cayley_graph(side='left', generators = S)
sage: A = Gamma.adjacency_matrix()
sage: Gamma1 = Graph(A, format = "adjacency_matrix")
sage: Gamma1.show(layout="circular", dpi = 300)
sage: Gamma1.characteristic_polynomial().factor()
(x - 6) * (x - 2)^6 * (x + 2)^9
```

2.6.2 Cayley graphs for non-abelian groups

Let $\alpha : G \rightarrow \mathbb{Z}$ be a given class function. Let $S \subset G$ be a subset which generates G satisfying $S = S^{-1}$ and $1 \notin S$. Let Γ denote the edge-weighted Cayley graph associated to (G, S, α), where edge (g_1, g_2) has weight $\alpha(g_2 g_1^{-1})$. In particular, we assume α is supported on S. In the notation above, $\Gamma = Cay(G, S, \alpha)$. We assume that the (weighted) adjacency matrix of Γ is the $|G| \times |G|$ matrix $A = (a_{g,h})$, where

$$a_{g,h} = \alpha(gh^{-1}).$$

Let G^* denote a complete set of inequivalent representations of G. We can write

$$\rho_i : G \rightarrow Aut(V_i),$$

where $V_i \cong \mathbb{C}^{d_i}$ and d_i is the degree of ρ_i, for $i = 1, \ldots, |G^*|$.

Example 2.6.4. While this is a very long example, we hope it will be useful to illustrate the ideas in the theorem below.

Let $G = D_6$ denote the dihedral group of order 12, written in the following order:

$$g_1 = 1, g_2 = (2,6)(3,5), g_3 = (1,2)(3,6)(4,5), g_4 = (1,2,3,4,5,6),$$
$$g_5 = (1,3)(4,6), g_6 = (1,3,5)(2,4,6), g_7 = (1,4)(2,3)(5,6),$$
$$g_8 = (1,4)(2,5)(3,6), g_9 = (1,5)(2,4), g_{10} = (1,5,3)(2,6,4),$$
$$g_{11} = (1,6,5,4,3,2), g_{12} = (1,6)(2,5)(3,4).$$

Let

$$S = \{g_4, g_{11}, g_3, g_7, g_{12}\}$$

and note S generates G, and is conjugation-invariant and closed under taking inverses. The conjugacy classes $cl_G(x)$ of G are ordered as follows:

$$cl_G(g_1), cl_G(g_2), cl_G(g_3), cl_G(g_4), cl_G(g_6), cl_G(g_8).$$

```
──────────────────────────── Sage ────────────────────────────
sage: G = DihedralGroup(6); G
Dihedral group of order 12 as a permutation group
sage: g1 = G.gens()[0]
sage: g2 = G.gens()[0]^(-1)
sage: g3 = G.gens()[1]
sage: Cg3 = [x^(-1)*g3*x for x in G]
sage: g4 = Cg3[1]; g5 = Cg3[2]
sage: S = [g1, g2, g3, g4, g5]; S
[(1,2,3,4,5,6), (1,6,5,4,3,2), (1,6)(2,5)(3,4), (1,2)(3,6)(4,5), (1,4)(2,3)(5,6)]
```

The character table of G is

	g_1	g_2	g_3	g_4	g_6	g_8
χ_1	1	1	1	1	1	1
χ_2	1	-1	-1	1	1	1
χ_3	1	-1	1	-1	1	-1
χ_4	1	1	-1	-1	1	-1
χ_6	2	0	0	1	-1	-2
χ_8	2	0	0	-1	-1	2

where χ_i is the i-th irreducible character of G (in the ordering given by Sage and GAP).

Define the class function $\alpha : G \to \mathbb{Z}$ by

$$\alpha(x) = \begin{cases} 1, & x \in \{(1,2,3,4,5,6), (1,6,5,4,3,2)\}, \\ 2, & x \in \{(1,2)(3,6)(4,5), (1,2)(3,6)(4,5), (1,4)(2,3)(5,6)\}, \\ 0, & \text{otherwise.} \end{cases}$$

The associated (weighted) adjacency matrix is

$$A = \begin{pmatrix} 0 & 0 & 2 & 1 & 0 & 0 & 2 & 0 & 0 & 0 & 1 & 2 \\ 0 & 0 & 1 & 2 & 0 & 0 & 0 & 2 & 0 & 0 & 2 & 1 \\ 2 & 1 & 0 & 0 & 1 & 2 & 0 & 0 & 0 & 2 & 0 & 0 \\ 1 & 2 & 0 & 0 & 2 & 1 & 0 & 0 & 2 & 0 & 0 & 0 \\ 0 & 0 & 1 & 2 & 0 & 0 & 1 & 2 & 0 & 0 & 2 & 0 \\ 0 & 0 & 2 & 1 & 0 & 0 & 2 & 1 & 0 & 0 & 0 & 2 \\ 2 & 0 & 0 & 0 & 1 & 2 & 0 & 0 & 1 & 2 & 0 & 0 \\ 0 & 2 & 0 & 0 & 2 & 1 & 0 & 0 & 2 & 1 & 0 & 0 \\ 0 & 0 & 0 & 2 & 0 & 0 & 1 & 2 & 0 & 0 & 2 & 1 \\ 0 & 0 & 2 & 0 & 0 & 0 & 2 & 1 & 0 & 0 & 1 & 2 \\ 1 & 2 & 0 & 0 & 2 & 0 & 0 & 0 & 2 & 1 & 0 & 0 \\ 2 & 1 & 0 & 0 & 0 & 2 & 0 & 0 & 1 & 2 & 0 & 0 \end{pmatrix}.$$

──────── Sage ────────

```
sage: def alpha(x):
          if x==S[0] or x==S[1]:
              return 1
          if x==S[2] or x==S[3] or x==S[4]:
              return 2
          return 0
sage: A = [[alpha(x*y^(-1)) for x in G] for y in G]; matrix(A)
```

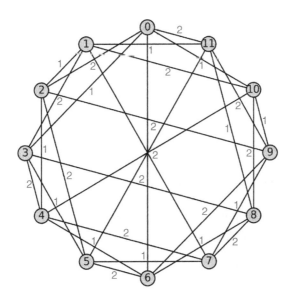

Figure 2.8: The undirected edge-weighted Cayley graph of D_6 with respect to S.

```
[0 0 2 1 0 0 2 0 0 0 1 2]
[0 0 1 2 0 0 0 2 0 0 2 1]
[2 1 0 0 1 2 0 0 0 2 0 0]
[1 2 0 0 2 1 0 0 2 0 0 0]
[0 0 1 2 0 0 1 2 0 0 2 0]
[0 0 2 1 0 0 2 1 0 0 0 2]
[2 0 0 0 1 2 0 0 1 2 0 0]
[0 2 0 0 2 1 0 0 2 1 0 0]
[0 0 0 2 0 0 1 2 0 0 2 1]
[0 0 2 0 0 0 2 1 0 0 1 2]
[1 2 0 0 2 0 0 0 2 1 0 0]
[2 1 0 0 0 2 0 0 1 2 0 0]
sage: Gamma = Graph(matrix(A), format = "adjacency_matrix", weighted=True)
sage: Gamma.show(layout="circular", dpi = 300, edge_labels=True)
sage: Gamma.automorphism_group(edge_labels=True).order()
144
sage: Gamma.automorphism_group().order()
1440
```

The weighted Cayley graph of (G, S, α) is shown in Figure 2.8.
The eigenvalues and eigenvectors of A are

$$\lambda_0 = 8, x_0 = (1,1,1,1,1,1,1,1,1,1,1,1),$$

$$\lambda_1 = -4, x_1 = (1,-1,-1,1,-1,1,-1,1,-1,1,1,-1),$$

$$\lambda_2 = 4, x_2 = (1,-1,1,-1,-1,1,1,-1,-1,1,-1,1),$$

$$\lambda_3 = -8, x_3 = (1,1,-1,-1,1,1,-1,-1,1,1,-1,-1),$$

each occurring with multiplicity 1. Note that, for $i = 0, 1, 2, 3$, we have

$$x_i = (\chi_i(g_1), \ldots, \chi_i(g_{12})),$$

where χ_i is the i-th character in the character table of G. Here they are in
Sage [2]:

──────────────────────────── Sage ────────────────────────────

```
sage: A*x1
(8, 8, 8, 8, 8, 8, 8, 8, 8, 8, 8, 8)
sage: x2 = vector([1,-1,-1,1,-1,1,-1,1,-1,1,1,-1])
sage: A*x2
(-4, 4, 4, -4, 4, -4, 4, -4, 4, -4, -4, 4)
sage: x3 = vector([1,-1,1,-1,-1,1,1,-1,-1,1,1,-1,1])
sage: A*x3
(4, -4, 4, -4, -4, 4, 4, -4, -4, 4, -4, 4)
sage: x4 = vector([1,1,-1,-1,1,1,-1,-1,1,1,-1,1])
sage: A*x4
(-4, -6, 8, 8, -8, -4, 8, 8, -6, -4, 8, 8)
sage: x4 = vector([1,1,-1,-1,1,1,-1,-1,1,1,-1,-1])
sage: A*x4
(-8, -8, 8, 8, -8, -8, 8, 8, -8, -8, 8, 8)
sage: x5 = vector([2,0,0,1,0,-1,0,-2,0,-1,1,0])
```

──

[2]Some care must be taken to record the entries of x_i consistent with the way the
elements of G are listed.

```
sage: A*x5
(2, 0, 0, 1, 0, -1, 0, -2, 0, -1, 1, 0)
```

Indeed, all these characters have degree 1. The remaining eigenvalues each occur with multiplicity 4. The eigenvalue $\lambda_4 = 1$, has eigenspace

$$
\begin{aligned}
E_{\lambda_4} = \operatorname{Span}((&1, 0, 0, 0, 0, -1, 0, -1, 0, 0, 1, 0), \\
(&0, 1, 0, 0, -1, 0, -1, 0, 0, 0, 0, 1), \\
(&0, 0, 1, 0, 1, 0, 0, 0, -1, 0, 0, -1), \\
(&0, 0, 0, 1, 0, 1, 0, 0, 0, -1, -1, 0)).
\end{aligned}
$$

It is easy to see that

$$
(\chi_4(g_1), \ldots, \chi_4(g_{12})) = (2, 0, 0, 1, 0, -1, 0, -2, 0, -1, 1, 0)
$$

is an element of E_{λ_4}. The eigenvalue $\lambda_5 = -1$, has eigenspace

$$
\begin{aligned}
E_{\lambda_5} = \operatorname{Span}((&1, 0, 0, 0, 0, -1, 0, 1, 0, 0, -1, 0), \\
(&0, 1, 0, 0, -1, 0, 1, 0, 0, 0, 0, -1), \\
(&0, 0, 1, 0, -1, 0, 0, 0, 1, 0, 0, -1), \\
(&0, 0, 0, 1, 0, -1, 0, 0, 0, 1, -1, 0)).
\end{aligned}
$$

It is easy to see that

$$
(\chi_5(g_1), \ldots, \chi_5(g_{12})) = (2, 0, 0, -1, 0, -1, 0, 2, 0, \ \ 1, -1, 0)
$$

is an element of E_{λ_5}.

The following well-known result describes a way to generalize the above example. Roughly speaking, it says that if Γ is a weighted Cayley graph attached to a group G and if the weight function of Γ is given by a class function of G, then the spectrum of Γ is determined by the representations of G.

Theorem 2.6.5. *Let α be a class function, and let $\Gamma = Cay(G, S, \alpha)$, as above (so, in particular, α is supported on S). If $n = |G|$, write*

$$
G = \{g_1 = 1, g_2, \ldots, g_n\}.
$$

Each eigenvector of the adjacency matrix A has the form

$$
(\chi(g_1), \ldots, \chi(g_n)),
$$

where $\chi = \mathrm{tr}(\rho)$, *for some* $\rho \in G^*$, *with eigenvalue*

$$\lambda = \lambda_\rho = \frac{1}{d_\rho} \sum_{s \in S} \alpha(s) \mathrm{tr}(\rho)(s),$$

where d_ρ *is the degree of* ρ. *Moreover, the multiplicity of* λ *is* $\chi(1)^2$.

The proof below follows Brouwer and Haemers [BH11], §6.3, and Kaski [KA02], §5, and is included for the reader's convenience. See also Rockmore, Kostelec, Hordijk, and Stadler [RKHS02].

Proof. Suppose Γ has vertex set V, and $W = C^0(\Gamma, \mathbb{R}) \cong \mathbb{R}^V$ is the $\mathbb{R}$-vector space spanned by the vertices of Γ. Part of this proof applies to any matrix which commutes with the action of G, such as the (unweighted) adjacency matrix A^* or the Laplacian Q. By Schur's Lemma[3], A^* acts as a scalar on each irreducible G-invariant subspace of W. In other words, the irreducible G-invariant subspaces are eigenspaces of A^*. If A^* acts like θI on the irreducible G-invariant subspace $U = W_\chi$ with character χ, then $tr(A^*g|_U) = \theta\chi(g)$.

Since S is a union of conjugacy classes of G, the weighted adjacency matrix A commutes with the elements of G, and the previous discussion applies. The regular representation of G decomposes into a direct sum of irreducible subspaces, where for each irreducible character χ there are $\chi(1)$ copies of W_χ. To be explicit, W_χ is spanned by $v_\chi = (\chi(g_1), \ldots, \chi(g_n))$ and all its images under the G-action. On each copy A acts like θI, for some $\theta \in \mathbb{R}$, and $\dim(W_\chi) = \chi(1)$, so θ has multiplicity $\chi(1)^2$. We saw that

$$tr(Ag|_{W_\chi}) = \theta\chi(g).$$

The first entry of Av_χ is equal to $\sum_{s \in S} \alpha(s)\chi(s)$. Since this is also $\theta\chi(1)$, we have

$$\theta = \frac{1}{\chi(1)} \sum_{s \in S} \alpha(s)\chi(s) = \frac{1}{\chi(1)} tr(A|_{W_\chi}).$$

$\square$

---

[3] If π is an irreducible n-dimensional representation of G and if $B \in GL(n, \mathbb{C})$ commutes with all matrices $\pi(g)$, $g \in G$, then B is a scalar matrix. A proof can be found in many textbooks on abstract algebra, e.g., [DF99], page 337.

Example 2.6.6. This is an extension of (the already very long) Example 2.6.4.

Let G and Γ be as in Example 2.6.4. The unweighted adjacency matrix is

$$
A_0 = \begin{pmatrix}
0 & 0 & 1 & 1 & 0 & 0 & 1 & 0 & 0 & 0 & 1 & 1 \\
0 & 0 & 1 & 1 & 0 & 0 & 0 & 1 & 0 & 0 & 1 & 1 \\
1 & 1 & 0 & 0 & 1 & 1 & 0 & 0 & 0 & 1 & 0 & 0 \\
1 & 1 & 0 & 0 & 1 & 1 & 0 & 0 & 1 & 0 & 0 & 0 \\
0 & 0 & 1 & 1 & 0 & 0 & 1 & 1 & 0 & 0 & 1 & 0 \\
0 & 0 & 1 & 1 & 0 & 0 & 1 & 1 & 0 & 0 & 0 & 1 \\
1 & 0 & 0 & 0 & 1 & 1 & 0 & 0 & 1 & 1 & 0 & 0 \\
0 & 1 & 0 & 0 & 1 & 1 & 0 & 0 & 1 & 1 & 0 & 0 \\
0 & 0 & 0 & 1 & 0 & 0 & 1 & 1 & 0 & 0 & 1 & 1 \\
0 & 0 & 1 & 0 & 0 & 0 & 1 & 1 & 0 & 0 & 1 & 1 \\
1 & 1 & 0 & 0 & 1 & 0 & 0 & 0 & 1 & 1 & 0 & 0 \\
1 & 1 & 0 & 0 & 0 & 1 & 0 & 0 & 1 & 1 & 0 & 0
\end{pmatrix}.
$$

--------- Sage ---------

```
sage: A0 = Gamma.adjacency_matrix()
sage: A0*v1==(5)*v1
True
sage: A0*v2==(-1)*v2
True
sage: A0*v3==(1)*v3
True
sage: A0*v4==(-5)*v4
True
sage: A0*v5==(1)*v5
True
sage: A0*v6==(-1)*v6
True
```

In other words, in this example, Sage helps us verify that, even if we use the unweighted adjacency matrix, the conclusion of the above Theorem 2.6.5 still holds.

2.7 Additive Cayley graphs

An *additive Cayley graph* (or *Cayley sum graph*) Γ with sum set S in a finite abelian group G has vertex set $V_\Gamma = G$, and two elements $g, h \in G$ are adjacent (i.e., connected by a edge) if and only if $g + h \in S$.

Let G^* denote the *dual group* of G, that is, the multiplicative group of multiplicative characters $\chi : G \to \mathbb{C}^\times$. Let $G^*_{\mathbb{R}}$ denote the subgroup of real-valued characters $\chi : G \to \mathbb{R}^\times$. For convenience of notation, we fix an indexing of the elements of these groups,

$$G = \{g_1, \ldots, g_n\},$$

and

$$G^* = \{\chi_1, \ldots, \chi_n\}.$$

Example 2.7.1. Consider the group $G = \mathbb{Z}/10\mathbb{Z}$ and the set $S = \{3, 5, 7\}$. The adjacency matrix of the associated additive Cayley graph is

$$\begin{pmatrix}
0 & 0 & 1 & 1 & 0 & 1 & 0 & 1 & 0 & 0 \\
0 & 0 & 1 & 0 & 1 & 0 & 1 & 0 & 0 & 0 \\
1 & 1 & 0 & 1 & 0 & 1 & 0 & 0 & 0 & 0 \\
1 & 0 & 1 & 0 & 1 & 0 & 0 & 0 & 0 & 1 \\
0 & 1 & 0 & 1 & 0 & 0 & 0 & 0 & 1 & 1 \\
1 & 0 & 1 & 0 & 0 & 0 & 0 & 1 & 1 & 0 \\
0 & 1 & 0 & 0 & 0 & 0 & 0 & 1 & 0 & 1 \\
1 & 0 & 0 & 0 & 0 & 1 & 1 & 0 & 1 & 0 \\
0 & 0 & 0 & 0 & 1 & 1 & 0 & 1 & 0 & 1 \\
0 & 0 & 0 & 1 & 1 & 0 & 1 & 0 & 1 & 0
\end{pmatrix}$$

and the graph itself is depicted in Figure 2.9.

If $\chi(S) = \sum_{s \in S} \chi(s)$, the values of $\chi(S)$, for $\chi \in G^*$, are listed as follows:

$$\chi_1(S) = 3, \quad \chi_2(S) = e^{\frac{7}{5}i\pi} + e^{\frac{3}{5}i\pi} - 1, \quad \chi_3(S) = e^{\frac{14}{5}i\pi} + e^{\frac{6}{5}i\pi} + 1,$$
$$\chi_4(S) = e^{\frac{21}{5}i\pi} + e^{\frac{9}{5}i\pi} - 1, \quad \chi_5(S) = e^{\frac{28}{5}i\pi} + e^{\frac{12}{5}i\pi} + 1, \quad \chi_6(S) = -3,$$
$$\chi_7(S) = e^{\frac{42}{5}i\pi} + e^{\frac{18}{5}i\pi} + 1, \quad \chi_8(S) = e^{\frac{49}{5}i\pi} + e^{\frac{21}{5}i\pi} - 1,$$
$$\chi_9(S) = e^{\frac{56}{5}i\pi} + e^{\frac{24}{5}i\pi} + 1, \quad \chi_{10}(S) = e^{\frac{63}{5}i\pi} + e^{\frac{27}{5}i\pi} - 1.$$

The following result can be found in Brouwer and Haemers [BH11].

Proposition 2.7.2. *Let Γ be the additive Cayley graph with sum set S in the finite abelian group G. The spectrum of Γ consists of*

$$\{\chi(S) \mid \chi \in G^*_{\mathbb{R}}\} \cup \{\pm|\chi(S)| \mid \chi \in G^* - G^*_{\mathbb{R}}\}.$$

The eigenvector of $\chi(S)$, for $\chi \in G^*_{\mathbb{R}}$, is

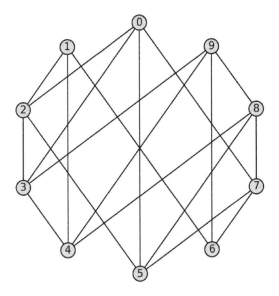

Figure 2.9:   The undirected additive Cayley graph of $\mathbb{Z}/10\mathbb{Z}$ with respect to S.

$$x_\chi = (\chi(g_1), \ldots, \chi(g_n)).$$

Pick $\alpha = \alpha_{S,\chi} \in \mathbb{C}^\times$ so that $|\chi(S)| = \alpha^2 \chi(S)$. The eigenvector of $|\chi(S)|$, for $\chi \in G^* - G_{\mathbb{R}}^*$, is

$$x_{\chi,+} = (\mathit{Re}\,(\alpha\chi(g_1)), \ldots, \mathit{Re}\,(\alpha\chi(g_n))).$$

The eigenvector of $-|\chi(S)|$, for $\chi \in G^* - G_{\mathbb{R}}^*$, is

$$x_{\chi,-} = (\mathit{Im}\,(\alpha\chi(g_1)), \ldots, \mathit{Im}\,(\alpha\chi(g_n))).$$

Proof. For $x, y \in G$, define $x \sim y$ if and only if $x + y \in S$. If the vertices of Γ are denoted $1, \ldots, n$, then the (unweighted) adjacency matrix A acts on $\mathbb{R}^n$ by sending a vector $x = (x_1, \ldots, x_n)$ to $x' = (x_1', \ldots, x_n')$, where

$$x_j' = \sum_{(i,j)\in E_\Gamma} x_i.$$

If $\chi : G \to \mathbb{C}^\times$ is a character of G, then

$$\sum_{y \sim x} \chi(y) = \sum_{s \in S} \chi(s - x) = \left(\sum_{s \in S} \chi(s)\right) \chi(-x) = \chi(S)\overline{\chi(x)}.$$

Since Γ is undirected, the spectrum is real.

If χ is a real character, then the computation displayed above tells us that $\mathbf{v}_\chi$ is an eigenvector of A with eigenvalue $\chi(S)$. If χ is nonreal, then $x_{\chi,+}$ and $x_{\chi,-}$ are eigenvectors with eigenvalues $|\chi(S)|$ and $-|\chi(S)|$, respectively. Indeed, the j-th component of $Ax_{\chi,+}$ is

$$
\begin{aligned}
(Ax_{\chi,+})_j &= \sum_{(i,j)\in E_\Gamma} Re(\alpha\chi(g_i)) \\
&= Re(\alpha \sum_{(i,j)\in E_\Gamma} \chi(g_i)) \\
&= Re(\alpha \sum_{s\in S} \chi(s - g_j)) \\
&= Re(\alpha^{-1}\chi(g_j)\alpha^2\chi(S)) \\
&= |\chi(S)| \cdot Re(\alpha\chi(g_j)).
\end{aligned}
$$

$\square$

An analogous definition of an additive Cayley graph associated to a subset $S \subset G$ is a graph Γ which has vertex set $V_\Gamma = G$, and two elements $g, h \in G$ are adjacent (i.e., connected by a edge) if and only if $g - h \in S$. In this case, we also require that S is symmetric: $S = -S$.

The next section is devoted to a class of graphs of this type.

2.7.1 Cayley graphs and p-ary functions

This section describes a type of Cayley graph attached to a p-ary function. More details are in Chapter 6 and in Celerier, Joyner, Melles, Phillips, and Walsh [CJMPW15].

Fix $n \geq 1$ and let $V = GF(p)^n$, where p is a prime.

If $f \colon V \to GF(p)$, then we let $f_{\mathbb{C}} \colon V \to \mathbb{C}$ be the function whose values are those of f but regarded as integers (i.e., we select the congruence class residue representative in the interval $\{0, 1, \ldots, p-1\}$). We sometimes abuse notation and often write f in place of $f_{\mathbb{C}}$.

Let f be a $GF(p)$-valued function on V such that $f(0) = 0$.

The *Cayley graph of f* is defined to be the edge-weighted directed graph

$$
\Gamma_f = (GF(p)^n, E_f), \tag{2.11}
$$

whose vertex set is $V = V(\Gamma_f) = GF(p)^n$ and whose set of edges is defined by

$$
E_f = \{(u, v) \in GF(p)^n \mid f(u - v) \neq 0\},
$$

where the edge $(u, v) \in E_f$ has weight $f(u-v)$. However, if f is even then we can (and do) regard Γ_f as a weighted (undirected) graph. We assume from this point on that f is even.

The adjacency matrix $A = A_f$ is the matrix whose entries are

$$A_{i,j} = f_{\mathbb{C}}(\eta(i) - \eta(j)), \tag{2.12}$$

where $\eta(k)$ is the p-ary representation as in (6.5). Ignoring edge weights, we let

$$A_{i,j}^* = \begin{cases} 1, & f_{\mathbb{C}}(\eta(i) - \eta(j)) \neq 0, \\ 0, & \text{otherwise.} \end{cases} \tag{2.13}$$

Let $\operatorname{supp}(f) = \{v \in V \mid f(v) \neq 0\}$ be the *support* of f, let

$$\omega_f = |\operatorname{supp}(f)|,$$

and let

$$\sigma_f = \sum_{v \in V} f_{\mathbb{C}}(v).$$

Clearly, the vertices in Γ_f connected to $0 \in V$ are in natural bijection with the elements of $\operatorname{supp}(f)$.

Recall that, given a graph Γ and its adjacency matrix A, the spectrum

$$\sigma(\Gamma) = \{\lambda_1, \lambda_2, \ldots, \lambda_N\},$$

is the multi-set of real eigenvalues of A. When Γ is the Cayley graph of a p-ary function on $GF(p)^n$, we have $N = p^n$. Following a standard convention, we index the elements $\lambda_i = \lambda_i(A)$ of the spectrum in such a way that they are monotonically increasing. Because Γ_f is regular, the row sums of A are all σ_f, hence the all 1's vector is an eigenvector of A with eigenvalue σ_f. We will see later (Corollary 2.7.6) that $\lambda_N(A) = \sigma_f$.

Let Δ denote the identity matrix multiplied by σ_f. The Laplacian of Γ_f is the matrix $Q = \Delta - A$.

Lemma 2.7.3. *Assume f is even. As an edge-weighted graph, Γ_f is connected if and only if $\lambda_{N-1}(A) < \lambda_N(A) = \sigma_f$, where A is the adjacency matrix of Equation (2.12). If we ignore edge weights, then Γ_f is connected if and only if $\lambda_{N-1}(A^*) < \lambda_N(A^*) = \omega_f$, where A^* is the unweighted adjacency matrix in (2.13).*

Proof. We only prove the statement for the edge-weighted case.

Note that for $i = 1, \ldots, N$, $\lambda_i(Q) = \sigma_f - \lambda_{N-i+1}(A)$, since $\det(Q - \lambda I) = \det(\sigma_f I - A - \lambda I) = (-1)^n \det(A - (\sigma_f - \lambda)I)$. Thus, $\lambda_i(Q) \geq 0$, for all i. By a theorem of Fiedler [Fi73], $\lambda_2(Q) > 0$ if and only if Γ_f is connected. But $\lambda_2(Q) > 0$ is equivalent to $\sigma_f - \lambda_{N-1}(A) > 0$. $\qquad\square$

Recall a circulant matrix is a square matrix where each row vector is a cyclic shift one element to the right relative to the preceding row vector.

Our Fourier transform matrix F is not circulant, but is "block circulant." Like circulant matrices, it has the property that $\mathbf{v}_a = (\zeta^{-\langle a, x \rangle} \mid x \in V)$ is an eigenvector with eigenvalue $\lambda_a = \hat{f}(-a)$ (something related to a value of the Hadamard transform of f). Thus, the proposition below shows that it "morally" behaves like a circulant matrix in some ways.

Proposition 2.7.4. *The eigenvalues $\lambda_a = \hat{f}(-a)$ of this matrix F are values of the Fourier transform of the function $f_\mathbb{C}$,*

$$\hat{f}(y) = \sum_{x \in V} f_\mathbb{C}(x)\zeta^{-\langle x, y \rangle},$$

and the eigenvectors are the vectors of p-th roots of unity,

$$\mathbf{v}_a = (\zeta^{-\langle a, x \rangle} \mid x \in V).$$

Proof. In $F = (F_{i,j})$, we have $F_{i,j} = f_\mathbb{C}(\eta(i) - \eta(j))$ for $i, j \in \{0, 1, \ldots, p^n - 1\}$. For each $a \in GF(p)^n$, let

$$\mathbf{v}_a = (\zeta^{-\langle a, \eta(i) \rangle} \mid i \in \{0, 1, \ldots, p^n - 1\}).$$

Then

$$F\mathbf{v}_a = (\sum_{y \in V} f_\mathbb{C}(x - y)\zeta^{-\langle a, y \rangle} \mid x \in V).$$

The entry in the i-th coordinate, where $x = \eta(i)$ is given by

$$\sum_{y \in V} f_\mathbb{C}(x - y)\zeta^{-\langle a, y \rangle} = \sum_{y \in V} f_\mathbb{C}(-y)\zeta^{-\langle a, y + x \rangle}$$

$$= \zeta^{-\langle a, x \rangle} \sum_{y \in V} f_\mathbb{C}(-y)\zeta^{-\langle a, y \rangle}$$

$$= \zeta^{-\langle a, x \rangle} \sum_{y \in V} f_\mathbb{C}(y)\zeta^{\langle a, y \rangle}$$

$$= \zeta^{-\langle a, x \rangle} \hat{f}(-a).$$

Therefore, the coordinates of the vector $F\mathbf{v}_a$ are the same as those of $\mathbf{v}_a$, up to a scalar factor. Thus $\lambda_a = \hat{f}(-a)$ is an eigenvalue and $\mathbf{v}_a = (\zeta^{-\langle a, x \rangle} \mid x \in V)$ is an eigenvector. $\qquad\square$

Corollary 2.7.5. *The matrix F is invertible if and only if none of the values of the Fourier transform of $f_\mathbb{C}$ vanish.*

Corollary 2.7.6. *The spectrum of the graph Γ_f is precisely the set of values of the Fourier transform of $f_\mathbb{C}$.*

Example 2.7.7. We take $V = GF(3)^2$ and consider an even function $f :$ $V \to GF(3)$ given by

$$f(x_0, x_1) = -x_0^2 x_1^2 + x_0^2 + x_0 x_1 - x_1^2.$$

Its Cayley graph Γ_f has weighted adjacency matrix

$$A_w = \begin{pmatrix} 0 & 1 & 1 & 2 & 0 & 1 & 2 & 1 & 0 \\ 1 & 0 & 1 & 1 & 2 & 0 & 0 & 2 & 1 \\ 1 & 1 & 0 & 0 & 1 & 2 & 1 & 0 & 2 \\ 2 & 1 & 0 & 0 & 1 & 1 & 2 & 0 & 1 \\ 0 & 2 & 1 & 1 & 0 & 1 & 1 & 2 & 0 \\ 1 & 0 & 2 & 1 & 1 & 0 & 0 & 1 & 2 \\ 2 & 0 & 1 & 2 & 1 & 0 & 0 & 1 & 1 \\ 1 & 2 & 0 & 0 & 2 & 1 & 1 & 0 & 1 \\ 0 & 1 & 2 & 1 & 0 & 2 & 1 & 1 & 0 \end{pmatrix}.$$

As a weighted graph, the adjacency spectrum is

$$8, 2, 2, -1, -1, -1, -1, -4, -4,$$

while the Laplacian spectrum is

$$12, 12, 9, 9, 9, 9, 6, 6, 0.$$

The unweighted adjacency matrix is

$$A = \begin{pmatrix} 0 & 1 & 1 & 1 & 0 & 1 & 1 & 1 & 0 \\ 1 & 0 & 1 & 1 & 1 & 0 & 0 & 1 & 1 \\ 1 & 1 & 0 & 0 & 1 & 1 & 1 & 0 & 1 \\ 1 & 1 & 0 & 0 & 1 & 1 & 1 & 0 & 1 \\ 0 & 1 & 1 & 1 & 0 & 1 & 1 & 1 & 0 \\ 1 & 0 & 1 & 1 & 1 & 0 & 0 & 1 & 1 \\ 1 & 0 & 1 & 1 & 1 & 0 & 0 & 1 & 1 \\ 1 & 1 & 0 & 0 & 1 & 1 & 1 & 0 & 1 \\ 0 & 1 & 1 & 1 & 0 & 1 & 1 & 1 & 0 \end{pmatrix}.$$

As an unweighted graph, its Laplacian has eigenvalues

$$9, 9, 6, 6, 6, 6, 6, 6, 0,$$

and the adjacency matrix has eigenvalues

$$6, 0, 0, 0, 0, 0, 0, -3, -3.$$

This function is not bent (in the sense of §6.3), yet the unweighted version of the graph Γ_f, shown in Figure 2.10, is a strongly regular graph with parameters $(9, 6, 3, 6)$. (Strongly regular is defined in §6.6.1 below.) However, the complement of Γ_f is disconnected, so Γ_f is not a primitive strongly regular graph[4].

Consider the subgraph Γ_1 of weight one edges. This has adjacency matrix

$$A_1 = \begin{pmatrix} 0 & 1 & 1 & 0 & 0 & 1 & 0 & 1 & 0 \\ 1 & 0 & 1 & 1 & 0 & 0 & 0 & 0 & 1 \\ 1 & 1 & 0 & 0 & 1 & 0 & 1 & 0 & 0 \\ 0 & 1 & 0 & 0 & 1 & 1 & 0 & 0 & 1 \\ 0 & 0 & 1 & 1 & 0 & 1 & 1 & 0 & 0 \\ 1 & 0 & 0 & 1 & 1 & 0 & 0 & 1 & 0 \\ 0 & 0 & 1 & 0 & 1 & 0 & 0 & 1 & 1 \\ 1 & 0 & 0 & 0 & 0 & 1 & 1 & 0 & 1 \\ 0 & 1 & 0 & 1 & 0 & 0 & 1 & 1 & 0 \end{pmatrix},$$

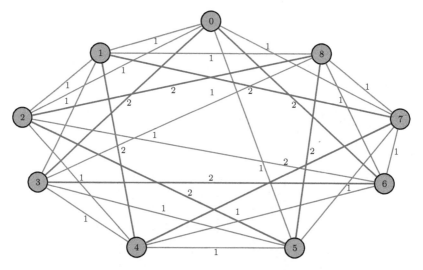

Figure 2.10: The undirected Cayley graph of an even $GF(3)$-valued function of two variables from Example 2.7.7. (The vertices are ordered as in the example.)

and is depicted in Figure 2.11.

Clearly, this is a 4-regular graph on 9 vertices. In fact, Sage allows us to verify that it is isomorphic to the Paley graph on 9 vertices.

---

[4] An SRG is *primitive* if both the graph and its complement are connected.

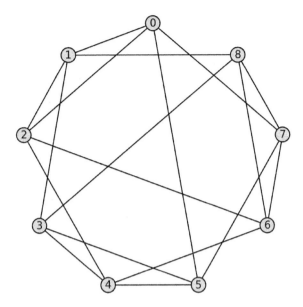

Figure 2.11: The subgraph Γ_1 of Γ is isomorphic to the Paley graph on 9 vertices.

```
─────────────────────────── Sage ───────────────────────────

sage: F = GF(3)
sage: V = F^2
sage: f = lambda x:-x[0]^2*x[1]^2+x[0]^2+x[0]*x[1]-x[1]^2
sage: Gamma = boolean_cayley_graph(f, V)
sage: Alist = [[ZZ(f(x-y)) for x in V] for y in V]; matrix(ZZ, Alist)
[0 1 1 2 0 1 2 1 0]
[1 0 1 1 2 0 0 2 1]
[1 1 0 0 1 2 1 0 2]
[2 1 0 0 1 1 2 0 1]
[0 2 1 1 0 1 1 2 0]
[1 0 2 1 1 0 0 1 2]
[2 0 1 2 1 0 0 1 1]
[1 2 0 0 2 1 1 0 1]
[0 1 2 1 0 2 1 1 0]
sage: A = matrix(GF(2), Alist); A
[0 1 1 0 0 1 0 1 0]
[1 0 1 1 0 0 0 0 1]
[1 1 0 0 1 0 1 0 0]
[0 1 0 0 1 1 0 0 1]
[0 0 1 1 0 1 1 0 0]
[1 0 0 1 1 0 0 1 0]
[0 0 1 0 1 0 0 1 1]
[1 0 0 0 0 1 1 0 1]
[0 1 0 1 0 0 1 1 0]
sage: Gamma1 = Graph(A)
sage: Gamma2 = graphs.PaleyGraph(9)
sage: Gamma1.is_isomorphic(Gamma2)
True
```

2.8 Graphs of group quotients

If $\Gamma = (V, E)$ is a graph and G a subgroup of its automorphism group, we define the *quotient graph* by G, denoted Γ/G, as follows:

1. The vertices of Γ/G are the G-orbits in V.

2. Distinct vertices $\overline{v_1}$, $\overline{v_2}$ of Γ/G are connected by an edge if and only if there is a vertex v_1 in V belonging to the orbit $\overline{v_1}$, and a vertex v_2 in V belonging to the orbit $\overline{v_2}$, for which (v_1, v_2) belongs to E.

3. Γ/G is simple.

For example, if Γ is any graph and G is any group that acts regularly[5] on Γ then Γ/G is the empty graph with one vertex.

Example 2.8.1. The graph Γ_2 depicted in Figure 2.12 is a 2-fold cover[6] of the diamond graph Γ_1 (isomorphic to the graph depicted in Figure 1.2).

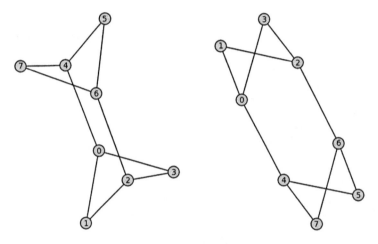

Figure 2.12:   Two depictions of a twofold cover of the diamond graph.

Moreover, the automorphism group G of Γ_2 is a cyclic group of order 16 and the quotient map $\Gamma_2 \to \Gamma_1 = \Gamma_2/G_0$ is harmonic[7], where G_0 is the subgroup of order 2 generated by $(0, 6)(1, 5)(2, 4)(3, 7)$. The Laplacian of Γ_1 is

---

[5]The adjective "regular" is over-used in mathematics. In this case, it means that G acts transitively on Γ (i.e., each vertex can be sent to any other by some element of G) and no vertex is fixed by any element of $G - \{1\}$.

[6]A cover is defined in §3.4.

[7]In the sense of Definition 3.3.5 below.

$$Q_1 = \begin{pmatrix} 3 & -1 & -1 & -1 \\ -1 & 2 & -1 & 0 \\ -1 & -1 & 3 & -1 \\ -1 & 0 & -1 & 2 \end{pmatrix},$$

and the Laplacian of Γ_2 is

$$Q_2 = \begin{pmatrix} 3 & -1 & 0 & -1 & -1 & 0 & 0 & 0 \\ -1 & 2 & -1 & 0 & 0 & 0 & 0 & 0 \\ 0 & -1 & 3 & -1 & 0 & 0 & -1 & 0 \\ -1 & 0 & -1 & 2 & 0 & 0 & 0 & 0 \\ -1 & 0 & 0 & 0 & 3 & -1 & 0 & -1 \\ 0 & 0 & 0 & 0 & -1 & 2 & -1 & 0 \\ 0 & 0 & -1 & 0 & 0 & -1 & 3 & -1 \\ 0 & 0 & 0 & 0 & -1 & 0 & -1 & 2 \end{pmatrix}.$$

```
─────────────────────────────── Sage ───────────────────────────────

sage: V1 = [0,1,2,3]
sage: E1 = [(0,1),(0,2),(0,3),(1,2),(2,3)]
sage: V2 = [0,1,2,3,4,5,6,7]
sage: E2 = [(0,1),(0,3),(0,4),(1,2),(2,3),(2,6),(4,5),(4,7),(6,5),(6,7)]
sage: Gamma1 = Graph([V1,E1], format='vertices_and_edges')
sage: Gamma2 = Graph([V2,E2], format='vertices_and_edges')
sage: AG = Gamma2.automorphism_group()
sage: AG.cardinality()
16
sage: AG.sylow_subgroup(2).cardinality()
16
sage: Q2 = Gamma2.laplacian_matrix()
sage: Q1 = Gamma1.laplacian_matrix()
sage: factor(Q2.charpoly())
x * (x - 4)^2 * (x - 2)^3 * (x^2 - 6*x + 4)
sage: factor(Q1.charpoly())
(x - 2) * x * (x - 4)^2
```

Recall that if Γ is a graph, we denote by $\sigma(\Gamma)$ the spectrum of Γ, i.e., the multi-set of eigenvalues of the adjacency matrix of A. Recall also that if G is a finite multiplicative group, S is a subset of G such that $S = S^{-1}$, and S does not contain the identity 1, then we denote by $Cay(G, S)$ the Cayley graph of the pair (G, S). The next proposition states that if S is also a generating set of G, and if H is a "good" subgroup of G, then the spectrum of the Cayley graph of $(H, H \cap S)$ is contained in a translate of the spectrum of the Cayley graph of (G, S).

Proposition 2.8.2. *Let G be a finite multiplicative group. Let $H \subset G$ be a normal subgroup. Assume that $S \subset G$ is a symmetric (i.e., $S = S^{-1}$) generating set with $1 \neq S$ such that (1) $H \cap S$ generates H, (2) $S - H \cap S$ generates a subgroup K disjoint from H (i.e., $H \cap K = \{1\}$), and (3) G acts by permutations on $H \cap S$ via conjugations. Then*

$$|S - H \cap S| + \sigma(Cay(H, H \cap S)) \subset \sigma(Cay(G, S)).$$

Proof. If G is abelian then this follows from Proposition 2.6.2. Indeed, in this case

$$\sigma(Cay(H, H \cap S)) = \{\chi(H \cap S) \mid \chi \in H^*\},$$

and

$$\sigma(Cay(G, S)) = \{\chi(S) \mid \chi \in G^*\}.$$

Since G is a direct product of cyclic subgroups of prime power order, H is a sub-product. By induction, we may assume without loss of generality that G is cyclic and so is H. For $\chi \in G^*$ such that $\chi|_K = 1$, we have

$$\chi(S) = \sum_{s \in S} \chi(s) = \sum_{s \in H \cap S} \chi(s) + \sum_{s \in S - H \cap S} \chi(s) = \chi(H \cap S) + |S - H \cap S|,$$

by our hypothesis on $K = G/H$.

Now, assume G is non-abelian. By Theorem 2.6.5,

$$\sigma(Cay(H, H \cap S)) = \left\{ \frac{1}{\deg(\chi)} \chi(H \cap S) \mid \chi \in H^* \right\},$$

and

$$\sigma(Cay(G, S)) = \left\{ \frac{1}{\deg(\chi)} \chi(S) \mid \chi \in G^* \right\}. \tag{2.14}$$

We need to understand the behavior of characters under restriction. The following fact can be found in Weintraub [Wi03] (Corollary 1.12, p. 105): Let $H \subset G$ be normal, $\rho : G \to Aut(V)$ an irreducible representation, $\tau : H \to Aut(W)$ an irreducible component of $Res_H^G(\rho) = \rho|_H$. Then

$$Res_H^G(\rho) = m \oplus_j \tau_j,$$

for some integer $m \geq 1$, where $\{\tau_j\}$ is a complete set of representatives of conjugates of $\tau = \tau_1$ under the action of G. It is easy to compute m if the number of conjugates $N = |\{\tau_j\}|$ is known:

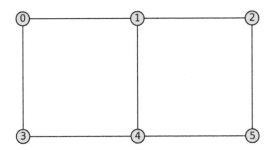

Figure 2.13: The ladder graph L_3.

$$m = \frac{1}{N}\frac{\deg(\rho)}{\deg(\tau)}.$$

We assumed that $S - H \cap S$ generates a group $K \subset G$ for which $H \cap K = \{1\}$. This implies that if $\chi = tr(\rho)$ and $\psi = tr(\tau)$, then

$$
\begin{aligned}
\chi(S) &= \sum_{s \in S} \chi(s) \\
&= \sum_{s \in H \cap S} \chi(s) + \sum_{s \in S - H \cap S} \chi(s) \\
&= \chi(H \cap S) + |S - H \cap S|\deg(\chi) \\
&= m\sum_{j} tr(\tau_j)(H \cap S) + |S - H \cap S|\deg(\chi) \\
&= \frac{1}{N}\frac{\deg(\rho)}{\deg(\tau)} \cdot N \cdot \psi(H \cap S) + |S - H \cap S|\deg(\chi) \qquad \text{by condition (3)} \\
&= \frac{\deg(\rho)}{\deg(\tau)} \cdot \psi(H \cap S) + |S - H \cap S|\deg(\chi).
\end{aligned}
$$

Dividing by $\deg(\chi) = \deg(\rho)$ and using Equation (2.14) gives the result. □

Example 2.8.3. We consider next the ladder graph of 6 vertices L_3, depicted in Figure 2.13. The automorphism group G of L_3 is the permutation group with generators $\{(0,2)(3,5),(0,3)(1,4)(2,5)\}$. Because G has only 4 elements, it can't be vertex transitive or edge transitive.

The quotient L_3/G is the connected graph with 2 vertices.

Example 2.8.4. We continue with the weighted Cayley graph Γ of $G = D_6$ in Example 2.6.4 above. The following Sage computation uses the Sage algebraic graph theory module written for this book[8]. It must be loaded before the following computations are possible.

```
─────────────────────────── Sage ───────────────────────────
sage: G = DihedralGroup(6); G
Dihedral group of order 12 as a permutation group
sage: G3 = G.sylow_subgroup(3)
sage: G3.order()
3
sage: A = [[alpha(x*y^(-1)) for x in G] for y in G]
sage: Gamma = Graph(matrix(A), format = "adjacency_matrix", weighted=True)
sage: Gamma.automorphism_group(edge_labels=True)
Permutation Group with generators [(3,10)(5,9), (2,11)(4,8),
   (1,4)(6,11), (0,1)(2,3)(4,5)(6,7)(8,9)(10,11), (0,2,5,6,9,11)(1,3,4,7,8,10)]
sage: AutGammaD12a = Gamma.automorphism_group()
sage: AutGammaD12b = Gamma.automorphism_group(edge_labels=True)
sage: AutGammaD12b.order()
144
sage: AutGammaD12a.order()
1440
sage: G9 = AutGammaD12b.sylow_subgroup(3)
sage: G9.order()
9
```

Let G_9 denote the Sylow subgroup of G. This group acts on the edge-weighted graph Γ (in the sense of §2.6). The orbits of G_9 on Γ are

$$\bar{0} = \{0,5,9\}, \bar{1} = \{1,4,8\}, \bar{2} = \{2,6,11\}, \bar{3} = \{3,7,10\},$$

and the edges in the cycle graph Γ/G_9 are $(\bar{0},\bar{2})$ (weight 2), $(\bar{1},\bar{3})$ (weight 2), $(\bar{1},\bar{2})$ (weight 1), $(\bar{0},\bar{3})$ (weight 1). The orbits of G_3 on Γ are

$$\bar{0} = \{0\}, \bar{1} = \{1,3,5\}, \bar{2} = \{2,4,6\}, \bar{7} = \{7\},$$
$$\bar{8} = \{8\}, \bar{9} = \{9\}, \overline{10} = \{10\}, \overline{11} = \{11\},$$

and the edges in Γ/G_9 are as in Figure 2.14, where 0 corresponds to $\bar{0}$, 1 corresponds to $\bar{1}$, 2 corresponds to $\bar{2}$, 3 corresponds to $\bar{7}$, 4 corresponds to $\bar{8}$, 5 corresponds to $\bar{9}$, 6 corresponds to $\overline{10}$, and 7 corresponds to $\overline{11}$.

Example 2.8.5. Consider the Paley graph on 9 vertices, Γ, depicted in Figure 1.12.

---

[8]This module is available from the github site for this book.

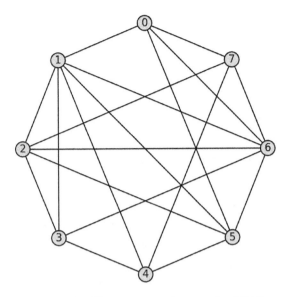

Figure 2.14:   The group quotient graph of Γ/G_3.

The automorphism group G of Γ has order 72. The action of G on the vertices of Γ is transitive. The action of G on the edges of Γ is also transitive. Thus, the quotient graph Γ/G is the empty graph with one vertex.

The 2-Sylow subgroup G_2 of G is order 8. The quotient graph Γ/G_2 is described as follows. Vertices of Γ:

$$0, 1, 2, a, a+1, a+2, 2a, 2a+1, 2a+2,$$

where $a \in GF(9) - GF(3)$ is a root of the generating polynomial, $x^2 + 2x + 2$. The orbits under the G_2-action:

$$\bar{0} = \{0\}, \bar{1} = \{1, 2, a+1, 2a+2\}, \bar{a} = \{a, a+2, 2a, 2a+1\}.$$

The quotient graph therefore has three vertices, $\bar{0}$, $\bar{1}$, and $\bar{a}$, with edges

$$(\bar{0}, \bar{1}), (\bar{1}, \bar{a}).$$

The quotient graph on 3 vertices, Γ/G_2, is depicted in Figure 2.15.

Figure 2.15: A quotient of a Paley graph created using Sage.

Definition 2.8.6. The *Cartesian graph product* of two graphs $\Gamma_1 = (V_1, E_1)$ and $\Gamma_2 = (V_2, E_2)$ is a graph $\Gamma_3 = \Gamma_1 \square \Gamma_2$ with the following properties:

- The vertex set of Γ_3 is the Cartesian product $V_1 \times V_2$.
- Two vertices (u_1, u_2) and (v_1, v_2) of Γ_3 are connected by an edge if and only if $u_1 = v_1$ and u_2 is a neighbor of v_2 in Γ_2 or $u_2 = v_2$ and u_1 is a neighbor of v_1 in Γ_1.

Exercise 2.13. Let G denote the cyclic group of order 3, Γ_1 denote the cycle graph on 3 vertices and let Γ_2 denote the cycle graph on 4 vertices. Take G to act on Γ_1 in the obvious way and to act on Γ_2 trivially. Let $\Gamma = \Gamma_1 \square \Gamma_2$ denote the graph product of them, depicted in Figure 2.16. Show that $\Gamma/G \cong \Gamma_2$.

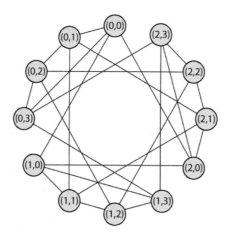

Figure 2.16: A graph product of two cycle graphs.

Exercise 2.14. Let G be a finite group, and let S be a subset of G such that $S = S^{-1}$ and $1 \notin S$. Let Γ_1 denote the Cayley graph of (G, S), let Γ_2 denote an arbitrary graph, and let $\Gamma = \Gamma_1 \square \Gamma_2$ denote the graph product of them. Show that there is a natural action of G on Γ such that $\Gamma/G \cong \Gamma_2$.

Example 2.8.7. Consider the symmetric group on $\{1, 2, 3\}$, G generated by $S = \{(1, 2, 3), (1, 2), (1, 3, 2)\}$. The Cayley graph Γ_2 of (G, S), depicted in Figure 4.8, has automorphism group of order 12 (with G as a normal subgroup). In particular, there is an action of G on Γ_2.

Let Γ_1 denote the cycle graph on 5 vertices and let Γ_3 denote the product graph $\Gamma_3 = \Gamma_1 \Box \Gamma_2$. This has an automorphism group of order 120, with the automorphism group of Γ_1 (a cyclic group of order 5) as a normal subgroup. The following Sage code shows there is a copy of G in the automorphism group of Γ_3 such that $\Gamma_3/G \cong \Gamma_1$.

```
                         ─────── Sage ───────

sage: G = SymmetricGroup(3)
sage: S = G.gens()+[G.gens()[0]^(-1)]; S
[(1,2,3), (1,2), (1,3,2)]
sage: A = G.cayley_graph(generators=S,simple=True).adjacency_matrix()
sage: Gamma1 = Graph(A)
sage: Gamma1.show(layout="spring", dpi = 300)
Launched png viewer for Graphics object consisting of 16 graphics primitives
sage: AG1 = Gamma1.automorphism_group()
sage: AG1.cardinality()
12
sage: Gamma2 = graphs.CycleGraph(5)
sage: Gamma3 = Gamma1.cartesian_product(Gamma2)
sage: AG3 = Gamma3.automorphism_group()
sage: AG3.cardinality()
120
sage: AG33 = AG3.sylow_subgroup(3)
sage: AG33.cardinality()
3
sage: AG32 = AG3.sylow_subgroup(2)
sage: AG32.cardinality()
8
sage: G0 = AG3.subgroup([AG32.list()[2],AG33.list()[1]])
sage: G0.cardinality()
6
sage: G0.is_normal(AG3)
True
sage: G0.is_abelian()
False
sage: Gamma4 = quotient_graph(Gamma3, G0)
sage: Gamma4.is_circulant()
True
sage: Gamma4.is_connected()
True
sage: len(Gamma4.vertices())
5
```

Next we give a construction of a graph with a given graph quotient.

Let G be a finite group, and let S be a subset of G such that $S = S^{-1}$ and $1 \notin S$. Let Γ_1 denote the Cayley graph of (G, S), and let Γ_2 denote an arbitrary graph with a distinguished vertex v_0. Let $\Gamma_3 = \Gamma_1 \Box \Gamma_2$ denote the Cartesian graph product of Γ_1 and Γ_2, and let Γ be the result of deleting from Γ_3 all edges of the form $((g, v), (g, v_0))$ where $v \neq v_0$.

Exercise 2.15. Show that there is a natural action of G on Γ such that $\Gamma/G \cong \Gamma_2$, where Γ is as in the construction above.

2.8.1 Example of the Biggs–Smith graph

Consider the Biggs–Smith graph Γ, encountered in Chapter 5 on graph examples. It is a 3-regular graph with 102 vertices and 153 edges, having an automorphism group $G \cong PSL(2, 17)$, of order 2448, which acts regularly on it. Using Sage, it can be shown that there is an edge e_0 of Γ such that the stabilizer of the edge e_0 is a subgroup G_0 of order 16. The quotient graph $\Gamma_0 = \Gamma/G_0$, shown in Figure 2.17, is a connected graph having 10 vertices and 10 edges, which itself has an automorphism group of order 2.

```
——————————————————————— Sage ———————————————————————
sage: Gamma = graphs.BiggsSmithGraph()
sage: G = Gamma.automorphism_group(); G.order()
2448
sage: E = Gamma.edges(); len(E)
153
sage: V = Gamma.vertices(); len(V)
102
sage: Gamma.delete_edge(0,1)
sage: E = Gamma.edges(); len(E)
152
sage: Gamma.add_edge((0,1), label="label")
sage: E = Gamma.edges(); len(E)
153
sage: G0 = Gamma.automorphism_group(edge_labels=True); G0.order()
16
```

The stabilizer of the vertex 0 is a subgroup G_1 of order 24. The quotient graph $\Gamma_1 = \Gamma/G_1$, shown in Figure 2.18, is a tree having 8 vertices and 7 edges, which itself has an automorphism group of order 2.

```
——————————————————————— Sage ———————————————————————
sage: Gamma = graphs.BiggsSmithGraph()
sage: G = Gamma.automorphism_group(); G.order()
2448
sage: G1 = Gamma.automorphism_group(partition=[[0],range(1,102)])
sage: G1.order()
24
```

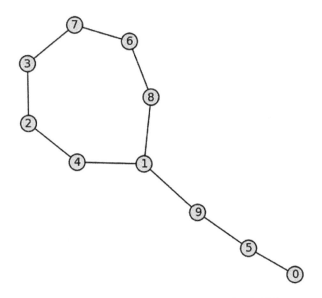

Figure 2.17: The group quotient graph of Γ/G_0.

The orbits of G_1 acting on Γ are:

$$\{0\}, \{7, 41, 10, 44, 50, 88, 94, 31\}, \{16, 1, 101\},$$
$$\{17, 2, 100, 25, 36, 15\}, \{64, 65, 67, 4, 98, 70, 71, 73, 76, 13, 77,$$
$$82, 19, 20, 23, 27, 28, 34, 38, 54, 55, 84, 61, 62\},$$
$$\{66, 35, 37, 72, 83, 14, 99, 18, 3, 24, 26, 63\},$$
$$\{68, 5, 97, 74, 75, 12, 78, 81, 21, 22, 90, 91, 86, 29, 09,$$
$$33, 39, 46, 47, 53, 56, 57, 60, 85\}, \{96, 6, 8, 9, 11, 79, 80, 87, 89, 92,$$
$$93, 30, 95, 32, 40, 42, 43, 45, 48, 49, 51, 52, 58, 59\}.$$

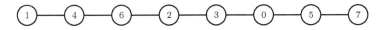

Figure 2.18: The group quotient graph of Γ/G_1.

Chapter 3
Graphs as Manifolds

3.1 Motivation

In this chapter, we are interested in exploring questions such as the following.

Question 3.1. *Do harmonic maps on a Riemann surface have an analog for graphs?*

Question 3.2. *Does the Riemann–Roch theorem for Riemann surfaces have an analog for graphs?*

Question 3.3. *If there is a harmonic morphism $\Gamma_2 \to \Gamma_1$, how are the Laplacians of Γ_2, Γ_1 related?*

There are several similarities between graphs and manifolds. This chapter explores some of these connections.

An excellent reference for Riemann surfaces can be found in Miranda's text [Mi95].

3.2 Calculus on graphs

If $\Gamma = (V, E)$ is a graph and F is a field such as $\mathbb{R}$ or $GF(q)$ or a ring such as $\mathbb{Z}$, recall

$$C^0(\Gamma, F) = \{f : V \to F\}, \quad C^1(\Gamma, F) = \{f : E \to F\},$$

are the sets of F-valued functions defined on V and E, respectively. The notation suggests we think of the functions on vertices as continuous functions on Γ and the functions on edges as continuously differentiable functions on

© Springer International Publishing AG 2017
W.D. Joyner and C.G. Melles, *Adventures in Graph Theory*,
Applied and Numerical Harmonic Analysis,
https://doi.org/10.1007/978-3-319-68383-6_3

Γ. Recall from (1.3) that if F is a field then these are F-inner product spaces and

$$\dim C^0(\Gamma, F) = |V|, \quad \dim C^1(\Gamma, F) = |E|.$$

The difference operator

$$B : C^1(\Gamma, F) \to C^0(\Gamma, F),$$
$$(Bf)(v) = \sum_{h(e)=v} f(e) - \sum_{t(e)=v} f(e)$$

can be regarded as the "derivative operator[1] on Γ." This operator is closely related to the incidence matrix. Indeed, with respect to the bases $\mathcal{F}$ and $\mathcal{G}$ in Lemma 1.1.20, we have the following result.

Lemma 3.2.1. The matrix representing the linear transformation

$$B : C^1(\Gamma, F) \to C^0(\Gamma, F)$$

with respect to $\mathcal{F}$ and $\mathcal{G}$ is the incidence matrix.

Proof. The proof is left as an exercise.   □

Think of $f \in C^1(\Gamma, F)$ as a set of edge weights on the graph Γ. In some sense, $(Bf)(v)$ measures the total of the edge weights incident to $v \in V$. Sometimes, a function f on E for which $(Bf)(v) = 0$, for all $v \in V$, is called a *flow*.

Recall, from Definition 1.1.25, for any nontrivial partition

$$V = V_1 \cup V_2, \quad V_i \neq \emptyset, \quad V_1 \cap V_2 = \emptyset,$$

the set of all edges $e = (v_1, v_2) \in E$, with $v_i \in V_i$ ($i = 1, 2$), is called a cocycle of Γ. A cocycle with a minimal set of edges is a bond of Γ. The set of cycles of Γ is denoted $Z(\Gamma)$ and the set of cocycles is denoted $Z^*(\Gamma)$. Recall, by Lemma 1.1.24, the F-vector space spanned by the vector representations of all the cycles is the cycle space of Γ,

$$\mathcal{Z}(\Gamma) = \mathcal{Z}(\Gamma, F).$$

This is the kernel of the incidence matrix of Γ and may be regarded as a subspace of $C^1(\Gamma, F)$.

Also, recall the F-vector space spanned by the vector representations of all the cocycles (in the sense of Definition 1.1.25) is the cocycle space of Γ, $\mathcal{Z}^*(\Gamma) = \mathcal{Z}^*(\Gamma, F)$. This is the column space of the transpose of the

---

[1]Indeed, it is often denoted by D but we shall have to use D for several other things, so we denote the difference operator by B instead.

incidence matrix of Γ and may be regarded as a subspace of $C^1(\Gamma, F)$ of dimension $rank(\Gamma)$.

Lemma 3.2.2. Under the inner product (1.3) on $C^1(\Gamma, F)$, the cycle space is orthogonal to the cocycle space.

Proof. See Proposition 4.7 in Biggs [Bi93] or Theorem 8.3.1 in Godsil and Royle [GR01]. □

Consider a spanning tree T of a graph Γ and its complementary subgraph $\overline{T}$. For each edge e of $\overline{T}$, the graph $T \cup e$ contains a unique cycle. The cycles which arise in this way are called the *fundamental cycles* of Γ, denoted $cyc(T, e)$.

First, we define a *circulation* on $\Gamma = (V, E)$ to be a function

$$f : E \to F,$$

satisfying[2]

- $\sum_{u \in V,\ (u,v) \in E} f(u,v) = \sum_{w \in V,\ (v,w) \in E} f(v,w)$.

Suppose Γ has a subgraph H and f is a circulation of Γ such that f is a constant function on H and 0 elsewhere. We call such a circulation a *characteristic function* of H. For example, if Γ has a cycle C and if f is the characteristic function on C, then f is a circulation.

The *circulation space* $\mathcal{C}$ is the F-vector space of circulation functions. The cycle space is "clearly" a subspace of the circulation space, since the F-vector space spanned by the characteristic functions of cycles may be identified with the cycle space of Γ. Under the inner product (1.3), i.e.,

$$(f, g) = \sum_{e \in E} f(e)g(e), \qquad (3.1)$$

this vector space is an inner product space.

3.3 Harmonic morphisms on a graph

We will define harmonic morphisms of graphs, derive some properties, and give some examples. We will state a Riemann–Hurwitz formula due to Baker and Norine [BN09]. Many of the key properties of harmonic morphisms were also originally established by Baker and Norine [BN09]. Where possible, we have tried to describe harmonic morphisms in terms of their associated matrices, for ease in computation.

---

[2]Note: Some authors add the condition $f(e) \geq 0$—see e.g., Chung [Ch05].

Let Γ_1 and Γ_2 be finite connected graphs which may have multi-edges but do not have loops.

We will assume that we have arbitrary orientations assigned to the edges of Γ_1 and Γ_2, so that for each edge e, one vertex is the head of e and the other vertex is the tail of e. The head and tail cannot be the same vertex, since there are no loops.

Definition 3.3.1. A nonconstant *graph morphism* $\phi : \Gamma_2 \to \Gamma_1$ consists of a map $\phi_V : V_2 \to V_1$ and a map $\phi_E : E_2 \to E_1 \cup V_1$ satisfying the following conditions: if e_i is any edge in E_2 and if v_a is the tail vertex of e_i and v_b is the head vertex of e_i, then one of the following holds:

(1) $\phi_E(e_i) = f_l$, for some edge f_l in E_1, $\phi_V(v_a)$ is the tail vertex of f_l, and $\phi_V(v_b)$ is the head vertex of f_l, in which case we say that ϕ *preserves orientation on e_i*, or

(2) $\phi_E(e_i) = f_l$, for some edge f_l in E_1, $\phi_V(v_a)$ is the head vertex of f_l, and $\phi_V(v_b)$ is the tail vertex of f_l, in which case we say that ϕ *reverses orientation on e_i*, or

(3) $\phi_E(e_i) = \phi_V(v_a) = \phi_V(v_b)$.

Definition 3.3.2. Given a map $\phi_E : E_2 \to E_1 \cup V_1$, an edge e_i is called *horizontal* if $\phi_E(e_i) \in E_1$ and is called *vertical* if $\phi_E(e_i) \in V_1$.

Definition 3.3.3. We say that a graph morphism $\phi : \Gamma_2 \to \Gamma_1$ is a *graph homomorphism* if $\phi_E(E_2) \subset E_1$. Thus, a graph morphism is a homomorphism if it has no vertical edges.

Definition 3.3.4. For any vertex v of a graph Γ, we define the *star* $St_\Gamma(v)$ to be the subgraph of Γ defined by the edges incident to v.

Definition 3.3.5. Suppose that Γ_1 has at least one edge. A graph morphism $\phi : \Gamma_2 \to \Gamma_1$ is called *harmonic* if for all vertices $v \in V(\Gamma_2)$, the quantity

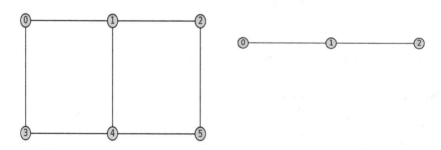

Figure 3.1: There is a harmonic morphism from the ladder graph to the path graph.

$$|\phi^{-1}(f) \cap St_{\Gamma_2}(v)| \tag{3.2}$$

(the number of edges in Γ_2 adjacent to v and mapping to the edge f in Γ_1) is independent of the choice of edge f in $St_{\Gamma_1}(\phi(v))$. If Γ_1 consists of a single vertex and no edges, we will say that a morphism $\phi : \Gamma_2 \to \Gamma_1$ is harmonic.

An equivalent definition is given in Definition 3.3.16 below.

Exercise 3.1. Show that

$$|\phi^{-1}(f) \cap St_{\Gamma_2}(v)| = |\{e \in E_2 \mid v \in e \text{ and } \phi_E(e) = f\}|.$$

Exercise 3.2. Show that the composite of harmonic morphisms is harmonic.

Example 3.3.6. Let Γ_2 be the ladder graph of 6 vertices and Γ_1 the path graph of 3 vertices (depicted in Figure 3.1). Consider the morphism $\phi : \Gamma_2 \to \Gamma_1$ defined by the map on vertices

$$\phi_V : \{0, 3\} \mapsto 0, \qquad \{1, 4\} \mapsto 1, \qquad \{2, 5\} \mapsto 5.$$

The morphism ϕ is harmonic. The edges $(0, 3)$, $(1, 4)$, and $(2, 5)$ are all vertical, while the remaining edges of Γ_2 are all horizontal.

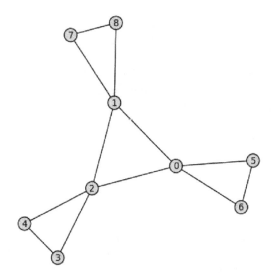

Figure 3.2: A covering graph of the cycle graph on 3 vertices.

Example 3.3.7. The graph morphism ϕ which sends two opposite vertices of the cube graph Γ_2 to a vertex of the tetrahedron graph $\Gamma_1 = K_4$ (depicted in Figure 1.1) is harmonic. We will return to this example later (see Example 3.3.12).

Example 3.3.8. Consider the graph morphism $\phi : \Gamma_2 \to \Gamma_1$ from the Paley graph on 9 vertices, $\Gamma_2 = (V_2, E_2)$ (depicted in Figure 1.12) to the cycle graph on 3 vertices, $\Gamma_1 = (V_1, E_1)$, defined by the map on vertices

$$\phi_V \; : \; \{0, 1, 2\} \mapsto 0, \{a, a+1, a+2\} \mapsto 2, \{2a, 2a+1, 2a+2\} \mapsto 1.$$

This morphism ϕ is harmonic. We examine this morphism in more detail in later examples, e.g., Example 3.3.34.

Example 3.3.9. Consider the graph Γ_2 having 9 vertices depicted in Figure 3.2.

Let Γ_1 denote the cycle graph on 3 vertices. Define the morphism $\phi_1 : \Gamma_2 \to \Gamma_1$ by

$$\{0, 5, 6\} \mapsto 0, \{1, 7, 8\} \mapsto 1, \{2, 3, 4\} \mapsto 2,$$

and define the morphism $\phi_2 : \Gamma_2 \to \Gamma_1$ by

$$\{0, 1, 2\} \mapsto 0, \{3, 5, 8\} \mapsto 1, \{4, 6, 7\} \mapsto 2.$$

Exercise 3.3. Show that both ϕ_1 and ϕ_2 are harmonic.

3.3.1 Matrix formulation of graph morphisms

Let Γ_1 and Γ_2 be connected graphs which may have multi-edges, but do not have loops. Let $\phi : \Gamma_2 \to \Gamma_1$ be a nonconstant graph morphism. We will assume, unless stated otherwise, that Γ_1 and Γ_2 each have at least one edge.

Suppose that we have orderings on the vertex sets and edge sets as follows:

$$
\begin{aligned}
V_1 &= \{w_1, w_2, \ldots, w_{n_1}\} \\
V_2 &= \{v_1, v_2, \ldots, v_{n_2}\} \\
E_1 &= \{f_1, f_2, \ldots, f_{m_1}\} \\
E_2 &= \{e_1, e_2, \ldots, e_{m_2}\}.
\end{aligned}
$$

Suppose also that we have arbitrary orientations assigned to the edges of Γ_1 and Γ_2. Let B_1 and B_2 be the incidence matrices of Γ_1 and Γ_2 with respect to the chosen orientations, i.e.,

$$
(B_1)_{ij} = \begin{cases} 1, & \text{if } w_i = \mathrm{h}(f_j), \\ -1, & \text{if } w_i = \mathrm{t}(f_j), \\ 0, & \text{otherwise}, \end{cases}
$$

and similarly for B_2.

We will now describe the maps ϕ_E and ϕ_V in terms of matrices Φ_E and Φ_V (with respect to the given orientations and orderings on edges and vertices) which satisfy certain conditions.

Definition 3.3.10. Given a graph morphism $\phi : \Gamma_2 \to \Gamma_1$ with an edge map $\phi_E : E_2 \to E_1 \cup V_1$, the *signed edge map matrix* with respect to the given orientations and orderings is defined to be the $m_2 \times m_1$ matrix Φ_E given by

$$(\Phi_E)_{il} = \begin{cases} 1, & \text{if } \phi_E(e_i) = f_l \text{ and } \phi \text{ preserves orientation on } e_i, \\ -1, & \text{if } \phi_E(e_i) = f_l \text{ and } \phi \text{ reverses orientation on } e_i, \\ 0, & \text{otherwise.} \end{cases}$$

Note that e_i is a horizontal edge in E_2 if and only if row i of Φ_E contains exactly one nonzero entry, and e_i is a vertical edge in E_2 if and only if row i of Φ_E consists entirely of zeros.

Definition 3.3.11. Given graphs Γ_1 and Γ_2 and a map $\phi_V : V_2 \to V_1$, the *vertex map matrix* Φ_V with respect to the given orientations and orderings is defined to be the $n_2 \times n_1$ matrix given by

$$(\Phi_V)_{ij} = \begin{cases} 1, & \text{if } \phi_V(v_i) = w_j, \\ 0, & \text{otherwise.} \end{cases}$$

Example 3.3.12. The automorphism which swaps two opposite vertices of the cube graph Γ_2 induces a graph morphism ϕ to its quotient, which can be identified with the tetrahedron graph Γ_1 (depicted in Figure 3.3).

Orient the edges so that $(0,1)$, $(0,2)$, $(0,3)$, $(1,2)$, $(1,3)$, $(2,3)$ have orientation $+1$. For a suitable labeling of the vertices of Γ_2 (for instance, see Example 3.3.35) the definition then gives us

$$\Phi_V = \begin{pmatrix} 1 & 0 & 0 & 0 \\ 0 & 1 & 0 & 0 \\ 0 & 0 & 1 & 0 \\ 0 & 0 & 0 & 1 \\ 0 & 0 & 0 & 1 \\ 0 & 0 & 1 & 0 \\ 0 & 1 & 0 & 0 \\ 1 & 0 & 0 & 0 \end{pmatrix},$$

and

$$\Phi_E = \begin{pmatrix} 1 & 0 & 0 & 0 & 0 & 0 \\ 0 & 1 & 0 & 0 & 0 & 0 \\ 0 & 0 & 1 & 0 & 0 & 0 \\ 0 & 0 & 0 & 0 & 1 & 0 \\ 0 & 0 & 0 & 1 & 0 & 0 \\ 0 & 0 & 0 & 0 & 0 & 1 \\ 0 & 0 & 0 & -1 & 0 & 0 \\ 0 & 0 & -1 & 0 & 0 & 0 \\ 0 & 0 & 0 & 0 & 0 & -1 \\ 0 & 0 & 0 & 0 & -1 & 0 \\ 0 & -1 & 0 & 0 & 0 & 0 \\ -1 & 0 & 0 & 0 & 0 & 0 \end{pmatrix}.$$

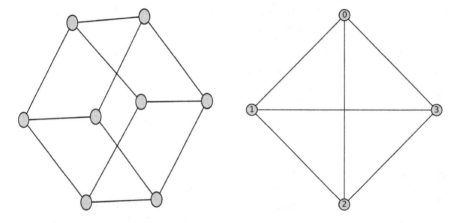

Figure 3.3: The cube graph is a 2-fold cover of the tetrahedron graph.

We will now give two exercises which lead to a graph morphism identity relating the incidence matrices and the vertex and edge map matrices.

Suppose that Γ_2 and Γ_1 are connected graphs and Γ_1 has at least one edge. Let B_1 and B_2 be the incidence matrices with respect to fixed orderings on the vertices and edges of the graphs and arbitrary orientations on their edges.

Exercise 3.4. Suppose that $\phi : \Gamma_2 \to \Gamma_1$ is a graph morphism with associated vertex map matrix Φ_V and signed edge map matrix Φ_E. Let e_i be a horizontal edge in Γ_2 and let w_j be a vertex in Γ_1. Show that

$$(\Phi_E B_1^t)_{ij} = \left\{ \begin{array}{l} 1, \quad \text{if } \phi_V(\text{head}(e_i)) = w_j, \\ -1, \quad \text{if } \phi_V(\text{tail}(e_i)) = w_j, \\ 0, \quad \text{otherwise,} \end{array} \right\} = (B_2^t \Phi_V)_{ij}.$$

Exercise 3.5. Suppose that $\phi : \Gamma_2 \to \Gamma_1$ is a graph morphism with associated vertex map matrix Φ_V and signed edge map matrix Φ_E. Let e_i be a vertical edge in Γ_2 and let w_j be a vertex in Γ_1. Show that

$$(\Phi_E B_1^t)_{ij} = 0 = (B_2^t \Phi_V)_{ij}.$$

The previous two exercises prove the following proposition.

Proposition 3.3.13 (Graph Morphism Identity). *Suppose that Γ_2 and Γ_1 are connected graphs and Γ_1 has at least one edge. Let B_2 and B_1 be the incidence matrices with respect to fixed orderings on the vertices and edges of the graphs and arbitrary orientations on their edges. If $\phi : \Gamma_2 \to \Gamma_1$ is a graph morphism, then the associated vertex map matrix Φ_V and signed edge map matrix Φ_E satisfy the identity*

$$B_2^t \Phi_V = \Phi_E B_1^t. \tag{3.3}$$

We can generalize the concept of an edge map matrix as follows.

Definition 3.3.14. If Γ_1 and Γ_2 are graphs and ψ is a map $\psi : E_2 \to E_1 \cup V_1$, (not necessarily coming from a graph morphism), we say an $m_2 \times m_1$ matrix Φ_E is a *signed edge map matrix* for ψ provided it satisfies

$$(\Phi_E)_{il} = \begin{cases} \pm 1, & \text{if } \psi(e_i) = f_l, \\ 0, & \text{otherwise,} \end{cases}$$

for some choice of signs.

The next exercise shows that identity 3.3 and condition (3) of Definition 3.3.1 characterize graph morphisms. Of course, if there are no vertical edges with respect to a given edge map, condition (3) is not needed, and identity 3.3 characterizes a graph homomorphism.

Exercise 3.6. Suppose that Γ_2 and Γ_1 are connected graphs with at least one edge each. Let B_1 and B_2 be the incidence matrices with respect to fixed orientations on their edges. Suppose that a map $\phi_V : V_2 \to V_1$ with vertex map matrix Φ_V and a map $\phi_E : E_2 \to E_1 \cup V_1$ with a signed edge map matrix Φ_E are given. Assume Γ_2 has at least one vertical edge with respect to this map. Show that (a) if Φ_V and Φ_E satisfy the identity 3.3, and (b) if ϕ_E and ϕ_V satisfy condition (3) of Definition 3.3.1 for graph morphisms, then ϕ_V and ϕ_E determine a graph morphism $\phi : \Gamma_2 \to \Gamma_1$.

In the next exercise, we reformulate the idea of Proposition 3.3.13.

Exercise 3.7. Let Γ_2 and Γ_1 be connected graphs with at least one edge each. Suppose that $\phi : \Gamma_2 \to \Gamma_1$ is a graph morphism. Let F be a field, such as $\mathbb{R}$, or a ring, such as $\mathbb{Z}$. We will use B_i to denote both the map $C^1(\Gamma_i, F) \to C^0(\Gamma_i, F)$ and its matrix with respect to the standard basis corresponding to some order on the vertices and edges of Γ_i (see Lemma 1.1.20 and 1.4). Let B_i^* be the dual to B_i. Show that there are maps Φ_V and Φ_E such that the following diagram is commutative:

$$
\begin{array}{ccc}
C^0(\Gamma_1, F) & \xrightarrow{B_1^*} & C^1(\Gamma_1, F) \\
\Phi_V \downarrow & & \Phi_E \downarrow \\
C^0(\Gamma_2, F) & \xrightarrow{B_2^*} & C^1(\Gamma_2, F)
\end{array}
$$

3.3.2 Identities for harmonic morphisms

In this section, we will describe a nonconstant harmonic morphism ϕ in terms of its associated matrices and establish some identities relating these matrices. We begin by discussing ways to count horizontal and vertical edges adjacent to a vertex in the domain. For a harmonic morphism ϕ, we define a multiplicity-vertex map matrix Φ_{mV} which is a product $M_\phi \Phi_V$ of a diagonal matrix of horizontal multiplicities with the vertex map matrix. Identities relating Φ_{mV} with the incidence and Laplacian matrices will be used in a later section to prove the existence of induced pushforward and pullback maps on the Jacobian groups (as defined in §4.6 below) of the graphs.

Let Γ_1 and Γ_2 be connected graphs. Let $\phi : \Gamma_2 \to \Gamma_1$ be a graph morphism. Unless stated otherwise, we will assume that Γ_1 has at least one edge.

Definition 3.3.15. Given a vertex v of Γ_2 and an edge f of Γ_1, we define the *local horizontal multiplicity of ϕ at v and f* to be

$$ \mu_\phi(v, f) = |\{e \in E_2 \mid v \in e \text{ and } \phi_E(e) = f\}|, $$

i.e., the local horizontal multiplicity of ϕ at v and f is the number of edges in Γ_2 which are adjacent to v and map to f. An element of $\{e \in E_2 \mid v \in e \text{ and } \phi_E(e) = f\}$ is called a *horizontal edge*. In other words, it is an edge of Γ_2 that does not get mapped to a vertex of Γ_1.

We now restate the definition of a harmonic morphism (see Definition 3.3.5) in terms of the local horizontal multiplicity.

Definition 3.3.16. Let $\phi : \Gamma_2 \to \Gamma_1$ be a morphism of connected graphs. If Γ_1 has at least one edge, we say that ϕ is *harmonic* if, for each vertex $v \in V_2$, the local horizontal multiplicity $\mu_\phi(v, f)$ is the same for all edges f such that $\phi_V(v) \in f$. If Γ_1 consists of a single vertex and no edges, we say that every graph morphism $\phi : \Gamma_2 \to \Gamma_1$ is harmonic.

Lemma 3.3.17. *The definition of harmonic in Definition 3.3.5 agrees with the definition of harmonic in Definition 3.3.16.*

Proof. This follows from the fact that $\mu_\phi(v, f) = |\phi^{-1}(f) \cap St_{\Gamma_2}(v)|$. $\qquad\square$

Definition 3.3.18. If $\phi : \Gamma_2 \to \Gamma_1$ is harmonic and Γ_1 has at least one edge, the *horizontal multiplicity of ϕ at $v \in V_2$* is defined to be

$$m_\phi(v) = \mu_\phi(v, f) \qquad \text{for any edge } f \text{ such that } \phi_V(v) \in f.$$

Thus $m_\phi(v)$ is the number of edges adjacent to v that map to any edge adjacent to $\phi_V(v)$. If Γ_1 has no edges, $m_\phi(v)$ is defined to be 0 for all $v \in V_2$.

Definition 3.3.19. If ϕ is harmonic, the *vertical multiplicity of ϕ at $v \in V_2$* is defined to be

$$\nu_\phi(v) = |\{e \in E_2 \mid v \in e \text{ and } \phi_E(e) = \phi_V(v)\}|.$$

Recall, an element of $\{e \in E_2 \mid v \in e \text{ and } \phi_E(e) = \phi_V(v)\}$ is called a vertical edge. In other words, it is an edge of Γ_2 which is sent to a vertex of Γ_1.

The following lemma is a direct consequence of the definitions.

Lemma 3.3.20. *For any vertex $v \in V_2$,*

$$deg(v) = m_\phi(v) \, deg(\phi_V(v)) + \nu_\phi(v).$$

The following exercise gives a way of calculating vertical multiplicities in terms of the signed edge map matrix and one of the incidence matrices.

Exercise 3.8. Suppose that $\phi : \Gamma_2 \to \Gamma_1$ is a harmonic morphism of connected graphs and Γ_1 has at least one edge. Let v_a be a vertex of Γ_2. Show that the vertical multiplicity of ϕ at v_a is given by

$$\nu_\phi(v_a) = \sum_{j=1}^{m_2} |(B_2)_{aj}| \left(1 - \prod_{k=1}^{m_1} |(\Phi_E)_{jk}| \right).$$

The next exercises and lemmas will lead to an identity relating the horizontal multiplicities of a harmonic morphism to the incidence, edge map, and vertex map matrices.

Exercise 3.9. Let $\phi : \Gamma_2 \to \Gamma_1$ be a graph morphism of connected graphs, where Γ_1 has at least one edge. Suppose that $v_a \in V_2$, $e_k \in E_2$, and $f_l \in E_1$. Show that

$$(B_2)_{ak}(\Phi_E)_{kl} = \begin{cases} 1, & \text{if } v_a \in e_k \text{ and } \phi_E(e_k) = f_l \text{ and } \phi_V(v_a) = \text{head}(f_l), \\ -1, & \text{if } v_a \in e_k \text{ and } \phi_E(e_k) = f_l \text{ and } \phi_V(v_a) = \text{tail}(f_l), \\ 0, & \text{otherwise.} \end{cases}$$

Exercise 3.10. Let $\phi : \Gamma_2 \to \Gamma_1$ be a graph morphism of connected graphs, where Γ_1 has at least one edge. Suppose that $v_a \in V_2$ and $f_l \in E_1$. Let $\tilde{a}$ be the index such that $w_{\tilde{a}} = \phi_V(v_a)$. Use Exercise 3.9 to show that

$$(B_2\Phi_E)_{al} = \mu_\phi(v_a, f_l)(B_1)_{\tilde{a}l}.$$

Example 3.3.21. We use the notation from Example 3.3.8. In this case, for each $v \in V_2$, we have
$$m_\phi(v) = 1.$$

Example 3.3.34 below discusses this in more detail.

In the case in which ϕ is harmonic, we can restate the result of Exercise 3.10 above as follows.

Lemma 3.3.22. Let $\phi : \Gamma_2 \to \Gamma_1$ be a harmonic graph morphism of connected graphs, where Γ_1 has at least one edge. Suppose that $v_a \in V_2$ and $f_l \in E_1$. Let $\tilde{a}$ be the index such that $w_{\tilde{a}} = \phi_V(v_a)$. Then

$$(B_2\Phi_E)_{al} = m_\phi(v_a)(B_1)_{\tilde{a}l}.$$

We now define the multiplicity-vertex map matrix and prove our first identity for harmonic morphisms.

Definition 3.3.23. Let $\phi : \Gamma_2 \to \Gamma_1$ be a harmonic graph morphism of connected graphs. Let n_i be the number of vertices of Γ_i and let Φ_V be the $n_2 \times n_1$ vertex map matrix (with respect to a given ordering on the vertices). We define M_ϕ to be the $n_2 \times n_2$ diagonal matrix of multiplicities of vertices of Γ_2, given by
$$(M_\phi)_{ii} = m_\phi(v_i).$$

The *multiplicity-vertex map matrix* is defined to be the product matrix

$$\Phi_{mV} = M_\phi \Phi_V.$$

Proposition 3.3.24 (Incidence matrix identity for harmonic morphisms). *Let $\phi : \Gamma_2 \to \Gamma_1$ be a harmonic graph morphism of connected graphs, where Γ_1 has at least one edge. Then*

$$B_2\Phi_E = \Phi_{mV} B_1, \tag{3.4}$$

where B_1 and B_2 are the incidence matrices of Γ_1 and Γ_2, Φ_E is the signed edge map matrix of ϕ, and Φ_{mV} is the multiplicity-vertex map matrix of ϕ.

Proof. Suppose that $v_a \in V_2$ and $f_l \in E_1$. Let $\tilde{a}$ be the index such that $w_{\tilde{a}} = \phi_V(v_a)$. If ϕ is harmonic then

$$
\begin{aligned}
(\Phi_{mV} B_1)_{al} &= \sum_{j=1}^{n_2} \sum_{k=1}^{n_1} (M_\phi)_{aj} (\Phi_V)_{jk} (B_1)_{kl} && \text{by the definition of } \Phi_{mV} \\
&= \sum_{k=1}^{n_1} m_\phi(v_a) (\Phi_V)_{ak} (B_1)_{kl} && \text{by the definition of } M_\phi \\
&= m_\phi(v_a) (B_1)_{\tilde{a}l} && \text{since } \phi_V(v_a) = w_{\tilde{a}} \\
&= (B_2 \Phi_E)_{al} && \text{by Lemma 3.3.22.}
\end{aligned}
$$

Thus, if ϕ is harmonic, $B_2 \Phi_E = \Phi_{mV} B_1$. $\qquad\square$

Our next exercise shows that the existence of a diagonal matrix M_ϕ such that $\Phi_{mV} = M_\phi \Phi_V$ satisfies identity 3.4 characterizes harmonic morphisms.

Exercise 3.11. Let $\phi : \Gamma_2 \to \Gamma_1$ be a graph morphism of connected graphs, where Γ_1 has at least one edge. If there exists an $n_2 \times n_2$ diagonal matrix M_ϕ such that the matrix product $\Phi_{mV} = M_\phi \Phi_V$ satisfies Equation (3.4), then ϕ is harmonic.

In the following exercise, we reformulate the idea of Proposition 3.3.24.

Exercise 3.12. Suppose that $\phi : \Gamma_2 \to \Gamma_1$ is a harmonic morphism of connected graphs and Γ_1 has at least one edge. Let F be a field, such as $\mathbb{R}$, or a ring, such as $\mathbb{Z}$. Suppose that $\phi : \Gamma_2 \to \Gamma_1$ is a graph morphism. We will use B_i to denote both the map $C^1(\Gamma_i, F) \to C^0(\Gamma_i, F)$ and its matrix with respect to the standard basis corresponding to some order on the vertices and edges of Γ_i (as in Lemma 1.1.20). Show that there are maps Φ_{mV} and Φ_E such that the following diagram is commutative:

$$
\begin{array}{ccc}
C^1(\Gamma_1, F) & \xrightarrow{\;\;B_1\;\;} & C^0(\Gamma_1, F) \\
\Big\downarrow{\scriptstyle \Phi_E} & & \Big\downarrow{\scriptstyle \Phi_{mV}} \\
C^1(\Gamma_2, F) & \xrightarrow{\;\;B_2\;\;} & C^0(\Gamma_2, F)
\end{array}
$$

The following identity involving Laplacian matrices will be used in a later section to establish properties of induced pushforward and pullback maps on the Jacobian groups of the graphs (see Propositions 4.8.4 and 4.8.2). Recall that the Laplacian matrices of Γ_1 and Γ_2 are

$$
Q_1 = B_1 B_1^t \text{ and } Q_2 = B_2 B_2^t.
$$

Proposition 3.3.25 (Laplacian matrix identity for harmonic morphisms).
If $\phi : \Gamma_2 \to \Gamma_1$ is a nonconstant harmonic morphism,

$$Q_2 \Phi_V = \Phi_{mV} Q_1. \tag{3.5}$$

Equivalently, since Q_1 and Q_2 are symmetric,

$$\Phi_V^t Q_2 = Q_1 \Phi_{mV}^t. \tag{3.6}$$

Proof. Recall from Proposition 3.3.13 that

$$B_2^t \Phi_V = \Phi_E B_1^t.$$

Multiplying on the left by B_2 gives

$$B_2 B_2^t \Phi_V = B_2 \Phi_E B_1^t.$$

Applying Proposition 3.3.24 to the right side of this equation gives

$$B_2 B_2^t \Phi_V = \Phi_{mV} B_1 B_1^t,$$

i.e.,

$$Q_2 \Phi_V = \Phi_{mV} Q_1.$$

$\square$

Exercise 3.13. Suppose that $\phi : \Gamma_2 \to \Gamma_1$ is a nonconstant harmonic morphism of connected graphs. Let F be a field, such as $\mathbb{R}$, or a ring, such as $\mathbb{Z}$. Suppose that $\phi : \Gamma_2 \to \Gamma_1$ is a graph morphism. We will use Q_i to denote both the Laplacian map $C^0(\Gamma_i, F) \to C^0(\Gamma_i, F)$ and the Laplacian matrix with respect to the standard basis corresponding to some order on the vertices Γ_i (see Lemma 1.1.20). Show that there are maps Φ_V and Φ_{mV} such that the following diagram is commutative:

$$
\begin{array}{ccc}
C^0(\Gamma_1, F) & \xrightarrow{\;\;Q_1\;\;} & C^0(\Gamma_1, F) \\
\Big\downarrow{\scriptstyle \Phi_V} & & \Big\downarrow{\scriptstyle \Phi_{mV}} \\
C^0(\Gamma_2, F) & \xrightarrow{\;\;Q_2\;\;} & C^0(\Gamma_2, F)
\end{array}
$$

Next we will show that if $\phi : \Gamma_2 \to \Gamma_1$ is a nonconstant harmonic morphism, then each edge in Γ_1 has the same number of pre-images in Γ_2.

Definition 3.3.26. Let $\phi : \Gamma_2 \to \Gamma_1$ be a graph morphism of connected graphs, where Γ_1 has at least one edge. Let f be an edge of Γ_1. We define the *local degree of ϕ at f* to be

$$d_\phi(f) = |\{e \in E_2 | \phi_E(e) = f\}|,$$

i.e., $d_\phi(f)$ is the number of horizontal edges in E_2 that are mapped to the edge $f \in \Gamma_1$ by ϕ_E.

Lemma 3.3.27. *Let $\phi : \Gamma_2 \to \Gamma_1$ be a graph morphism of connected graphs, where Γ_1 has at least one edge. Let f be an edge of Γ_1 and let w be a vertex adjacent to f. Then*

$$d_\phi(f) = \sum_{v \in \phi_V^{-1}(w)} \mu_\phi(v, f),$$

i.e., the local degree of ϕ at $f \in \Gamma_1$ is the sum of the local horizontal multiplicities of ϕ at v and f for all vertices $v \in \Gamma_2$ such that $\phi_V(v) = w$.

Proof. Every edge in Γ_2 that maps to f must be adjacent to a vertex that maps to w, by items (1) and (2) of the definition of a graph morphism. □

Corollary 3.3.28. *Let $\phi : \Gamma_2 \to \Gamma_1$ be a harmonic graph morphism of connected graphs, where Γ_1 has at least one edge. Let f be an edge of Γ_1 and let w be a vertex adjacent to f. Then*

$$d_\phi(f) = \sum_{v \in \phi_V^{-1}(w)} m_\phi(v).$$

Lemma 3.3.29. *Let $\phi : \Gamma_2 \to \Gamma_1$ be a harmonic graph morphism of connected graphs, where Γ_1 has at least one edge. Then the local degree $d_\phi(f)$ of ϕ at f is the same for all edges $f \in \Gamma_1$, i.e., each edge in Γ_1 has the same number of pre-images in Γ_2.*

Proof. Since Γ_1 is connected, it is enough to prove that for any two adjacent edges f_i and f_j, $d_\phi(f_i) = d_\phi(f_j)$. Suppose that w is a vertex in Γ_1 that is adjacent to both f_i and f_j. By Corollary 3.3.28,

$$d_\phi(f_i) = \sum_{v \in \phi_V^{-1}(w)} m_\phi(v) = d_\phi(f_j).$$

□

Lemma 3.3.30. *If $\phi : \Gamma_2 \to \Gamma_1$ is a nonconstant harmonic morphism of connected graphs, then all edges and vertices of Γ_1 are in the image of ϕ.*

Proof. If ϕ is nonconstant, then ϕ_E maps onto at least one edge in Γ_1. Since all edges in Γ_1 have the same number of pre-image edges in Γ_2, it follows that all edges in Γ_1, and hence all vertices in Γ_1, are in the image of ϕ. □

Thus we can make the following definition.

Definition 3.3.31. Let $\phi : \Gamma_2 \to \Gamma_1$ be a harmonic morphism of connected graphs. If Γ_1 has at least one edge, we define the degree of ϕ to be

$$\deg(\phi) = d_\phi(f)$$

for any edge $f \in \Gamma_1$, i.e., $\deg(\phi)$ is the number of pre-images of each edge in Γ_1. If Γ_1 consists of a single vertex and no edges, we define $\deg(\phi) = 0$.

Lemma 3.3.32. *If ϕ is harmonic and w_p is any vertex in V_1, then*

$$\deg(\phi) = \sum_{v \in \Phi_V^{-1}(w_p)} m_\phi(v) \tag{3.7}$$

and

$$\deg(\phi) = \sum_{k=1}^{n_2} (\Phi_{mV})_{kp}. \tag{3.8}$$

Proof. The first statement was shown in the proof of Lemma 3.3.29 and the second follows from the definition of Φ_{mV}. $\square$

Lemma 3.3.33. *If ϕ is harmonic then*

(1)

$$\Phi_V^t \Phi_{mV} = \deg(\phi) I_{n_1},$$

where I_{n_1} is the $n_1 \times n_1$ identity matrix, and

(2)

$$\Phi_E^t \Phi_E = \deg(\phi) I_{m_1},$$

where I_{m_1} is the $m_1 \times m_1$ identity matrix.

Proof. Note that

$$(\Phi_{mV})_{kp} = \begin{cases} m_\phi(v_k), & \text{if } \phi_V(v_k) = w_p, \\ 0, & \text{otherwise.} \end{cases}$$

Then

$$(\Phi_V^t \Phi_{mV})_{ij} = \sum_{k=1}^{n_2} (\Phi_V)_{ki}(\Phi_{mV})_{kj}$$

$$= \begin{cases} \sum_{k=1}^{n_2} (\Phi_V)_{ki}(\Phi_{mV})_{ki}, & \text{if } i = j, \\ 0, & \text{if } i \neq j, \end{cases}$$

$$= \begin{cases} \sum_{v_k \in \Phi_V^{-1}(w_p)} m_\phi(v_k), & \text{if } i = j, \\ 0, & \text{if } i \neq j, \end{cases}$$

$$= \begin{cases} \deg(\phi), & \text{if } i = j, \\ 0, & \text{if } i \neq j. \end{cases}$$

For the identity involving Φ_E, note that

$$(\Phi_E^t \Phi_E)_{ij} = \sum_{k=1}^{m_2} (\Phi_E)_{ki}(\Phi_E)_{kj}$$

$$= \begin{cases} \sum_{k=1}^{m_2} (\Phi_E)_{ki}^2, & \text{if } i = j, \\ 0, & \text{if } i \neq j. \end{cases}$$

Since $(\Phi_E)_{ki}$ is 1 or -1 if $\phi(e_k) = f_i$, and is 0 otherwise, the quantity $\sum_{k=1}^{m_2} (\Phi_E)_{ki}^2$ is the number of horizontal edges in Γ_2 mapping to the edge f_i in Γ_1, which is $\deg(\phi)$. $\qquad\square$

Example 3.3.34. There is a morphism ϕ from the Paley graph on 9 vertices, Γ_2 (depicted in Figure 1.12) to the cycle graph on 3 vertices, Γ_1, given by the map of vertices

$$\phi_V : \{0, 1, 2\} \mapsto 0, \{a, a+1, a+2\} \mapsto 2, \{2a, 2a+1, 2a+2\} \mapsto 1. \quad (3.9)$$

This is a harmonic morphism from this Paley graph to a graph with 3 vertices. We can check that in this case, the eigenvalues and eigenvectors of the Laplacians on these graphs are closely related, as Theorem 3.3.44 below predicts. However, this morphism does have vertical edges (so is not a cover in the sense of §3.4 but what we call it a quasicover).

```
──────────────────── Sage ────────────────────
sage: Gamma2 = graphs.PaleyGraph(9); Gamma2
Paley graph with parameter 9: Graph on 9 vertices
sage: V2 = Gamma2.vertices(); V2
[0, 1, 2, a, a + 1, a + 2, 2*a, 2*a + 1, 2*a + 2]
sage: Gamma1 = graphs.CycleGraph(3); Gamma1
Cycle graph: Graph on 3 vertices
sage: V1 = Gamma1.vertices(); V1
[0, 1, 2]
sage: Q2 = Gamma2.laplacian_matrix()
sage: E2 = Q2.right_eigenspaces()
sage: Q2.eigenvalues()
```

```
[0, 6, 6, 6, 6, 3, 3, 3, 3]
sage: Q1 = Gamma1.laplacian_matrix()
sage: E1 = Q1.right_eigenspaces()
sage: Q1.eigenvalues()
[0, 3, 3]
```

Next, using Sage, we compute the matrix Φ_V associated to ϕ. Sage tells us that

$$\Phi_V = \begin{pmatrix} 1 & 0 & 0 \\ 1 & 0 & 0 \\ 1 & 0 & 0 \\ 0 & 0 & 1 \\ 0 & 0 & 1 \\ 0 & 0 & 1 \\ 0 & 1 & 0 \\ 0 & 1 & 0 \\ 0 & 1 & 0 \end{pmatrix}.$$

Now look at the image of the eigenvectors under Φ_V. Sage computations show:

- the vector

$$v_0 = (1, 1, 1, 1, 1, 1, 1, 1, 1)$$

is an eigenvector of Q_2 having eigenvalue $\lambda = 0$ and

$$v_0 \Phi_V = (3, 3, 3).$$

- the vectors

$$x_1 = (1, 0, -1, 0, 0, 0, 0, 1, -1), \quad x_2 = (0, 1, -1, 0, 0, 0, -1, 1, 0),$$

$$x_3 = (0, 0, 0, 1, 0, -1, 1, -1, 0), \quad x_4 = (0, 0, 0, 0, 1, -1, 1, 0, -1),$$

are eigenvectors of Q_2 having eigenvalue $\lambda = 6$ and

$$x_i \Phi_V = (0, 0, 0), \quad i = 1, 2, 3, 4.$$

- the vectors

$$x_5 = (1, 0, 0, 0, 1, 0, -1, -1, 0), \quad x_6 = (0, 1, 0, 0, 0, 1, 0, -1, -1),$$

$$x_7 = (0, 0, 1, 0, -1, -1, 0, 1, 0), \quad x_8 = (0, 0, 0, 1, 1, 1, -1, -1, -1),$$

are eigenvectors of Q_2 having eigenvalue $\lambda = 3$ and

$$x_i \Phi_V = (1, -2, 1), \qquad i = 5, 6,$$

while

$$x_7 \Phi_V = (1, 1, -2), \qquad x_8 \Phi_V = (0, -3, 3).$$

While most are omitted, some of the computations supporting this are given below.

```
─────────────────────── Sage ───────────────────────
sage: v0 = E2[0][1].basis()[0]; v0
(1, 1, 1, 1, 1, 1, 1, 1, 1)
sage: Q2*v0 == 0*v0
True
sage: v0*PhiV
(3, 3, 3)
sage: v1 = E2[1][1].basis()[0]; v1
(1, 0, -1, 0, 0, 0, 0, 1, -1)
sage: Q2*v1 == 6*v1
True
sage: v1*PhiV
(0, 0, 0)
sage: v5 = E2[2][1].basis()[0]; v5
(1, 0, 0, 0, 1, 0, -1, -1, 0)
sage: Q2*v5
(3, 0, 0, 0, 3, 0, -3, -3, 0)
sage: v5*PhiV
(1, -2, 1)
```

It can be verified that this morphism ϕ is harmonic and satisfies the graph morphism identity of Proposition 3.3.13.

Give the edges of Γ_2,

$$(0, 1), (0, 2), (0, a + 1), (0, 2a + 2), (1, 2), (1, a + 2), (1, 2a), (2, a),$$
$$(2, 2a + 1), (a, a + 1), (a, a + 2), (a, 2a + 1), (a + 1, a + 2),$$
$$(a + 1, 2a + 2), (a + 2, 2a), (2a, 2a + 1), (2a, 2a + 2), (2a + 1, 2a + 2),$$

the orientation $+1$. Likewise, give the edges of Γ_2,

$$(0, 1), (0, 2), (1, 2),$$

the orientation $+1$. The signed incidence matrix of Γ_1 is

$$B_1 = \begin{pmatrix} 1 & 1 & 0 \\ -1 & 0 & 1 \\ 0 & -1 & -1 \end{pmatrix}$$

and the signed incidence matrix of Γ_2 is

$$B_2 = \begin{pmatrix}
1 & 1 & 1 & 1 & 0 & 0 & 0 & 0 & 0 & 0 & 0 & 0 & 0 & 0 & 0 & 0 & 0 & 0 \\
-1 & 0 & 0 & 0 & 1 & 1 & 1 & 0 & 0 & 0 & 0 & 0 & 0 & 0 & 0 & 0 & 0 & 0 \\
0 & -1 & 0 & 0 & -1 & 0 & 0 & 1 & 1 & 0 & 0 & 0 & 0 & 0 & 0 & 0 & 0 & 0 \\
0 & 0 & 0 & 0 & 0 & 0 & -1 & 0 & 0 & 1 & 1 & 1 & 0 & 0 & 0 & 0 & 0 & 0 \\
0 & 0 & -1 & 0 & 0 & 0 & 0 & 0 & -1 & 0 & 0 & 1 & 1 & 0 & 0 & 0 & 0 & 0 \\
0 & 0 & 0 & 0 & 0 & -1 & 0 & 0 & 0 & 0 & -1 & 0 & -1 & 0 & 1 & 0 & 0 & 0 \\
0 & 0 & 0 & 0 & 0 & 0 & -1 & 0 & 0 & 0 & 0 & 0 & 0 & -1 & 1 & 1 & 0 & 0 \\
0 & 0 & 0 & 0 & 0 & 0 & 0 & 0 & -1 & 0 & 0 & -1 & 0 & 0 & 0 & -1 & 0 & 1 \\
0 & 0 & 0 & -1 & 0 & 0 & 0 & 0 & 0 & 0 & 0 & 0 & -1 & 0 & 0 & -1 & -1 \\
\end{pmatrix}.$$

For the function in (3.9), we have

$$\Phi_V = \begin{pmatrix}
1 & 0 & 0 \\
1 & 0 & 0 \\
1 & 0 & 0 \\
0 & 0 & 1 \\
0 & 0 & 1 \\
0 & 0 & 1 \\
0 & 1 & 0 \\
0 & 1 & 0 \\
0 & 1 & 0 \\
\end{pmatrix},$$

and

$$\Phi_E = \begin{pmatrix}
0 & 0 & 0 \\
0 & 0 & 0 \\
0 & 1 & 0 \\
1 & 0 & 0 \\
0 & 0 & 0 \\
0 & 1 & 0 \\
1 & 0 & 0 \\
0 & 1 & 0 \\
1 & 0 & 0 \\
0 & 0 & 0 \\
0 & 0 & 0 \\
0 & 0 & -1 \\
0 & 0 & 0 \\
0 & 0 & -1 \\
0 & 0 & -1 \\
0 & 0 & 0 \\
0 & 0 & 0 \\
0 & 0 & 0 \\
\end{pmatrix}.$$

We leave it as an exercise to check that $\Phi_V B_1 = B_2 \Phi_E$.

Example 3.3.35 We return to the graph morphism in Example 3.3.12 sending the cube graph Γ_2 to the tetrahedron graph Γ_1. If the vertices of Γ_2 are placed at the coordinates of the unit cube in $\mathbb{R}^3$, the morphism $f : \Gamma_2 \to \Gamma_1$ is given by

$$f : \{000, 111\} \mapsto 0, \{001, 110\} \mapsto 1, \{010, 101\} \mapsto 2, \{100, 011\} \mapsto 3,$$

and the edges of Γ_1,

$$(0, 1), (0, 2), (0, 3), (1, 2), (1, 3), (2, 3)$$

have orientation $+1$. For this function, we have

$$\Phi_V = \begin{pmatrix} 1 & 0 & 0 & 0 \\ 0 & 1 & 0 & 0 \\ 0 & 0 & 1 & 0 \\ 0 & 0 & 0 & 1 \\ 0 & 0 & 0 & 1 \\ 0 & 0 & 1 & 0 \\ 0 & 1 & 0 & 0 \\ 1 & 0 & 0 & 0 \end{pmatrix},$$

and, for a suitable orientation of the edges of Γ_2,

$$\Phi_E = \begin{pmatrix} 1 & 0 & 0 & 0 & 0 & 0 \\ 0 & 1 & 0 & 0 & 0 & 0 \\ 0 & 0 & 1 & 0 & 0 & 0 \\ 0 & 0 & 0 & 0 & 1 & 0 \\ 0 & 0 & 0 & 1 & 0 & 0 \\ 0 & 0 & 0 & 0 & 0 & 1 \\ 0 & 0 & 0 & -1 & 0 & 0 \\ 0 & 0 & -1 & 0 & 0 & 0 \\ 0 & 0 & 0 & 0 & 0 & -1 \\ 0 & 0 & 0 & 0 & -1 & 0 \\ 0 & -1 & 0 & 0 & 0 & 0 \\ -1 & 0 & 0 & 0 & 0 & 0 \end{pmatrix}.$$

It is left as an exercise to verify $\Phi_E^t \Phi_E = 2I_6$ (the 6×6 scalar matrix with 2s on the diagonal) and $\Phi_V^t \Phi_V = 2I_4$.

3.3.3 Covering maps

In this section, we will define coverings and state our identities for this special case of harmonic morphisms. We will also prove an identity for edge Laplacians of coverings.

Definition 3.3.36. A harmonic morphism $\phi : \Gamma_2 \to \Gamma_1$ is called a *covering* if there are no vertical edges and all the horizontal multiplicities $m_\phi(v_i)$ at vertices v_i in Γ_2 are equal to 1. Thus $\Phi_{mV} = \Phi_V$ for a covering. The degree $d = \deg(\phi)$ of ϕ is called the degree of the covering.

We give two generalizations of coverings, for which many of our results for coverings also hold.

Definition 3.3.37. If ϕ is a harmonic morphism such that all horizontal multiplicities $m_\phi(w_i)$ at vertices w_i are equal to a constant m and Γ has no vertical edges, then we call ϕ a *multicover*.

Definition 3.3.38. If ϕ is a harmonic morphism, possibly with vertical edges, such that all horizontal multiplicities $m_\phi(w_i)$ at vertices w_i are equal to a constant m, then we call ϕ a *quasi-multicover*. If $m = 1$, then we call ϕ a *quasicover*.

Thus, a cover is a special case of a multicover, in which the horizontal multiplicity m is 1. A multicover is a special case of a quasi-multicover. A quasi-multicover may have vertical edges but a multicover does not. For a cover, the multiplicity-vertex map matrix Φ_{mV} and the vertex map matrix Φ_V are equal. For a multicover or quasi-multicover with horizontal multiplicity m, we have $\Phi_{mV} = m\Phi_V$.

Example 3.3.39. This is a continuation of Example 3.3.9.
  Let ϕ_1 and ϕ_2 denote the morphisms from Γ_2 to Γ_1 defined in Example 3.3.9. We know that they are both harmonic. Note that the cyclic group of order 3, G, acts on Γ_2 and that ϕ_2 is the quotient map $\Gamma_2 \to \Gamma_2/G \cong \Gamma_1$.

Exercise 3.14. Show that ϕ_1 and ϕ_2 are both quasicovers.

We will now state our previous identities for harmonic morphisms for the special case of coverings.

Proposition 3.3.40 (Identities for coverings). *If $\phi : \Gamma_2 \to \Gamma_1$ is a nonconstant covering of degree d, then the following three identities hold:*

$$B_2\Phi_E = \Phi_V B_1 \tag{3.10}$$

$$Q_2\Phi_V = \Phi_V Q_1 \tag{3.11}$$

$$\Phi_V^t \Phi_V = d I_{n_1},\tag{3.12}$$

where n_1 is the number of vertices of Γ_1 and I_{n_1} is the $n_1 \times n_1$ identity matrix.

Definition 3.3.41. The *edge Laplacian* matrix $\tilde{Q}$ of a graph Γ with incidence matrix B is defined to be

$$\tilde{Q} = B^t B.$$

Proposition 3.3.42. *Let* $\phi : \Gamma_2 \to \Gamma_1$ *be a nonconstant covering. Then*

$$\tilde{Q}_2 \Phi_E = \Phi_E \tilde{Q}_1.$$

Proof. Using the identities above, and the fundamental identity for graph morphisms,

$$\begin{aligned}
\tilde{Q}_2 \Phi_E &= B_2^t B_2 \Phi_E \\
&= B_2^t \Phi_V B_1 \\
&= \Phi_E B_1^t B_1 \\
&= \Phi_E \tilde{Q}_1.
\end{aligned}$$

$\square$

See also Corry [Co11], Baker-Norine [BN09], and [MM14] Mednykh and Mednykh for further material on graph coverings and harmonic morphisms.

3.3.4 Graph spectra for harmonic morphisms

In this section, we will relate the eigenvalues and eigenvectors of Q_1 and $\Phi_V^t Q_2 \Phi_V$. We will also prove a slightly stronger statement for covering maps.

Theorem 3.3.43. *Let* $\phi : \Gamma_2 \to \Gamma_1$ *be a nonconstant harmonic morphism. Let* Q_1 *and* Q_2 *be the Laplacian matrices of* Γ_1 *and* Γ_2 *and let* Φ_V *be the vertex map matrix (with respect to some orderings on the vertices and edges of the graphs). Let* $\mathbf{v}$ *be an eigenvector of* Q_1 *corresponding to an eigenvalue* λ. *Then* $\mathbf{v}$ *is an eigenvector of the matrix*

$$\Phi_V^t Q_2 \Phi_V$$

corresponding to an eigenvalue $deg(\phi)\lambda$.

Proof. If $\mathbf{v}$ is an eigenvector of Q_1 corresponding to an eigenvalue λ, then

$$\Phi_V^t Q_2 \Phi_V x = \Phi_V^t \Phi_{mV} Q_1 x \quad \text{by Proposition 3.3.25}$$
$$= \deg(\phi) \lambda x \quad \text{by Lemma 3.3.33}$$

□

Theorem 3.3.44. *Let $\phi : \Gamma_2 \to \Gamma_1$ be a nonconstant covering map. Let Q_1 and Q_2 be the Laplacian matrices of Γ_1 and Γ_2 and let Φ_V be the vertex map matrix (with respect to some orderings on the vertices and edges of the graphs). Let x be an eigenvector of Q_1 corresponding to an eigenvalue λ. Then $\Phi_V x$ is an eigenvector Q_2 corresponding to the eigenvalue λ.*

Proof. Recall that if ϕ is a covering map then all the horizontal multiplicities are 1 so that the multiplicity-vertex map matrix Φ_{mV} equals the vertex map matrix Φ_V. Let x be an eigenvector of Q_1 corresponding to the eigenvalue λ, and let $x' = \Phi_V x$. Then

$$Q_2 x' = Q_2 \Phi_V x$$
$$= \Phi_V Q_1 x \quad \text{by Proposition 3.3.25}$$
$$= \Phi_V \lambda x$$
$$= \lambda x'.$$

□

Exercise 3.15. Suppose that $\phi : \Gamma_2 \to \Gamma_1$ is a quasi-multicover, i.e., ϕ is a harmonic morphism such that all horizontal multiplicities $m_\phi(v_i)$ at vertices v_i are equal to a constant m. Show that if x is an eigenvector of Q_2 corresponding to an eigenvalue λ, then $\Phi_V x$ is an eigenvector of Q_1 corresponding to the eigenvalue $m\lambda$.

3.3.5 Riemann–Hurwitz formula

By a graph, we mean a finite connected multigraph without loops. Recall that we define the *genus* of a graph Γ as

$$\text{genus}(\Gamma) = |E(\Gamma)| - |V(\Gamma)| + 1.$$

Riemann–Hurwitz formula (classical)

The following statement is the *Riemann–Hurwitz formula for graphs* (see Baker and Norine [BN09]).

Theorem 3.3.45. *(Riemann–Hurwitz formula for graphs, Baker and Norine)* Let $\phi : \Gamma_2 \to \Gamma_1$ be a harmonic morphism. Then

$$\text{genus}(\Gamma_2) - 1 = \deg(\phi)(\text{genus}(\Gamma_1) - 1) + \sum_{v \in V_2} ((m_\phi(v) - 1) + \frac{1}{2}\nu_\phi(v)), \quad (3.13)$$

where $m_\phi(v)$ denotes the horizontal multiplicity and $\nu_\phi(v)$ denotes the vertical multiplicity of ϕ at $v \in V_2$.

Proof. The proof follows from Lemmas 3.3.20 and 3.3.32. For more details, see Baker and Norine [BN09], Corry [Co11]. □

Riemann–Hurwitz formula (with group action)

Let G be a finite group acting on a graph Γ.

Definition 3.3.46. An edge $\{v, w\} \in E(\Gamma)$ is said to be *fixed* by G if there is a nontrivial element $g \in G$ that fixes v and w. An edge $\{v, w\} \in E(\Gamma)$ is said to be *reversed* by G if there is a nontrivial element $g \in G$ that swaps v and w. We say that G acts on Γ *without edge reversal* if Γ has no edges swapped by G.

The following statement is the *Riemann–Hurwitz Formula for graphs with a group action* (see Mednykh [M15]).

Theorem 3.3.47. Let G be a finite group acting on graph Γ without fixed and invertible edges. We have

$$\text{genus}(\Gamma) - 1 = |G|(\text{genus}(\Gamma/G) - 1) + \sum_{v \in V_\Gamma} (|G_v| - 1), \quad (3.14)$$

where G_v denotes the stabilizer of v in G.

Example 3.3.48. Consider the example of the Paley graph having 9 vertices, Γ. The automorphism group G has order 72. The genus of Γ is 10. The quotient Γ/G is the trivial graph, so has genus 0. The group G satisfies

$$\sum_{v \in V_\Gamma} (|G_v| - 1) = 63.$$

Therefore, $\text{genus}(\Gamma) - 1 = 9$, and $|G|(\text{genus}(\Gamma/G) - 1) = -72$, so the right-hand side is -9 and the left-hand side is 9. Indeed, there are edges of Γ fixed by a nontrivial automorphism.

3.4 *G*-equivariant covering graphs

If $\Gamma = (V, E)$ is a graph and $v \in V$ is a vertex, recall that $N_\Gamma(v)$ denotes the set of neighbors of v in Γ, i.e., the set of all vertices adjacent to v. Other than v itself, these are the vertices of the star graph $St_\Gamma(v)$ (in Definition 3.3.4).

Let $\Gamma_1 = (V_1, E_1)$ and $\Gamma_2 = (V_2, E_2)$ be two graphs, and let $\phi : \Gamma_2 \to \Gamma_1$ be a surjection (so $\phi|_{V_2} : V_2 \to V_1$ and $\phi|_{E_2} : E_2 \to E_1$ are surjections). Then ϕ is a *covering map* from Γ_2 to Γ_1 if for each $v \in V_2$,

$$\phi|_{N_{\Gamma_2}(v)} : N_{\Gamma_2}(v) \to N_{\Gamma_1}(\phi(v))$$

is a bijection onto the neighborhood of $\phi(v) \in V_1$ in Γ_1. In other words, ϕ maps edges incident to v one-to-one onto edges incident to $\phi(v)$. In this case, we call Γ_2 a *covering graph*, or *lift*, of Γ_1. Such a covering graph has an *automorphism* (or *covering transformation*) if there is a graph automorphism $\alpha \in Aut(\Gamma_2)$ such that $\phi \circ \alpha = \phi$. In other words, the diagram

$$\begin{array}{ccc} \Gamma_2 & \xrightarrow{\ \alpha\ } & \Gamma_2 \\ {\scriptstyle \phi}\downarrow & & \downarrow{\scriptstyle \phi} \\ \Gamma_1 & \xrightarrow{\ id\ } & \Gamma_1 \end{array}$$

commutes. Such an automorphism of the cover is an element of $Aut(\Gamma_2)$ whose restriction determines an action on each of the fibers $\phi^{-1}(v)$, for $v \in V_1$. The group of all such automorphisms is denoted $Aut(\Gamma_2/\Gamma_1)$. If $G \subset Aut(\Gamma_2/\Gamma_1)$ is a subgroup, then such a covering graph is called a G-*cover*. If this action is transitive on each fiber, then we call the cover *Galois*. In this case, the group of all such automorphisms is denoted $Gal(\Gamma_2/\Gamma_1)$.

The following interesting example can be found in Terras [Te10].

Example 3.4.1. This is a continuation of Example 3.3.12. There is an action of $G = \mathbb{Z}/2\mathbb{Z}$ on the cube graph in $\mathbb{R}^3$ which allows it to be regarded as a G-cover of the tetrahedron graph. Indeed, each of the four vertices of the tetrahedron lifts to one of the four pairs of vertices of the cube graph of the form (vertex, antipode vertex). The G-action on the vertices of the cube graph is simply to swap each vertex with its antipode.

For the construction below, we recall a general graph-theoretical construction. Let $\Gamma_1 = (V_1, E_1)$ and $\Gamma_2 = (V_2, E_2)$ be arbitrary graphs. Fix a vertex v_0 of Γ_2 (the *root*). The *rooted graph product* of Γ_1 and Γ_2, denoted

$$\Gamma_1 \circ \Gamma_2,$$

is obtained by replacing each vertex of Γ_1 by v_0, along with the rest of Γ_2. An example of a rooted graph product is given in Figure 3.2.

The Jacobian, defined in (3.16), of such graphs is easily computed (see Exercise 4.12 below for a more general result).

Lemma 3.4.2. *The rooted graph product satisfies*

$$Jac(\Gamma_1 \circ \Gamma_2) = Jac(\Gamma_1) \times Jac(\Gamma_2)^n,$$

where n is the number of vertices of Γ_1.

Theorem 3.4.3. *Given a finite group G and a graph Γ, there is a graph $\overline{\Gamma}$ with the following properties:*

- *G acts on $\overline{\Gamma}$.*

- *The quotient graph $\overline{\Gamma}/G$ is isomorphic to Γ and the quotient map $\overline{\Gamma} \to \Gamma$ is harmonic.*

- *$\overline{\Gamma}$ is a quasicovering of Γ.*

- *The action of G on $\overline{\Gamma}$ induces an action on the Jacobian $Jac(\overline{\Gamma})$ and we have a natural isomorphism*

$$Jac(\overline{\Gamma})/G \cong Jac(\Gamma).$$

Proof. The proof, by construction, is very simple. Pick a set S of generators of G which does not contain the identity and which is symmetric (i.e., S is closed under taking inverses). Let

$$\overline{\Gamma} = Cay(G, S) \circ \Gamma, \tag{3.15}$$

denote the rooted graph product of the Cayley graph of G with the graph Γ, for any fixed root v_0 of Γ. In this case, we see that, by the Lemma above, the Jacobian satisfies

$$Jac(\overline{\Gamma}) \cong Jac(Cay(G, S)) \times Jac(\Gamma)^{|G|},$$

where G acts via the natural action on $Jac(Cay(G, S))$ and by permuting the coordinates of product $Jac(\Gamma)^{|G|}$. This implies $Jac(\overline{\Gamma})/G \cong Jac(\Gamma)$. $\square$

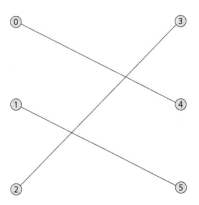

Figure 3.4:   A covering graph of $\Gamma = (\{0, 1\}, \{(0, 1)\})$.

Example 3.4.4. Let G denote the cyclic group on three letters and let $\Gamma = (\{0,1\},\{(0,1)\})$ denote the simple graph having two vertices and one edge connecting them. A G-cover $\overline{\Gamma} \to \Gamma$ is depicted in Figure 3.4.

Example 3.4.5. This is a continuation of Example 3.3.12.

There is a harmonic morphism $\phi : \Gamma_2 \to \Gamma_1$, between the cube graph Γ_2 and the tetrahedron graph Γ_1 (depicted in 3.3). The corresponding map on the edges Φ_E and the map Φ_V on the vertices, described above, can be computed explicitly.

For the Laplacian of Γ_2, the eigenvalues are $\lambda_1 = 0$, $\lambda_2 = 2$, $\lambda_3 = 4$, $\lambda_4 = 6$, and the corresponding (right) eigenspaces are

$$E_1 = \mathrm{span}\{(1,1,1,1,1,1,1,1)\},$$

$$E_2 = \mathrm{span}\{(1,0,0,-1,1,0,0,-1),$$
$$(0,1,0,1,-1,0,-1,0), (0,0,1,1,-1,-1,0,0)\},$$

$$E_3 = \mathrm{span}\{(1,0,0,-1,-1,0,0,1),$$
$$(0,1,0,-1,-1,0,1,0), (0,0,1,-1,-1,1,0,0)\},$$

and

$$E_4 = \mathrm{span}\{(1,-1,-1,1,-1,1,1,-1)\}.$$

For the Laplacian of Γ_1, the eigenvalues are $\lambda_1 = 0$, $\lambda_2 = 4$, and the corresponding (right) eigenspaces are

$$E_1 = \mathrm{span}\{(1,1,1,1)\},$$

$$E_2 = \mathrm{span}\{(1,0,0,-1), (0,1,0,-1), (0,0,1,-1)\}.$$

It is left as an exercise to verify that $\Phi_V(E_1(\Gamma_1)) \subset E_1(\Gamma_2)$.

Example 3.4.6. We continue Example 3.3.34, with ϕ denoting the morphism from the Paley graph on 9 vertices, Γ_2 (depicted in Figure 1.12), to the cycle graph on 3 vertices, Γ_1. The morphism ϕ defined by the map of vertices

$$\phi_V : \{0,1,2\} \mapsto 0, \quad \{\alpha, \alpha+1, \alpha+2\} \mapsto 2, \quad \{2\alpha, 2\alpha+1, 2\alpha+2\} \mapsto 1,$$

is harmonic and satisfies the property that all horizontal multiplicities $m_\phi(v_i)$ equal 1. However, it is not a multicover because there are vertical edges (for example $(0,1)$ is vertical). We call it a quasicover. The following computation shows that the statement in the above theorem holds in this case anyway.

We have

$$\Phi_V = \begin{pmatrix} 1 & 0 & 0 \\ 1 & 0 & 0 \\ 1 & 0 & 0 \\ 0 & 0 & 1 \\ 0 & 0 & 1 \\ 0 & 0 & 1 \\ 0 & 1 & 0 \\ 0 & 1 & 0 \\ 0 & 1 & 0 \end{pmatrix}.$$

This matrix sends the (one-dimensional) eigenspace of the Laplacian of Γ_1 with eigenvalue 0 to the eigenspace of the Laplacian of Γ_2 with eigenvalue 0. This matrix also sends the (two-dimensional) eigenspace of the Laplacian of Γ_1 with eigenvalue 3 to the (four-dimensional) eigenspace of the Laplacian of Γ_2 with eigenvalue 3.

Suppose that $G \subset Aut(\Gamma_2)$ is a group of automorphisms of a graph Γ_2. Then G *acts harmonically* on Γ_2 if for any subgroup $K \subset G$, the quotient map $\Gamma_2 \to \Gamma_2/K$ is a harmonic morphism. We say G *acts freely* on Γ_2 if $g \in G$ and $x \in V_2$ satisfy $gx = x$ then $g = 1$. We say G *acts transitively* on Γ_2 if for each pair $x, y \in V_2$ there exists a $g \in G$ such that $gx = y$. See Corry [Co11] for a proof of the following lemma.

Lemma 3.4.7. *(Corry and Nedela) A nontrivial group G acts harmonically on a graph Γ if and only if it acts freely[3] on the set of edges of Γ.*

For example, the cycle group C_n acts transitively without fixed points on the (edges of the) cycle graph having n vertices, so that action is harmonic.

To obtain a correct graph-theoretic analog of the "Galois theory" of coverings of Riemann surfaces, Terras works with non-simple graphs. The following result may be found in Terras [Te10] (see Chapters 13 and 14 in Terras [Te10]).

Theorem 3.4.8. *("Fundamental Theorem of Galois Theory") Suppose $\pi :$ $\Gamma_2 \to \Gamma_1$ is an unramified normal covering with Galois group $G = Gal$ (Γ_2/Γ_1). Assume the graphs Γ_i satisfy the conditions in Terras [Te10].*

1. *Given a subgroup H of G, there exists a unique graph Γ intermediate to $\Gamma_2 \to \Gamma_1$ such that $H = Gal(\Gamma_2/\Gamma)$. Define the map σ from subgroups of G to intermediate graphs by $\sigma(H) = \Gamma$.*

2. *Suppose Γ is intermediate to $\Gamma_2 \to \Gamma_1$. Then there is a unique subgroup H of G which equals $Gal(\Gamma_2/\Gamma)$. Define the map τ from intermediate graphs to subgroups of G by $\tau(\Gamma) = H$.*

---

[3]To be clear, if an edge is flipped by an element of G, then it is not fixed, according to the definition Corry gives.

3. *Two intermediate graphs Γ and Γ' are equal if and only if $\tau(\Gamma) = \tau(\Gamma')$.*

4. *We have $H = \tau(\sigma(H))$ and $\Gamma = \sigma(\tau(\Gamma))$, for each subgroup H of G and each Γ intermediate to $\Gamma_2 \to \Gamma_1$. Therefore, we have a bijective correspondence between intermediate graphs and subgroups of the Galois group.*

5. *If H and H' are subgroups of G, and if Γ and Γ' are the corresponding intermediate graphs, then Γ is intermediate to $\Gamma_2 \to \Gamma'$ if and only if $H \subset H'$.*

3.5 A Riemann–Roch theorem on graphs

In this section, we will discuss divisors on graphs and prove Baker and Norine's [BN07] Riemann–Roch theorem for graphs.

We will assume that Γ is a finite connected graph which has no loops but may have multiple edges between pairs of vertices. We will let V be the set of vertices and E the set of edges of Γ.

Recall that the genus of the graph Γ is defined to be

$$\text{genus}(\Gamma) = |E| - |V| + 1.$$

3.5.1 Divisors and the Jacobian

We define a *divisor* on Γ to be an expression of the form

$$D = \sum_{v \in V} D(v) v,$$

where $D(v) \in \mathbb{Z}$ for all $v \in V$. The set of all divisors on Γ is denoted $\text{Div}(\Gamma)$ and is the free abelian group on V.

The *degree* of a divisor D is defined to be

$$\deg(D) = \sum_{v \in V} D(v).$$

We denote by $\text{Div}^d(\Gamma)$ the set of all divisors of degree d on Γ. Note $\text{Div}^0(\Gamma)$ is an abelian subgroup of $\text{Div}(\Gamma)$.

If f is a function in $C^0(\Gamma, \mathbb{Z})$, i.e., a function $f : V \to \mathbb{Z}$, we define the *principal divisor* associated with f to be

$$\text{div}(f) = \sum_{v \in V} \left(f(v)\deg(v) - \sum_{w \neq v} f(w)A_{vw} \right) v$$

where A is the adjacency matrix of Γ. Note that if we fix an ordering $v_1, v_2, \ldots, v_n$ of the vertices of Γ and represent f and $\text{div}(f)$ as vectors, then we can write $\text{div}(f) = Qf$, where Q is the Laplacian matrix of Γ.

The set of principal divisors $\text{Prin}(\Gamma)$ forms a subgroup of $\text{Div}^0(\Gamma)$. The quotient group is the *Jacobian*

$$\text{Jac}(\Gamma) = \text{Div}^0(\Gamma)/\text{Prin}(\Gamma). \tag{3.16}$$

Suppose that $\phi : \Gamma_2 \to \Gamma_1$ is a nonconstant harmonic morphism. We will show later that ϕ determines a pushforward map

$$\phi_* : \text{Jac}(\Gamma_2) \to \text{Jac}(\Gamma_1)$$

which is surjective (Proposition 4.8.4) and a pullback map

$$\phi^* : \text{Jac}(\Gamma_1) \to \text{Jac}(\Gamma_2)$$

which is injective (Proposition 4.8.2).

3.5.2 Linear systems on graphs

We say that divisors D and D' are *linearly equivalent* and write $D \sim D'$ if $D' - D$ is a principal divisor, i.e., if $D' = D + \text{div}(f)$ for some function $f : V \to \mathbb{Z}$. Note that if D and D' are linearly equivalent, they must have the same degree, since the degree of every principal divisor is 0. Divisors of degree 0 are linearly equivalent if and only if they determine the same element of the Jacobian. If D is a divisor of degree 0, we denote by $[D]$ the element of the Jacobian determined by D.

A divisor D is said to be *effective* if $D(v) \geq 0$ for all vertices v. We write $D \geq 0$ to mean that D is effective.

The *linear system* associated to a divisor D is the set

$$|D| = \{D' \in \text{Div}(\Gamma) : D' \geq 0 \text{ and } D' \sim D\},$$

i.e., $|D|$ is the set of all effective divisors linearly equivalent to D. Note that if $D_1 \sim D_2$, then $|D_1| = |D_2|$. We note also that if $\deg(D) < 0$, then $|D|$ must be empty.

Example 3.5.1. Let Γ be the cycle graph with 4 vertices having vertices (arranged counterclockwise around the graph) 0, 1, 2, 3 and let D be the divisor $(1, 0, 1, 0)$ (supported on 0 and 2). In this case,

$$|D| = \{(0, 2, 0, 0), (0, 0, 0, 2), (1, 0, 1, 0)\}.$$

Definition 3.5.2. The *dimension* $r(D)$ of a linear system $|D|$ is defined as follows. If $|D|$ is empty, $r(D) = -1$. Otherwise, if $|D|$ is non-empty, $r(D)$ is the largest nonnegative integer d such that, for all effective divisors D' of degree d, $|D - D'|$ is non-empty. We will also refer to $r(D)$ as the *Riemann–Roch dimension* of D.

Note that the only effective divisor D' of degree $d = 0$ is $D' = 0$, so if $|D|$ is non-empty, then $r(D) \geq 0$. Also, if $\deg(D') > \deg(D)$, then $D - D'$ has negative degree and $|D - D'|$ is empty, so $r(D) \leq \deg(D)$. If $\deg(D) = 0$, then $r(D) = 0$ if $D \sim 0$, and $r(D) = -1$ otherwise.

3.5.3 Non-special divisors

The notion of a non-special divisor is central in Baker and Norine's proof of the Riemann–Roch theorem for graphs.

Definition 3.5.3. A divisor v on a graph Γ is called *non-special* if

1. $\deg(v) = g - 1$, where g is the genus of Γ, and

2. $|v| = \phi$, i.e., the linear system associated to v is empty.

The collection of all non-special divisors on Γ is denoted $\mathcal{N}$.

Note that if v is a non-special divisor, then so is v', for all divisors v' which are linearly equivalent to v.

We first show that non-special divisors always exist, using an algorithm of Baker and Norine [BN07].

Lemma 3.5.4. *Let $P = (v_1, v_2, \ldots, v_n)$ be an ordering of the vertices of Γ, and let A be the adjacency matrix of Γ with respect to this ordering, i.e., A_{ij} is the number of edges between vertices v_i and v_j. Let v_P be the divisor given by*

$$v_P(v_i) = \begin{cases} -1, & \text{if } i = 1, \\ -1 + \sum_{j=1}^{i-1} A_{ij}, & \text{if } i \geq 2. \end{cases}$$

Then v_P is non-special.

We will call v_P the *non-special divisor determined by the ordering P*.

Remark 3.5.5. We will usually abbreviate the formula for v_P, using the convention that $\sum_{j=1}^{0} A_{ij} = 0$, so that we write $v_P(v_i) = -1 + \sum_{j=1}^{i-1} A_{ij}$ for all i.

Proof. Let $v = v_P$. The degree of v is

$$\deg(v) = -n + \sum_{j<i} A_{ij}$$

$$= -n + \frac{1}{2} \sum_{i=1}^{n} \deg(v_i)$$

$$= -|V| + |E|$$

$$= g - 1.$$

Next, we will show that no divisor linearly equivalent to v can be effective, so that $|v|$ must be empty.

Suppose that v is linearly equivalent to an effective divisor D. Let Q be the Laplacian matrix with respect to the vertex ordering $v_1, v_2, \ldots, v_n$. Then D must satisfy the equation

$$D - Qx = v,$$

for some n-vector $x = (x_1, x_2, \ldots, x_n)$ of integers. We can assume that $x_i \geq 0$ for all i and $x_i = 0$ for at least one i, since if x' differs from x by a multiple of the all 1's vector, then $Qx' = Qx$.

Let v_k be the lowest index vertex such that $x_k = 0$. Then

$$v(v_k) = D(v_k) - (Qx)_k$$

$$= D(v_k) + \sum_{i \neq k} x_i A_{ki}$$

$$\geq A_{k1} + A_{k2} + \cdots + A_{k,k-1} \qquad \text{since } D(v_k) \geq 0.$$

This is a contradiction, because $v(v_k) = A_{k1} + A_{k2} + \cdots + A_{k,k-1} - 1$.

Thus, no divisor linearly equivalent to v can be effective, so v is non-special. $\qquad\square$

The next proposition is key to proving the Riemann–Roch theorem. It relies on a result which we will not prove until the next chapter. First, we state two definitions.

Definition 3.5.6. For any set $W \subset V$, and any vertex $v \in W$, we define the *outdegree* of W at v, denoted $\text{outdeg}_W(v)$ to be the number of edges from v to vertices in the complement of W.

Definition 3.5.7. Let q be any vertex in Γ. We say that a divisor D is *q-reduced* if $D(v) \geq 0$ for all $v \neq q$, and for each set $W \subset V \setminus \{q\}$, there exists a vertex v in W such that $D(v) < \text{outdeg}_W(v)$.

Reduced divisors are closely related to reduced configurations[4], studied in the next chapter. In fact, we will show there (see Proposition 4.5.7) that for each divisor D of degree 0 and for each vertex q, there is a unique q-reduced divisor linearly equivalent to D. It follows that for each divisor D of any degree and for each vertex q, there is a unique q-reduced divisor linearly equivalent to D.

Example 3.5.8. If $\Gamma = (V, E)$ denotes the tetrahedron graph on 4 vertices $v_1 (= q)$, v_2, v_3, v_4 (depicted in Figure 1.1), then there are exactly 16 q-reduced divisors of degree 0. If we denote the divisor $D = av_1 + bv_2 + cv_3 + dv_4$ by $[a, b, c, d]$, then these reduced divisors are

$$[-3, 0, 1, 2], [-3, 0, 2, 1], [-3, 1, 0, 2], [-3, 1, 2, 0],$$

$$[-3, 2, 0, 1], [-3, 2, 1, 0], [-2, 0, 0, 2], [-2, 0, 1, 1],$$

$$[-2, 0, 2, 0], [-2, 1, 0, 1], [-2, 1, 1, 0], [-2, 2, 0, 0],$$

$$[-1, 0, 0, 1], [-1, 0, 1, 0], [-1, 1, 0, 0], [0, 0, 0, 0].$$

In fact, as we will see later, this set of divisors has a natural abelian finite group structure—the critical group of Γ.

Proposition 3.5.9. *Let D be any divisor on Γ. Then*

1. *if the linear system $|D|$ is non-empty, then the linear system $|v - D|$ is empty for all non-special divisors v, and*

2. *if the linear system $|D|$ is empty, then there exists a non-special divisor v_P, associated to some ordering P of the vertices of Γ, such that the linear system $|v_P - D|$ is non-empty.*

Proof. Suppose that the linear system $|D|$ is non-empty, and that D' is an effective divisor linearly equivalent to D. Note that $|v - D| = |v - D'|$ since $v - D$ is linearly equivalent to $v - D'$. If there is a non-special divisor v such that $|v - D|$ is non-empty, then there is an effective divisor E such that $v - D'$ is linearly equivalent to E. Thus, v is linearly equivalent to the effective divisor $D' + E$, which is impossible, since v is non-special.

Now suppose that the linear system $|D|$ is empty. Pick a vertex v_1 and let D_r be the unique v_1-reduced divisor linearly equivalent to D. Note that $D_r(v_1) \leq -1$ since otherwise, D_r would be effective and $|D|$ would not be

---

[4]Reduced configurations are, essentially, reduced divisors of degree 0.

empty. We define an order on the remaining vertices as follows. Suppose that for some $k > 1$ we have selected vertices $v_1, v_2, \ldots, v_{k-1}$. Let $W_k = V \setminus \{v_1, v_2, \ldots, v_{k-1}\}$ and let v_k be an element of W_k such that $D_r(v_k) < \text{outdeg}_{W_k}(v_k)$. It is possible to pick such a vertex v_k, since D_r is v_1-reduced. Let A be the adjacency matrix of Γ with respect to this ordering of the vertices. Note that for $k > 1$ we have

$$\text{outdeg}_{W_k}(v_k) = \sum_{j=1}^{k-1} A_{kj}.$$

Now we apply the method of Lemma 3.5.4 to obtain the non-special divisor υ_P associated to the ordering $P = (x_1, x_2, \ldots, x_n)$, and given by

$$\upsilon_P(v_k) = \begin{cases} -1, & \text{if } k = 1, \\ -1 + \sum_{j=1}^{k-1} A_{kj}, & \text{if } k > 1, \end{cases}$$

$$= \begin{cases} -1, & \text{if } k = 1, \\ -1 + \text{outdeg}_{W_k}(v_k), & \text{if } k > 1. \end{cases}$$

Consider the divisor $\upsilon_P - D_r$. We have

$$\upsilon_P(v_1) - D_r(v_1) = -1 - D_r(v_1) \geq 0$$

since $D_r(v_1) \leq -1$, as noted above. For $k > 1$ we have $D(v_k) < \text{outdeg}_{W_k}(v_k)$ by the choice of vertices v_k. Thus, for $k > 1$, we have

$$\upsilon_P(v_k) - D_r(v_k) = -1 + \text{outdeg}_{W_k}(v_k) - D_r(v_k) \geq 0.$$

It follows that $|\upsilon_P - D|$ is non-empty, since $\upsilon_P - D_r$ is an effective divisor linearly equivalent to $\upsilon_P - D$. $\square$

Proposition 3.5.10. *If the degree of a divisor υ is $g - 1$, then υ is non-special if and only if υ is linearly equivalent to a non-special divisor υ_P associated to some ordering P of the vertices of Γ.*

Proof. If υ is a divisor of degree $g - 1$ which is linearly equivalent to υ_P, for some ordering P, then $|\upsilon| = |\upsilon_P|$ is empty, so υ is non-special.

Suppose that υ is non-special. Then $|\upsilon|$ is empty, so by Proposition 3.5.9 there exists a non-special divisor υ_P such that $|\upsilon_P - \upsilon|$ is not empty. But the degree of $\upsilon_P - \upsilon$ is 0, so $|\upsilon_P - \upsilon|$ is non-empty if and only if $\upsilon_P - \upsilon$ is linearly equivalent to the zero divisor. Therefore, υ is linearly equivalent to υ_P. $\square$

Definition 3.5.11. We define the *canonical divisor* K on Γ by

$$K = \sum_{v \in V} (\deg(v) - 2)\, v.$$

The canonical divisor has degree $\deg(K) = 2g - 2$, where g is the genus of the graph Γ.

Proposition 3.5.12. *Let v be a divisor of degree $g-1$. Then v is non-special if and only if $K - v$ is non-special, where K is the canonical divisor.*

Proof. Suppose that v is non-special. Choose an ordering $P = (v_1, v_2, \ldots, v_n)$ such that v is linearly equivalent to v_P.

Let $v_{P'}$ be the non-special divisor corresponding to the reverse ordering $P' = (v_n, v_{n-1}, \ldots, v_1)$.

Then

$$v_P(v_k) = \begin{cases} -1, & \text{if } k = 1, \\ -1 + \sum_{j=1}^{k-1} A_{kj}, & \text{if } k > 1, \end{cases}$$

and

$$v_{P'}(v_k) = \begin{cases} -1, & \text{if } k = n, \\ -1 + \sum_{j=k+1}^{n} A_{kj}, & \text{if } k < n. \end{cases}$$

Thus

$$v_P(v_k) + v_{P'}(v_k) = -2 + \sum_{j \neq k} A_{kj}$$
$$= -2 + \deg(v_k)$$
$$= K(v_k),$$

where K is the canonical divisor. Thus $v_P + v'_P = K$.

It follows that $K - v_P$ is non-special, and hence $K - v$ is non-special. The proof that if $K - v$ is non-special then v is also non-special is similar. $\square$

3.5.4 The Riemann–Roch theorem for graphs

We give a characterization of the dimension of a linear system in terms of non-special divisors. We then prove Baker and Norine's Riemann–Roch theorem for graphs [BN07]. Baker and Norine also give criteria under which a Riemann–Roch theorem holds in more general settings, but we will restrict our discussion here to graphs.

For any divisor $D = \sum a_v v$, we define

$$D^+ = \sum_{a_v > 0} a_v v \quad \text{and} \quad D^- = -\sum_{a_v < 0} a_v v$$

so that D^+ and D^- are effective divisors with disjoint supports, and

$$D = D^+ - D^-.$$

We define

$$\deg^+(D) = \deg(D^+) \text{ and } \deg^-(D) = \deg(D^-).$$

Thus

$$\deg(D) = \deg^+(D) - \deg^-(D). \tag{3.17}$$

Note also that

$$\deg^+(D) = \deg^-(-D). \tag{3.18}$$

Here's a natural question: Is it true that, for any harmonic morphism $\phi : \Gamma_2 \to \Gamma_1$, the identity

$$\deg^+(\Phi_V D) = \deg(\Phi_V(D^+)) = \deg(\phi)\deg^+(D)$$

holds? The following example addresses this.

Example 3.5.13. We use the notation of Examples 3.3.8, 3.3.34 and 3.4.6 above: $\phi : \Gamma_2 \to \Gamma_1$ is a harmonic morphism, where Γ_2 is the Paley graph on 9 vertices in $GF(9)$ (see Figure 1.12) and Γ_1 is the cycle graph on three vertices $0, 1, 2$. Let $v_0 = 0$, $v_1 = 1$, $v_2 = 2$, $v_3 = a$, $v_4 = a + 1$, $v_5 = a + 2$, $v_6 = 2a$, $v_7 = 2a + 1$, and $v_8 = 2a + 2$. The divisor

$$D = -4 \cdot v_0 - 3 \cdot v_1 - 2 \cdot v_2 - v_3 + 0 \cdot v_4 + v_5 + 2 \cdot v_6 + 3 \cdot v_7 + 3 \cdot v_8,$$

is represented by the vector $D = (-4, -3, -2, -1, 0, 1, 2, 3, 4)$. The 9×3 matrix Φ_V was computed in Example 3.4.6. We have[5] $\Phi_V^t D = (-9, 9, 0)$, so

$$(\Phi_V^t D)^+ = (0, 9, 0), \quad \deg^+(\Phi_V D) = 9.$$

On the other hand, $D^+ = (0, 0, 0, 0, 0, 1, 2, 3, 4)$ and we have $\Phi_V^t(D^+) = (0, 9, 1)$. Therefore,

$$9 = \deg^+(\Phi_V^t D) \neq \deg(\Phi_V^t(D^+)) = 10.$$

Since

---

[5]Note that $\Phi_V^t D = \phi_*(D)$ is the pushforward of the divisor, in the sense of §3.6.1.

$$\Phi_E = \begin{pmatrix} 0 & 0 & 0 \\ 0 & 0 & 0 \\ 0 & 1 & 0 \\ 1 & 0 & 0 \\ 0 & 0 & 0 \\ 0 & 1 & 0 \\ 1 & 0 & 0 \\ 0 & 1 & 0 \\ 1 & 0 & 0 \\ 0 & 0 & 0 \\ 0 & 0 & 0 \\ 0 & 0 & -1 \\ 0 & 0 & 0 \\ 0 & 0 & -1 \\ 0 & 0 & -1 \\ 0 & 0 & 0 \\ 0 & 0 & 0 \\ 0 & 0 & 0 \end{pmatrix}$$

and $\Phi_E^t \Phi_E = 3I_3$, where I_3 is the 3×3 identity matrix, it follows from Lemma 3.3.33 that $\deg(\phi) = 3$. Therefore,

$$9 = \deg^+(\Phi_V^t D) \neq \deg(\phi)\deg^+(D) = 3 \cdot 10 = 30.$$

The next proposition gives an alternate characterization of the dimension of a linear system in terms of non-special divisors.

Proposition 3.5.14. *The dimension $r(D)$ of the linear system of a divisor D satisfies*

$$r(D) = -1 + \min\{deg^+(D' - v) : D' \sim D \text{ and } v \in \mathcal{N}\}.$$

Proof. Let $m = \min\{\deg^+(D' - v) : D' \sim D \text{ and } v \in \mathcal{N}\}$ and let D' and v be divisors at which the minimum is achieved. Let E_m be the effective divisor of degree m given by $E_m = (D' - v)^+$. Consider the linear system

$$
\begin{aligned}
|D - E_m| &= |D' - E_m| \\
&= |D' - (D' - v)^+| \\
&= |D' - v - (D' - v)^+ + v| \\
&= |-(D' - v)^- + v|.
\end{aligned}
$$

But $|v|$ is empty, because v is non-special, so $|v - F|$ is also empty, for any effective divisor F. In particular, for $F = (D' - v)^-$, we have $|-(D' - v)^- + v|$ is empty. Therefore, $r(D) \leq m - 1$.

Next, suppose that $r(D) < m - 1$, where m is as above. Then there exists an effective divisor E_{m-1} of degree $m - 1$ such that $|D - E_{m-1}|$ is empty. By Proposition 3.5.9, there exists a non-special divisor v such that $|v - (D - E_{m-1})|$ is not empty. Thus, there exists a divisor D' linearly equivalent to D such that $v - D' + E_{m-1}$ is effective, i.e., $v - D' + E_{m-1} \geq 0$. Rewriting as $E_{m-1} \geq D' - v$ gives $\deg(E_{m-1}) \geq \deg^+(D' - v)$, i.e., $m - 1 \geq \deg^+(D' - v)$. This contradicts our definition of m above. Thus $r(D) = m - 1$. $\square$

Theorem 3.5.15. *(Riemann–Roch theorem for graphs, Baker and Norine [BN07]) Let D be a divisor on the graph Γ and let K be the canonical divisor. Let g be the genus of Γ. Then*

$$r(D) - r(K - D) = deg(D) + 1 - g.$$

Proof. We will show that the theorem follows directly from Proposition 3.5.14.

Let $m = \min\{\deg^+(D' - v) : D' \sim D \text{ and } v \in \mathcal{N}\}$ and let D' and v be divisors for which the minimum is achieved. Then

$$
\begin{aligned}
r(D) &= -1 + \deg^+(D' - v) \\
&= -1 + \deg(D' - v) + \deg^-(D' - v) && \text{by Equation (3.17)} \\
&= -1 + \deg(D' - v) + \deg^+(-D' + v) && \text{by Equation (3.18)} \\
&= -1 + \deg(D' - v) + \deg^+\left(K - D' - (K - v)\right).
\end{aligned}
$$

Let $m' = \min\{\deg^+(K - D'' - v') : D'' \sim D \text{ and } v' \in \mathcal{N}\}$. By the above calculation, this minimum is achieved by $D'' = D'$ and $v' = K - v$ (recalling that $K - v$ is also non-special, by Proposition 3.5.12). Thus,

$$
\begin{aligned}
r(D) &= \deg(D' - v) + r(K - D) \\
&= \deg(D) - (g - 1) + r(K - D),
\end{aligned}
$$

i.e., $r(D) - r(D - K) = \deg(D) - g + 1$. $\square$

Open Question 1. *Is there an analog of Theorem 3.5.15 for graphs with a group action, as with the Riemann–Hurwitz theorem?*

3.6 Induced maps on divisors

In this section, we explore the pushforward and pullback maps.

3.6.1 The pushforward map

Suppose that ϕ is a nonconstant harmonic morphism from Γ_2 to Γ_1. Let

$$D = \sum_{i=1}^{n_2} a_i v_i$$

be a divisor on Γ_2. We will let $a = (a_1, a_2, \ldots, a_{n_2})$ be the vector of coefficients.

Definition 3.6.1. We define the *pushforward* $\phi_*(D) \in \mathrm{Div}(\Gamma_1)$ to be the divisor

$$\phi_*(D) = \Phi_V^t D \qquad (3.19)$$

i.e.,

$$\phi_*(D) = D' = \sum_{j=1}^{n_1} b_j w_j$$

with vector of coefficients $b = (b_1, b_2, \ldots, b_{n_1})$ given by

$$b = \Phi_V^t a.$$

Lemma 3.6.2.
$$deg(\phi_*(D)) = deg(D)$$

Proof. Let $D' = \phi_*(D) = \sum_{j=1}^{n_1} b_j w_j$. We have

$$deg(D') = \sum_{j=1}^{n_1} b_j$$
$$= \sum_{j=1}^{n_1} \sum_{k=1}^{n_2} (\Phi_V)_{kj} a_k.$$

Note that for each k, there is exactly one j such that $(\Phi_V)_{kj} = 1$. Therefore,

$$deg(D') = \sum_{k=1}^{n_2} a_k = deg(D).$$

$\square$

Lemma 3.6.3. *If $\phi : \Gamma_2 \to \Gamma_1$ is a nonconstant harmonic morphism, then ϕ_* is a surjective map from the group of all divisors on Γ_2 onto the group of all divisors on Γ_1.*

Proof. Since a nonconstant harmonic morphism surjects onto all vertices of Γ_1, it follows that ϕ_* is a surjective map onto the group of all divisors on Γ_1. $\square$

3.6.2 The pullback map

Definition 3.6.4. Let

$$D' = \sum_{i=1}^{n_1} b_i w_i$$

be a divisor on Γ_1. We define the pullback $\phi^*(D') \in \operatorname{Div}(\Gamma_2)$ to be the divisor

$$\phi^*(D') = \Phi_{mV}(D'), \tag{3.20}$$

i.e., $\phi^*(D')$ is the divisor

$$\phi^*(D') = D = \sum_{j=1}^{n_2} a_j v_j$$

with vector of coefficients $a = (a_1, a_2, \ldots, a_{n_2})$ given by

$$a = \Phi_{mV} b.$$

Lemma 3.6.5.
$$deg(\phi^* D') = deg(\phi)deg(D')$$

Proof. In the notation above,

$$
\begin{aligned}
\deg(\phi^* D) &= \sum_{j=1}^{n_2} a_j \\
&= \sum_{j=1}^{n_2} \sum_{k=1}^{n_1} (\Phi_{mV})_{jk} b_k \\
&= \begin{cases} m_\phi(v_j) b_k, & \text{if } \phi(v_j) = w_k, \\ 0, & \text{otherwise}, \end{cases} \\
&= \deg(\phi)\deg(D').
\end{aligned}
$$

$\square$

Lemma 3.6.6. *Suppose that ϕ is a harmonic morphism and D' is a divisor on Γ_1. Then*

$$\phi_*(\phi^*(D')) = deg(\phi)D'.$$

Proof. If b is the vector of coefficients of D', then the vector of coefficients of $\phi_*(\phi^*(D'))$ is

$$\Phi_V^t \Phi_{mV} b = \deg(\phi)b$$

by Lemma 3.3.33.   □

We will discuss the induced pullback map on Jacobians later, after introducing the energy pairing.

3.6.3 Dimensions of linear systems and harmonic morphisms

Proposition 3.6.7. *Let $\phi : \Gamma_2 \to \Gamma_1$ be a nonconstant harmonic morphism of graphs. Let D be a divisor on Γ_2. Then $r(D) \leq r(\phi_* D)$.*

Proof. Recall that if the linear system $|D|$ of D is empty then $r(D) = -1$, and if $|D|$ is not empty, then $r(D)$ is the largest nonnegative integer d such that $|D - E|$ is non-empty, for all effective divisors E of degree d. If $|D|$ is empty, the statement of the proposition holds trivially. Suppose that $r(D) = d \geq 0$. Recall that the pushforward map ϕ_* is a surjective map on vertices. Therefore, if E' is an effective divisor of degree d on Γ_1, then E' is of the form $\phi_* E$ for some effective divisor E of degree d on Γ_2. Since $r(D) = d$, the linear system $|D - E|$ is non-empty, so there exists an effective divisor F on Γ_2 and an n_2-vector of integers x such that

$$F = D - E + Q_2 x.$$

Then, using our identity for Laplacians of harmonic morphisms,

$$
\begin{aligned}
\phi_*(F) &= \phi_*(D - E + Q_2 x) \\
&= \phi_*(D) - \phi_*(E) + \Phi_V^t Q_2 x \\
&= \phi_*(D) - \phi_*(E) + Q_1 \Phi_{mV}^t x.
\end{aligned}
$$

Thus $\phi_*(F)$ is linearly equivalent to $\phi_*(D) - \phi_*(E)$. Furthermore, $\phi_*(F)$ is effective, since F is effective. Therefore, $r(\phi_*(D)) \geq d = r(D)$.   □

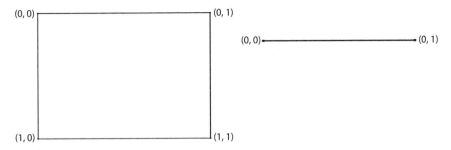

Figure 3.5: There is a harmonic morphism from the grid graph to the path graph.

Example 3.6.8. Define ϕ to be the morphism which "collapses" the square graph $\Gamma_2 = (V_2, E_2)$ with 4 vertices

$$V_2 = \{(0,0), (0,1), (1,0), (1,1)\},$$

to the subgraph $\Gamma_1 = (V_1, E_1)$ consisting of two vertices

$$V_1 = \{(0,0), (0,1)\},$$

and one edge:

$$\phi : \{(0,0), (1,0)\} \to (0,0), \ \{(0,1), (1,1)\} \to (0,1).$$

This morphism ϕ is harmonic. The graphs are depicted in Figure 3.5.
   Let D be the divisor on Γ_2,

$$D = 1 \cdot (0,0) + 1 \cdot (1,1).$$

The map Φ_V induces a map between the linear system $|D|$ of a divisor D on Γ_2, and the corresponding linear system $|\phi_*(D)|$ of the pushforward divisor $\phi_*(D)$ on Γ_1.

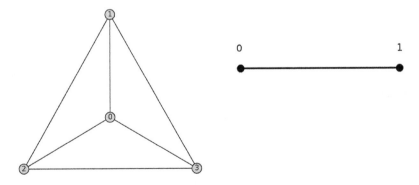

Figure 3.6: There is a harmonic morphism from the tetrahedron graph to the path graph.

We have

$$\Phi_V = \begin{pmatrix} 1 & 0 \\ 0 & 1 \\ 1 & 0 \\ 0 & 1 \end{pmatrix},$$

$$|D| = \{2 \cdot (0,1), 1 \cdot (0,0) + 1 \cdot (1,1), 2 \cdot (1,0)\},$$

$$|\phi_*(D)| = \{2 \cdot (0,0), 1 \cdot (0,0) + 1 \cdot (1,0), 2 \cdot (0,1)\}.$$

In the following example, we compare the Riemann–Roch dimension (the dimension $r(D)$ of $|D|$) of a divisor D on Γ_2, and the corresponding (possibly larger) Riemann–Roch dimension of the pushforward divisor $\phi_*(D)$ on Γ_1.

Example 3.6.9. Consider the wheel graph on 4 vertices, $\Gamma_2 = (V_2, E_2)$ (also known as the tetrahedron graph), and the path graph on 2 vertices, $\Gamma_1 = (V_1, E_1)$. The morphism $f : \Gamma_2 \to \Gamma_1$ defined by

$$\{0\} \mapsto 0, \ \{1,2,3\} \mapsto 1,$$

is harmonic. The graphs are depicted in Figure 3.6.

The associated Φ_V is

$$\Phi_V = \begin{pmatrix} 1 & 0 \\ 0 & 1 \\ 0 & 1 \\ 0 & 1 \end{pmatrix}.$$

In general, if $V = \{v_1, \ldots, v_n\}$ is the set of vertices of the graph, we identify a divisor $a_1 v_1 + a_2 v_2 + \cdots + a_n v_n$ with the vector of coefficients $(a_1, a_2, \ldots, a_n)$.

With the help of a computer algebra program, we compute the following Riemann–Roch dimension values on Γ_2:

$$r((1,0,0,1)) = 0,$$
$$r((2,0,0,1)) = 0,$$
$$r((3,0,0,1)) = 1,$$
$$r((2,1,1,-1)) = 0, \text{and}$$
$$r((3,1,1,-1)) = 1.$$

Likewise, we compute the following Riemann–Roch dimension values on Γ_1:

$$r(1,1) = r(\Phi_V^t \cdot (1,0,0,1)) = 2,$$
$$r(2,1) = r(\Phi_V^t \cdot (2,0,0,1)) = 3,$$
$$r(3,1) = r(\Phi_V^t \cdot (3,0,0,1)) = 4,$$
$$r(2,1) = r(\Phi_V^t \cdot (2,1,1,-1)) = 3, \text{and}$$
$$r(3,1) = r(\Phi_V^t \cdot (3,1,1,-1)) = 4.$$

Suppose that we have fixed a distinguished vertex q, designated the sink vertex. We assume that we have an algorithm for finding the q-reduced form of a degree 0 divisor and that we have a list of all q-reduced divisors of degree 0.

Lemma 3.6.10. *If D is a divisor of degree d and $red(D)$ is the q-reduced divisor linearly equivalent to D, then $red(D) = r + dq$ for some degree 0 reduced divisor r.*

Proof. We need only note that $r + dq$ is q-reduced if and only if r is q-reduced, which follows directly from the definition of q-reduced divisors.   □

Proposition 3.6.11. *Let D be a divisor on Γ and let $red(D)$ be the q-reduced divisor linearly equivalent to D. The linear system $|D|$ is non-empty if and only the q-component of $red(D)$ is nonnegative.*

Proof. If the q-component of $red(D)$ is nonnegative, then $red(D)$ is effective, so $|D|$ is non-empty. Conversely, suppose that $|D|$ is non-empty and that D' is an effective divisor linearly equivalent to D. Then all components of D', including the q-component, are nonnegative. The reduced divisor $red(D)$ may be obtained from D' by firing sets that do not include q, so the q-component of $red(D)$ is at least as large as the q-component of D'.   □

Corollary 3.6.12. *If E is an effective divisor of degree s, then $E \sim r + sq$ for some reduced degree 0 divisor r such that the q-component $r(q)$ of r satisfies*

$$r(q) \geq -s.$$

Suppose now that the q-reduced degree 0 divisors on Γ are $r_1, r_2, \ldots, r_K$. We denote by $red(r_i - r_j)$ the reduced divisor linearly equivalent to $r_i - r_j$.

Proposition 3.6.13. *Suppose that $D \sim r_i + dq$ is a degree d divisor whose linear system $|D|$ is non-empty. Let s be a nonnegative integer. Then $r(D) \geq s$ if and only if $red(r_i - r_j)(q) + d - s \geq 0$ for all j such that $r_j(q) \geq -s$.*

Proof. By the previous corollary, a degree s divisor $E \sim r_j + sq$ is effective if any only if $r_j(q) \geq -s$. By Lemma 3.6.11, the linear system $|D - E|$ is non-empty if and only if $red(D - E)(q) \geq 0$. Now

$$red(D - E) = red(r_i + dq - r_j - sq) = red(r_i - r_j) + (d - s)q.$$

Thus, $|D - E|$ is non-empty if and only if $red(r_i - r_j)(q) + d - s \geq 0$.   □

Example 3.6.14. Consider the theta graph Γ with two vertices, v_1 and v_2, and three edges between v_1 and v_2. The genus of Γ is 2. The canonical divisor of Γ is $K = v_1 + v_2$.

The Laplacian matrix of Γ is

$$Q = \begin{pmatrix} 3 & -3 \\ -3 & 3 \end{pmatrix}.$$

If f is the integer-valued function on vertices of Γ given by $f(v_1) = 1$ and $f(v_2) = 0$, then $\mathrm{div}(f) = 3v_1 - 3v_2$. Every principal divisor is a multiple of $\mathrm{div}(f)$. The Jacobian of Γ is isomorphic to $\mathbb{Z}/3\mathbb{Z}$.

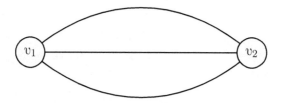

Figure 3.7:   A theta graph.

There are three linear equivalence classes of degree 0 divisors, the classes determined by $D_{01} = 0$, $D_{02} = -v_1 + v_2$, and $D_{03} = -2v_1 + 2v_2$. The corresponding linear systems are

$$|D_{01}| = \{0\}, \ |D_{02}| = \phi, \ \text{ and } \ |D_{03}| = \phi,$$

and their dimensions are

$$r(D_{01}) = 0, \ r(D_{02}) = -1, \ \text{ and } \ r(D_{03}) = -1.$$

Consider a divisor of degree $d > 0$. We may write D as $D = (d-a)v_1 + av_2$, where a is an integer. Let $a' = a \pmod 3$ and let k be the integer such that $a' = a + 3k$. Let $D' = (d - a')v_1 + a'v_2$. Then $D' \sim D$, since

$$\begin{aligned} D' &= (d - a')v_1 + a'v_2 \\ &= (d - a - 3k)v_1 + (a + 3k)v_2 \\ &= D - k(3v_1 - 3v_2). \end{aligned}$$

Thus, there are three linear equivalence classes of degree d divisors, the classes determined by $D_{d1} = dv_1$, $D_{d2} = (d-1)v_1 + v_2$, and $D_{d3} = (d-2)v_1 + 2v_2$.
For $d = 1$, the corresponding linear systems are

$$|D_{11}| = \{v_1\}, \ |D_{12}| = \{v_2\}, \ \text{ and } \ |D_{13}| = \psi,$$

and their dimensions are

$$r(D_{11}) = 0, \ r(D_{12}) = 0, \ \text{and} \ r(D_{13}) = -1.$$

For $d = 2$, the corresponding linear systems are

$$|D_{21}| = \{2v_1\}, \ |D_{22}| = \{K = v_1 + v_2\}, \ \text{and} \ |D_{33}| = \{2v_2\},$$

and their dimensions are

$$r(D_{21}) = 0, \ r(D_{22}) = 1, \ \text{and} \ r(D_{23}) = 0.$$

For $d = 3$, the corresponding linear systems are

$$|D_{31}| = \{3v_1, 3v_2\}, \ |D_{32}| = \{(2v_1 + v_2\}, \ \text{and} \ |D_{33}| = \{v_1 + 2v_2\},$$

and their dimensions are

$$r(D_{31}) = 1, \ r(D_{32}) = 1, \ \text{and} \ r(D_{33}) = 1.$$

Note that for any divisor D of degree $d \geq 3$, we have $\deg(K - D) = 2 - d < 0$, so $r(K - D) = -1$. Thus, by the Riemann–Roch theorem,

$$
\begin{aligned}
r(D) &= r(K - D) + \deg(D) + 1 - g \\
&= -1 + d + 1 - 2 \\
&= d - 2
\end{aligned}
$$

for all divisors D of degree $d \geq 3$.

For further reading on the Riemann–Roch theorem for graphs and related topics, see Backman [Ba14], Baker and Norine [BN07] and [BN09], Luo [Lu11], and Manjunath [Ma11].

Chapter 4
Chip-Firing Games

A chip-firing game on a vertex-weighted graph Γ is a one-player game where a move amounts to selecting a vertex and firing it, i.e., redistributing the weight (or chips) of that vertex among its neighbors. There are a number of variations in this game and we shall discuss some of them below.

Chip-firing games on graphs relate to a surprising variety of mathematical models used in physics, theoretical computer science, and dynamical systems.

4.1 Motivation

In this chapter, we are interested in exploring questions such as the following.

Question 4.1. *To what extent does the symmetry of a graph affect the chip-firing game on that graph?*

Question 4.2. *To what extent does the Laplacian determine the chip-firing game on a graph?*

Question 4.3. *If there is a harmonic morphism $\Gamma_2 \to \Gamma_1$ between graphs Γ_1 and Γ_2, how are the critical groups of graphs related? In other words, what functorial properties do critical groups have?*

4.2 Introduction

Consider a vertex-weighted undirected graph $\Gamma = (V, E)$. The weights are assumed to be integers, sometimes measured in units called chips or dollars.

© Springer International Publishing AG 2017
W.D. Joyner and C.G. Melles, *Adventures in Graph Theory*,
Applied and Numerical Harmonic Analysis,
https://doi.org/10.1007/978-3-319-68383-6_4

The literature contains many variations of the rules defining a chip-firing game. We mention a few of them.

(a) In the simplest version, no weight can be negative. In this version, a chip-firing game on Γ is a one-player game where a move amounts to selecting a vertex $v \in V$ with "enough chips" and then redistributing the weight of that vertex to its $\deg_\Gamma(v)$ neighbors, one chip per edge incident to v. This is referred to as *firing* v. If the weight of v was w before firing, it is $w - \deg_\Gamma(v)$ afterwards. Since no weight can be negative, v can be fired if and only if $w \geq \deg_\Gamma(v)$.

This version seems to have been introduced in A. Björner, L. Lovász, P. Shor, [BLoS91].

(b) In another version of the chip-firing game, we fix a distinguished vertex $q \in V$. The vertex q can have a negative weight, but all other weights must be non-negative. In this case, a chip-firing game on Γ is a one-player game where a move amounts to selecting a non-distinguished vertex $v \in V$ having weight at least $\deg_\Gamma(v)$ and firing it, if possible. If no vertex $v \neq q$ can be fired then the vertex q can be fired.

Biggs [Bi99] calls this the *dollar game*.

(c) In a related version of the chip-firing game, we fix a distinguished vertex $q \in V$. The vertex q can have a negative weight, but all other weights must be non-negative. In this version, a chip-firing game on Γ is a one-player game where a move amounts to selecting a subset W of non-distinguished vertices $W \subset V - \{q\}$ and firing them all simultaneously, provided no negative weights are created in the end. In this version, each vertex in W sends a chip to each of its neighbors in $V - W$.

This version was introduced in M. Baker and F. Shokrieh, [BS13].

Chip-firing games on graphs relate to "abelian sandpile models" from physics, "rotor-routing models" from theoretical computer science (designing efficient computer multiprocessor circuits), "self-organized criticality" (a subdiscipline of dynamical systems), "algebraic potential theory" on a graph (see Biggs [Bi97]), and cryptography (via the Biggs cryptosystem). Moreover, the analysis of chip-firing games relates the concepts of the Laplacian of a graph, the tree number, the circulation space, the incidence matrix, and many other ideas. Some good references are Baker and Norine [BN07]; Biggs [Bi99]; Durgin [Du09]; Holroyd, Levine, Meszaros, Peres, Propp, and Wilson [HLMPPW08]; Perkinson, Perlman, and Wilmes [PPW11]; and Perlman [Pe09].

4.2.1 The Laplacian

Let Γ be a finite connected graph with vertex set V and edge set E. We will assume that Γ may be a multigraph (i.e., there may be more than one edge between two distinct vertices) but Γ has no loops (i.e., there are no edges from a vertex to itself). Let $n = |V|$ be the number of vertices and $m = |E|$ the number of edges. The Laplacian matrix Q of Γ, as defined in (2.3), is the $n \times n$ matrix indexed by vertices of Γ, whose (u, v)-th entry is the negative of the number of edges from u to v, if $u \neq v$, and whose (u, u)-th entry is the degree or valency of the vertex u.

Let $x = (x_0, x_1, \ldots, x_{n-1})$ be a variable vector indexed by the vertices of Γ and let $y = Qx$. Roughly speaking, each coordinate y_j is a linear combination of the x_is, and the only x_i that occur in this relation are those indexed by vertices connected to j by an edge in Γ.

Throughout this chapter, unless stated otherwise, we will treat Q as a map $Q : \mathbb{Z}^n \to \mathbb{Z}^n$. By the *image* of Q, we mean the set of all vectors Qx, where $x \in \mathbb{Z}^n$.

Example 4.2.1. The Laplacian matrix of the complete graph on 4 vertices $K_4 = (V, E)$ (shown in Figure 1.1 and in Figure 1.10) is

$$Q = \begin{pmatrix} 3 & -1 & -1 & -1 \\ -1 & 3 & -1 & -1 \\ -1 & -1 & 3 & -1 \\ -1 & -1 & -1 & 3 \end{pmatrix}.$$

As predicted by Proposition 2.2.6, the kernel of Q is $(1, 1, 1, 1)$ and the rank of Q is 3.

Using the indexing in Figure 1.10, let (w_0, w_1, w_2, w_3) denote the vertex weights on Γ, where w_v is the weight of vertex $v \in V$. If, for example, we fire vertex 2 then the resulting vertex weights will be $(w_0 + 1, w_1 + 1, w_2 - 3, w_3 + 1)$. For another interpretation, note that if the rows of Q are denoted Q_0, Q_1, Q_2, Q_3, then firing vertex 2 results in the vertex weights $(w_0, w_1, w_2, w_3) - Q_2$.

Example 4.2.2. The Dürer graph on 12 vertices, Γ, depicted in Figure 5.6, has Laplacian

$$Q = \begin{pmatrix} 3 & -1 & 0 & 0 & 0 & -1 & -1 & 0 & 0 & 0 & 0 & 0 \\ -1 & 3 & -1 & 0 & 0 & 0 & 0 & -1 & 0 & 0 & 0 & 0 \\ 0 & -1 & 3 & -1 & 0 & 0 & 0 & 0 & -1 & 0 & 0 & 0 \\ 0 & 0 & -1 & 3 & -1 & 0 & 0 & 0 & 0 & -1 & 0 & 0 \\ 0 & 0 & 0 & -1 & 3 & -1 & 0 & 0 & 0 & 0 & -1 & 0 \\ -1 & 0 & 0 & 0 & -1 & 3 & 0 & 0 & 0 & 0 & 0 & -1 \\ -1 & 0 & 0 & 0 & 0 & 0 & 3 & 0 & -1 & 0 & -1 & 0 \\ 0 & -1 & 0 & 0 & 0 & 0 & 0 & 3 & 0 & -1 & 0 & -1 \\ 0 & 0 & -1 & 0 & 0 & 0 & -1 & 0 & 3 & 0 & -1 & 0 \\ 0 & 0 & 0 & -1 & 0 & 0 & 0 & -1 & 0 & 3 & 0 & -1 \\ 0 & 0 & 0 & 0 & -1 & 0 & -1 & 0 & -1 & 0 & 3 & 0 \\ 0 & 0 & 0 & 0 & 0 & -1 & 0 & -1 & 0 & -1 & 0 & 3 \end{pmatrix}.$$

As the Sage code below shows, the rank of Q is indeed 11 and it does, in fact, have the all 1's vector in its kernel.

```
                                    ───── Sage ─────

sage: Gamma = graphs.DurerGraph()
sage: Gamma
Durer graph: Graph on 12 vertices
sage: Q = Gamma.laplacian_matrix()
sage: Q.rank()
11
sage: Q.right_kernel()
Free module of degree 12 and rank 1 over Integer Ring
Echelon basis matrix:
[1 1 1 1 1 1 1 1 1 1 1 1]
```

Exercise 4.1. Verify the following: If the rows of Q are denoted $Q_0, Q_1, \ldots, Q_{11}$, then firing vertex 2 results in the vertex weights $(w_0, w_1, \ldots, w_{11}) - Q_2$.

4.3 Configurations on graphs

In this section, we describe chip-firing games on a graph with a distinguished vertex.

Let $\Gamma = (V, E)$ denote a vertex-weighted undirected graph with a distinguished vertex, $q \in V = \{v_1, \ldots, v_n\}$. If the weight of v_i is denoted $w_i \in \mathbb{Z}$, we may consider a distribution of weights as either a vector $(w_1, w_2, \ldots, w_n) \in \mathbb{Z}^n$, or a formal linear combination of the vertices,

$$w_1 v_1 + \cdots + w_n v_n \in \mathbb{Z}[V].$$

In either case, such a distribution of weights is called a *configuration* on Γ. A more formal version is stated in Definition 4.3.1 below.

We are interested in certain legal configurations of chips on the vertices and legal chip-firing moves which transform one configuration into another. We define two complementary types of configurations: critical configurations and reduced configurations. In §4.5 we will define an equivalence relation on configurations which produces a group, called the critical group. Each equivalence class has a unique critical representative and a unique reduced representative (with respect to a given fixed sink vertex q).

4.3.1 Legal configurations

Let Γ be a finite connected graph and let q be a distinguished vertex of Γ, sometimes called the *sink* or the *government*. In a chip-firing game, each vertex is assigned an integer number of chips (possibly negative). As indicated above, an assignment of chips is called a configuration. We will consider games in which a *legal configuration* is a configuration $(w_1, w_2, \ldots, w_n) \in \mathbb{Z}^n$ which has non-negative numbers of chips on all non-sink vertices (so $w_i \geq 0$ if $v_i \neq q$) and the total number of chips (the *degree* of the configuration) is zero (so $w_1 + \cdots + w_n = 0$).

Definition 4.3.1. We define a *configuration* on Γ to be any integer-valued function $s : V \to \mathbb{Z}$ on the vertex set V of Γ. The *degree* of a configuration is

$$\deg(s) = \sum_{v \in V} s(v).$$

If $\phi : \Gamma_2 \to \Gamma_1$ is a morphism between graphs and if $s : V_2 \to \mathbb{Z}$ is a configuration on $\Gamma_2 = (V_2, E_2)$, then the *pushforward configuration* $\phi_*(s)$ on $\Gamma_1 = (V_1, E_1)$ is given by

$$\phi_*(s)(w) = \sum_{v \in V, \ \phi(v) = w} s(v). \tag{4.1}$$

Exercise 4.2. Check that the following are true.

- Each degree 0 configuration on Γ_2 maps to a degree 0 configuration on Γ_1.

- The pushforward configuration defined by (4.1) agrees with the pushforward of the associated divisor, as defined in §3.6.1 (and therefore can be computed using the matrix Φ_V^t).

Definition 4.3.2. A *q-legal configuration* of a chip-firing game on Γ is a degree 0 configuration s which takes non-negative values on every vertex other than the sink q, i.e., a configuration s such that

$$s(v) \geq 0 \text{ for } v \neq q$$

and

$$s(q) = -\sum_{v \neq q} s(v).$$

Note that some authors use the terminology "configuration" to refer only to q-legal configurations.

We think of $s(v)$ as representing the number of chips or dollars at the vertex v. Only the vertex q is allowed to be "in debt". When we think of the chips as dollars at each vertex, the game is also called a dollar game. The sum of all chips on all vertices is assumed to be 0. (If the sum were different from 0, it would not affect the allowed moves, described below.)

4.3.2 Chip-firing and set-firing moves

In chip games, certain allowable or legal moves on configurations are defined. We seek to reach certain special types of configurations through legal moves. In some games, we will allow only chip-firing moves, in which a single vertex is fired at a time. In other games, we will allow set-firing moves, in which all vertices in a set are fired simultaneously.

Definition 4.3.3. A *chip-firing move* applied to a configuration s consists of firing a vertex v by moving a chip along each edge from v to an adjacent vertex. After a vertex v is fired, it has $\deg(v)$ fewer chips and each adjacent vertex w has one additional chip for each edge from v to w. A chip-firing move of a vertex $v \neq q$ is *q-legal* if v has at least as many chips on it as the number of incident edges, i.e., if $s(v) \geq \deg(v)$. A chip-firing move firing q is *q-legal* if no other vertex may be fired q-legally.

Definition 4.3.4. Suppose that s is a q-legal configuration and we fire a sequence of vertices $v_1, v_2, \ldots, v_k$ (not necessarily distinct), obtaining a sequence of configurations $s = s_0, s_1, s_2, \ldots, s_k$. We say the sequence of vertex-firings is a *q-legal sequence* if firing v_i is q-legal for configuration s_{i-1}, for $i = 1, 2, \ldots, k$.

Example 4.3.5. Consider the diamond graph shown in Figure 4.1 with vertices $\{q = v_1, v_2, v_3, v_4\}$.

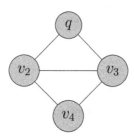

Figure 4.1: A diamond graph.

Let s be the configuration $s = (-6, 2, 1, 3)$. Then the sequence of vertex-firings $\{v_4, v_2, v_4, v_3, q\}$ is q-legal, and results in the sequence of q-legal configurations

$$s_1 = (-6, 3, 2, 1)$$
$$s_2 = (-5, 0, 3, 2)$$
$$s_3 = (-5, 1, 4, 0)$$
$$s_4 = (-4, 2, 1, 1)$$
$$s_5 = (-6, 3, 2, 1).$$

Suppose now that we start with a configuration s on a graph Γ and apply a sequence of firings in which vertex v is fired $x(v)$ times, for v in V. The new configuration s' satisfies

$$s'(v) = s(v) - x(v)\deg(v) + \sum_{w \neq v} x(w)A_{vw}, \tag{4.2}$$

where A is the adjacency matrix of Γ, defined by setting A_{vw} to be the number of edges between v and w. We will usually assume that an ordering $q = v_1, v_2, \ldots, v_n$ of the vertices of Γ has been specified, and that Q is the Laplacian matrix with respect to this ordering. We treat x as a vector with the same ordering. Using these conventions, we can rewrite Equation (4.2) in the shorter form

$$s' = s - Qx. \tag{4.3}$$

Exercise 4.3. Consider the diamond graph of Example 4.3.5 with configuration $s = (-6, 2, 1, 3)$. Let $x = (1, 1, 1, 2)$ (corresponding to the sequence of chip-firings of Example 4.3.5), and let Q be the Laplacian matrix

$$Q = \begin{pmatrix} 2 & -1 & -1 & 0 \\ -1 & 3 & -1 & -1 \\ -1 & -1 & 3 & -1 \\ 0 & -1 & -1 & 2 \end{pmatrix}.$$

Check that $s - Qx = (-6, 3, 2, 1)$, which is the final configuration of Example 4.3.5.

Next we look at sets of vertices and set-firing moves.

Definition 4.3.6. Suppose that W is a non-empty set of vertices. The *outdegree* of a vertex v in W, denoted $\text{outdeg}_W(v)$, is defined to be the number of edges from v to vertices in the complement W^c of W. (Note that it is possible for the outdegree of a vertex v in W to be zero, which occurs if and only if all neighbors of v are in W.) The *outdegree of W* is defined to be

$$\text{outdeg}(W) = \sum_{v \in W} \text{outdeg}_W(v).$$

Example 4.3.7. Consider the diamond graph of Example 4.3.5 and let W be the vertex set $W = \{v_2, v_3, v_4\}$. Then

$$\text{outdeg}_W(v_2) = 1, \quad \text{outdeg}_W(v_3) = 1, \quad \text{and} \quad \text{outdeg}_W(v_4) = 0.$$

It follows that $\text{outdeg}(W) = 2$.

Before defining set-firing moves, we will establish some useful results about outdegrees of sets.

Definition 4.3.8. Let W be a non-empty set of vertices in a graph Γ. The *characteristic vector* of W is the vector x_W given by

$$x_W(v) = \begin{cases} 1, & \text{if } v \in W, \\ 0, & \text{if } v \notin W. \end{cases}$$

Lemma 4.3.9. *Let x_W be the characteristic vector of a non-empty set W of the vertices of a graph Γ and let Q denote the Laplacian matrix of Γ. Let W^c be the complement of W in the set of vertices of Γ. Then*

1. *if $v \in W$ we have $Qx_W(v) = outdeg_W(v)$,*
2. *if $v \notin W$ we have $Qx_W(v) = -outdeg_{W^c}(v)$, and*
3. *$x_W^t Q x_W = \sum_{v \in W} outdeg_W(v) = outdeg(W)$.*

Proof. If $v \in W$,

$$\begin{aligned}
outdeg_W(v) &= \sum_{w \notin W} A_{vw} \\
&= \sum_{w \neq v} (1 - x_W(w)) A_{vw} \\
&= \deg(v) - \sum_{w \neq v} x_W(w) A_{vw} \\
&= Q x_W(v).
\end{aligned}$$

The third statement follows immediately.

If $v \notin W$, then $x_W(v) = 0$ and

$$\begin{aligned}
Q x_W(v) &= x_W(v)\deg(v) - \sum_{w \neq v} x_W(w) A_{vw} \\
&= - \sum_{w \in W} A_{vw} \\
&= -outdeg_{W^c}(v). \qquad \square
\end{aligned}$$

Exercise 4.4. Verify conditions (1) - (3) of Lemma 4.3.9 for the diamond graph of Figure 4.1, the set $W = \{v_2, v_3, v_4\}$, and its characteristic vector $x_W = (0, 1, 1, 1)$.

Exercise 4.5. Let W be the set of vertices $\{2, 3\}$ of the graph of Figure 4.3 and let Q be the Laplacian matrix of this graph. Calculate $x_W^t Q x_W$.

Definition 4.3.10. If W is a non-empty set of vertices in $V \setminus \{q\}$, and s is a configuration of chips, we define the *relative degree of s on W* to be

$$\deg_W(s) = \sum_{v \in W} s(v).$$

Note that if W can be fired q-legally, then $s(v) \geq \text{outdeg}_W(v)$ for all vertices v in W. Consequently,

$$\text{deg}_W(s) \geq \text{outdeg}(W).$$

Example 4.3.11. Consider the cycle graph on 4 vertices having configuration $(-8, 4, 3, 1)$, as depicted in Figure 4.2.

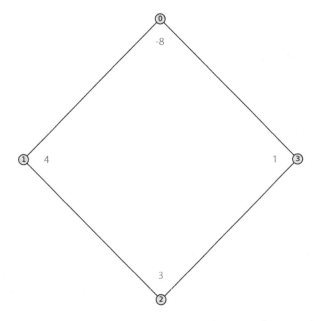

Figure 4.2: The cycle graph with legal configuration $(-8, 4, 3, 1)$.

Can you find a sequence of vertex-firings that gives the configuration $(-2, 1, 1, 1)$?

The next lemma gives a formula for the relative degree of a configuration s on a non-empty set W.

Lemma 4.3.12. *Suppose that s is a configuration on a graph Γ and W is a non-empty set of vertices of Γ. Then the relative degree of s on W is*

$$deg_W(s) = \sum_{v \in W} s(v) = x_W^t s.$$

Proof. The formula follows immediately from the definition of the relative degree $deg_W(s)$.   □

Definition 4.3.13. Let s be a configuration on a graph Γ and let W be a non-empty set of vertices which do not contain q. We say that the set W may be fired *q-legally* (on s) if each vertex $v \in W$ has at least $\mathrm{outdeg}_W(v)$ chips on it, i.e., if $s(v) \geq \mathrm{outdeg}_W(v)$ for all $v \in W$. A *set-firing move*, firing the set W, consists of moving a chip from W along each edge from a vertex in W to a vertex in the complement W^c, or equivalently, by firing every vertex in W at the same time.

If s is a configuration on a graph Γ, and an ordering $q = v_1, v_2, \ldots, v_n$ has been chosen for the vertices of Γ, we will sometimes treat s as a vector, so that equations involving s may be written in matrix form.

The configuration obtained from s after firing all the chips in W does not depend on the order in which they are fired, by Equation (4.3) and is given by

$$s' = s - Qx_W, \tag{4.4}$$

where Q is the Laplacian matrix of the graph.

Exercise 4.6. Let Γ be the diamond graph of Figure 4.1, let $s = (-5, 2, 2, 1)$, and let $W = \{v_2, v_3, v_4\}$. Show that W may be fired q-legally and that the resulting configuration is $s' = (-3, 1, 1, 1)$.

4.3.3 Stable, recurrent, and critical configurations

We next define stable, recurrent, and critical configurations on a graph. A stable recurrent (degree 0) configuration is critical. There are only finitely many stable configurations, and hence only finitely many critical configurations.

Definition 4.3.14. Let s be a q-legal configuration. A vertex $v \neq q$ is called *q-stable* for s if v cannot be fired q-legally, i.e., if $s(v) < \deg(v)$. A vertex $v \neq q$ that can be fired q-legally, i.e., such that $s(v) \geq \deg(v)$, is called an *active vertex*. The vertex q is defined to be *active* if and only if all other vertices are stable.

Definition 4.3.15. A q-legal configuration s is called *q-stable* if it is q-stable for all vertices $v \neq q$, i.e., no vertex other than q may be fired q-legally.

Note, that since we have defined q-legal configurations to be of degree 0, q-stable configurations are also of degree 0.

Remark 4.3.16. There are only finitely many q-stable configurations on Γ. It is easy, in principle, to list all q-stable configurations. There are

$$\prod_{v \neq q} \deg(v)$$

of them.

Definition 4.3.17. A non-empty (finite) sequence of chip-firings is called *q-recurrent* for a configuration s if it is q-legal and takes s to s. A configuration for which there exists a q-recurrent sequence called a *q-recurrent* configuration.

Definition 4.3.18. A q-stable configuration which is q-recurrent is called *q-critical*.

Note, that since we have defined q-stable configurations to be of degree 0, q-critical configurations are also of degree 0.

Note also, that there are only finitely many q-critical configurations, since there are only finitely many q-stable configurations, by Remark 4.3.16.

Example 4.3.19. Consider the graph of Figure 4.3.

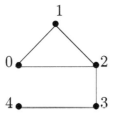

Figure 4.3: A graph with 5 vertices.

This graph has Laplacian

$$Q = \begin{pmatrix} 2 & -1 & -1 & 0 & 0 \\ -1 & 2 & -1 & 0 & 0 \\ -1 & -1 & 3 & -1 & 0 \\ 0 & 0 & -1 & 2 & -1 \\ 0 & 0 & 0 & -1 & 1 \end{pmatrix}.$$

Suppose the initial configuration is $s = (3, 1, 0, 1, -5)$, i.e.,

- vertex 0 has 3 chips,
- vertex 1 has 1 chip,
- vertex 2 has no chips,
- vertex 3 has 1 chip, and
- vertex 4 is the sink vertex q.

Notice that vertex 0 is active. If we fire 0 then we get the new configuration $s' = (1, 2, 1, 1, -5)$. Indeed, if we compute $s' = s - Qx_{\{0\}}$, we get:

$$s' = \begin{pmatrix} 3 \\ 1 \\ 0 \\ 1 \\ -5 \end{pmatrix} - \begin{pmatrix} 2 & -1 & -1 & 0 & 0 \\ -1 & 2 & -1 & 0 & 0 \\ -1 & -1 & 3 & -1 & 0 \\ 0 & 0 & -1 & 2 & -1 \\ 0 & 0 & 0 & -1 & 1 \end{pmatrix} \begin{pmatrix} 1 \\ 0 \\ 0 \\ 0 \\ 0 \end{pmatrix} = \begin{pmatrix} 1 \\ 2 \\ 1 \\ 1 \\ -5 \end{pmatrix}.$$

This can be written more concisely as

$$(3, 1, 0, 1, -5) \xrightarrow{0} (1, 2, 1, 1, -5).$$

The legal sequence of chip-firings $(0, 1, 0, 2, 1, 0, 3, 2, 1, 0)$ leads to the stable configuration $(0, 1, 2, 1, -4)$. If q is fired then the configuration $(0, 1, 2, 2, -5)$ is achieved. This is recurrent since we have the cyclic legal sequence

$$(0, 1, 2, 2, -5) \xrightarrow{3} (0, 1, 3, 0, -4) \xrightarrow{2} (1, 2, 0, 1, -4)$$
$$\xrightarrow{1} (2, 0, 1, 1, -4) \xrightarrow{0} (0, 1, 2, 1, -4) \xrightarrow{q} (0, 1, 2, 2, -5).$$

In particular, the configuration $(0, 1, 2, 1, -4)$ is also recurrent. Since this configuration is both stable and recurrent, it is critical.

Exercise 4.7. Show that the configurations $(1, 0, 2, 1, -4)$ and $(1, 1, 2, 1, -5)$ are also critical for the graph of Figure 4.3 with $q = 4$ as the sink vertex.

4.3.4 Identifying critical configurations

In this section, we will show that a q-stable configuration on a graph Γ is q-critical if and only if there is an ordering $q = v_1, v_2, \ldots, v_n$ of the vertices of Γ such that the corresponding sequence of chip-firings is q-legal and q-recurrent (takes s to s).

Remark 4.3.20. Note that, by Lemma 2.2.6 on the kernel of the Laplacian, a q-recurrent sequence of chip-firings applied to a configuration must fire each vertex of the graph the same number of times. Conversely, any q-legal sequence in which each vertex is fired the same number of times must be q-recurrent.

Definition 4.3.21. A q-legal sequence of chip-firings is called q-*proper* if the vertex q is not fired. A q-legal firing of q followed by a q-proper sequence of chip-firings is called q-*pointed*. A q-pointed sequence of chip-firings in which each vertex is fired exactly once will be called *complete*.

Thus, a complete q-pointed sequence of chip-firings corresponds to an ordering $q = v_1, v_2, \ldots, v_n$ of the vertices of Γ such that the corresponding sequence of chip-firings is q-legal. This sequence is necessarily q-recurrent, since each vertex is fired the same number of times.

Proposition 4.3.22. *If s is a q-stable configuration, then any q-recurrent sequence of chip-firings of s must consist of a finite sequence of complete q-pointed sequences.*

Proof. Suppose that s is a q-stable configuration with a q-recurrent sequence $\mathcal{S}$. By Remark 4.3.20, each vertex of Γ is fired the same number of times in the sequence $\mathcal{S}$. Since s is q-stable, the first vertex fired must be q. If $\mathcal{S}$ does not consist of complete q-pointed sequences, then there must either be q-pointed subsequences of $\mathcal{S}$ of length longer and shorter than the number of vertices n of Γ, or there must be a q-pointed subsequence of length n which is not complete. In either case, there exists a q-pointed subsequence of $\mathcal{S}$ in which at least one vertex is fired twice. Consider such a subsequence. Let w be the first vertex which is fired twice, and let $\mathcal{S}'$ consist of the part of this subsequence that starts with q and ends with the second firing of w. Thus $\mathcal{S}'$ is a q-legal sequence chip-firings of the form $q = v_1, v_2, \ldots, v_k, w$, where the vertices $v_2, \ldots, v_k$ are distinct vertices, not equal to q, and $w = v_j$ for some j with $2 \leq j \leq k$. Let s' be the configuration which has been reached by the sequence of chip-firings preceding $\mathcal{S}'$ in $\mathcal{S}$. Note that s' must be q-stable, since it is now legal to fire q, so $s'(w) < \deg(w)$. Let W consist of the complement of the vertices $q, v_2, \ldots, v_{j-1}, v_{j+1}, \ldots, v_k$, i.e., W is the complement of all vertices in $\mathcal{S}'$ except for $w = v_j$.

Now we calculate the number of chips on w just before w is fired for the second time in $\mathcal{S}'$. This number must be at least $\deg(w)$, since it is now legal to fire w. Thus

$$s'(w) - \deg(w) + \operatorname{outdeg}_W(w) \geq \deg(w),$$

i.e.,

$$s'(w) + \operatorname{outdeg}_W(w) \geq 2\deg(w).$$

This is impossible, since $s'(w) < \deg(w)$, as observed above (since s' is q-stable), and $\operatorname{outdeg}_W(w) \leq \deg(w)$. Thus the q-recurrent sequence $\mathcal{S}$ must consist of a finite number of complete q-pointed subsequences. $\square$

Corollary 4.3.23. *A q-stable configuration s is q-critical, if and only if there is an ordering $q = v_1, v_2, \ldots, v_n$ of the vertices of Γ, such that the corresponding sequence of chip-firings is q-legal and takes s to s, i.e., if and only if there exists a q-legal complete q-pointed sequence of chip-firings.*

Proof. The result follows immediately from Proposition 4.3.22, since any q-legal complete q-pointed sequence of chip-firings is necessarily q-recurrent. $\square$

Exercise 4.8. Let Γ be the diamond graph of Figure 4.1. Show that the configuration $(-4, 2, 2, 0)$ is q-critical by showing that the sequence of vertex-firings $\{q, v_3, v_2, v_4\}$ is q-legal.

4.3.5 Reduced configurations

We introduce the notion of reduced configurations and show that they are complementary to critical configurations, with respect to the maximal degree 0 q-stable configuration. There are only finitely many reduced configurations, since there are only finitely many critical configurations. Reduced configurations may also be described in terms of the combinatorial notion of a G-parking function on a graph.

We begin by extending the notion of a stable vertex $v \neq q$, to a stable set $W \subset V \setminus \{q\}$.

Definition 4.3.24. Let s be a configuration on a graph Γ. A non-empty set $W \subset V \setminus \{q\}$ is said to be q-stable for s if W cannot be fired q-legally, i.e., if $s(v) < \mathrm{outdeg}_W(v)$ for some $v \in W$.

We now define a q-reduced configuration as one for which no non-empty set may be fired q-legally.

Definition 4.3.25. A q-legal configuration s is called q-reduced if every non-empty set $W \subset V \setminus \{q\}$ is q-stable, i.e., no non-empty set can be fired q-legally.

Example 4.3.26. Let Γ be the diamond graph of Figure 4.1 and let s be the configuration $(-3, 1, 1, 1)$. This configuration is not q-reduced, since the set $W = \{v_2, v_3, v_4\}$ may be fired q-legally. After firing W, we obtain the configuration $(-1, 0, 0, 1)$, which is q-reduced.

Definition 4.3.27. The *maximal q-stable configuration* is the configuration M_q given by

$$M_q(v) = \deg(v) - 1 \text{ for } v \neq q$$

and

$$M_q(q) = -\sum_{v \neq q} M_q(v).$$

Example 4.3.28. The maximal q-stable configuration of the diamond graph of Figure 4.1 is $M_q = (-5, 2, 2, 1)$.

We will show that the notions of being q-critical and q-reduced are complementary with respect to the maximal q-stable configuration.

Definition 4.3.29. Let s be a q-stable configuration and let M_q be the maximal q-stable configuration. We define the q-*complementary configuration* s^* by $s + s^* = M_q$, i.e.,

$$s^*(v) = \deg(v) - 1 - s(v) \text{ for } v \neq q$$

and

$$s^*(q) = -\sum_{v \neq q} s^*(v).$$

Example 4.3.30. Let Γ be the diamond graph of Figure 4.1 and let $s = (-1, 0, 0, 1)$. The configuration s is q-reduced. Using Example 4.3.28, we compute its complement

$$s^* = (-5, 2, 2, 1) - (-1, 0, 0, 1) = (-4, 2, 2, 0).$$

Note that s^* is the q-critical configuration of Example 4.8.

Remark 4.3.31. Note that the complement s^* of a q-stable configuration s is also q-stable and that $s^{**} = s$. If a configuration s is not q-stable, then s^* is not necessarily q-legal.

Remark 4.3.32. It is important to note that if s_1 and s_2 are legal configurations, the sum $s_1^* + s_2^*$ of the complements is not necessarily the same as the complement $(s_1 + s_2)^*$ of the sum.

For example, consider the q-critical configurations on the C_4, the cycle graph on 4 vertices, described in Example 4.7.3 and Remark 4.7.4.

The following result is due to Baker and Norine (see [BN07]).

Lemma 4.3.33. *A configuration s is q-critical if and only if its complementary configuration s^* is q-reduced.*

Proof. Suppose that a configuration s is q-critical and the sequence $q = v_1, v_2, \ldots, v_n$ is an ordering of the vertices of Γ giving a q-legal sequence of chip-firings taking s to s. Let W be any non-empty set of vertices which do not contain q. We wish to show that W is q-reduced with respect to s^*, i.e., we wish to show that there is a vertex w in W such that $s^*(w) < \text{outdeg}_W(w)$. Let k be the least integer such that v_k is in W and let $w = v_k$. Set $\overline{W} = \{v_k, v_{k+1}, \ldots, v_n\}$ and note that since $\overline{W} \supset W$, $\text{outdeg}_{\overline{W}}(w) \leq \text{outdeg}_W(w)$. The complement of $\overline{W}$ contains the vertices $q, v_2, \ldots, v_{k-1}$ used in the first $k - 1$ firings for s. The number of chips added to vertex w in the first $k - 1$ chip-firings of s is exactly $\text{outdeg}_{\overline{W}}(w)$. After the first $k - 1$ vertices have been fired, there are at least $\deg(w)$ chips on vertex w, so

$$s(w) + \text{outdeg}_{\overline{W}}(w) \geq \deg(w).$$

Equivalently,

$$\deg(w) - 1 - s(w) - \mathrm{outdeg}_{\overline{W}}(w) \leq -1,$$

i.e., in terms of s^*,

$$s^*(w) \leq \mathrm{outdeg}_{\overline{W}}(w) - 1$$
$$\leq \mathrm{outdeg}_W(w) - 1,$$

which means that W is q-reduced with respect to s^*.

Next suppose that s^* is q-reduced. We wish to show that there is an ordering $q = v_1, v_2, \ldots, v_n$ of the vertices of Γ which gives a q-legal sequence of chip-firings of s. First we note that s is q-stable, since s^* is, so that q may be fired legally in the configuration s. Next, suppose that we have found a q-legal sequence $q, v_2, \ldots, v_k$ of firings of s, for some $k < n$. Let W be the complement of $\{q, v_2, \ldots, v_k\}$. Since s^* is q-reduced and W is a non-empty set of vertices which does not contain q, there is a vertex w in W such that $s^*(w) < \mathrm{outdeg}_W(w)$. Rewriting in terms of s, we have

$$\deg(w) - 1 - s^*(w) > \deg(w) - 1 - \mathrm{outdeg}_W(w)$$

i.e.,

$$s(w) + \mathrm{outdeg}_W(w) > \deg(w) - 1.$$

Since $s(w) + \mathrm{outdeg}_W(w)$ is the number of chips on w after the sequence of chip-firings $q, v_2, \ldots v_k$ has been applied to s, it follows that w can now be fired q-legally, so we set $v_{k+1} = w$.   □

It follows immediately from the previous result that there are only finitely many q-reduced configurations.

In a later section, we will show that for every q-legal configuration s, there is a unique q-reduced configuration which can be reached from s by a sequence of q-legal set-firing moves (see Proposition 4.5.7).

Similarly, we will show that for every q-legal configuration s, there is a unique q-critical configuration which can be reached from s by a sequence of q-legal chip-firing moves (see Proposition 4.5.8).

G-parking functions are closely related to reduced configurations.

Definition 4.3.34. An integer-valued function on the set of vertices other than q, $f : V \setminus \{q\} \to \mathbb{Z}$, is called a G-parking function based at q if $f(v) \geq 0$ for all vertices v in $V \setminus \{q\}$, and for every non-empty set of vertices $W \subset V \setminus \{q\}$, there exists a vertex v in W such that $f(v) < \mathrm{outdeg}_W(v)$.

Alternatively, f is a G-parking function based at q if and only if the configuration s defined by

$$s(v) = \begin{cases} f(v), & \text{if } v \neq q, \\ -\sum_{v \neq q} f(v), & \text{if } v = q, \end{cases}$$

is a q-reduced configuration.

For more on G-parking functions, see Benson, Chakrabarty, and Tetali [BCT10]; Chebikin and Pylyavskyy [CP05]; Postnikov and Shapiro [PS04]; and Stanley [St99].

4.4 Energy pairing on degree 0 configurations

In this section, we define an energy pairing on degree 0 configurations, using the Moore–Penrose pseudoinverse (see Definition 2.3.5). The energy pairing takes values in the rational numbers $\mathbb{Q}$. After defining the critical group (Jacobian) of a graph in §4.6, we will show that the energy pairing determines a non-degenerate pairing on the critical group, with values in $\mathbb{Q}$ mod $\mathbb{Z}$.

Definition 4.4.1. Let s_1 and s_2 be degree 0 configurations on Γ and let Q^+ be the Moore–Penrose pseudoinverse. The *energy pairing* of s_1 and s_2 is given by
$$\langle s_1, s_2 \rangle = s_1^t Q^+ s_2.$$

Example 4.4.2. The q-reduced configurations on C_4 are

$$\begin{aligned}
s_1 &= (0, 0, 0, 0) \\
s_2 &= (-1, 1, 0, 0) \\
s_3 &= (-1, 0, 1, 0) \\
s_4 &= (-1, 0, 0, 1)
\end{aligned}$$

and the corresponding q-critical configurations are

$$\begin{aligned}
s_1^* &= (-3, 1, 1, 1) \\
s_2^* &= (-2, 0, 1, 1) \\
s_3^* &= (-2, 1, 0, 1) \\
s_4^* &= (-2, 1, 1, 0).
\end{aligned}$$

The pseudoinverse of the Laplacian of C_4 is

$$Q^+ = \begin{pmatrix} \frac{5}{16} & -\frac{1}{16} & -\frac{3}{16} & -\frac{1}{16} \\ -\frac{1}{16} & \frac{5}{16} & -\frac{1}{16} & -\frac{3}{16} \\ -\frac{3}{16} & -\frac{1}{16} & \frac{5}{16} & -\frac{1}{16} \\ -\frac{1}{16} & -\frac{3}{16} & -\frac{1}{16} & \frac{5}{16} \end{pmatrix}.$$

From this, we compute

$$\langle s_1^*, s_2^* \rangle = (s_1^*)^t Q^+ s_2^* = 7/2.$$

Definition 4.4.3. The *energy* of a degree 0 configuration s is defined to be

$$\mathcal{E}(s) = \langle s, s \rangle = s^t Q^+ s.$$

Exercise 4.9. Consider the cycle graph on 3 vertices C_3 of Example 2.3.9. The Moore–Penrose pseudoinverse of the Laplacian matrix of C_3 is

$$Q^+ = \begin{pmatrix} \frac{2}{9} & -\frac{1}{9} & -\frac{1}{9} \\ -\frac{1}{9} & \frac{2}{9} & -\frac{1}{9} \\ -\frac{1}{9} & -\frac{1}{9} & \frac{2}{9} \end{pmatrix}.$$

Label the vertices v_1, v_2, and v_3 and let $q = v_1$ be the sink vertex. Let $s_1 = (-1, 1, 0)$, $s_2 = (-3, 1, 2)$, and $s_3 = (-2, 1, 1)$.

(a) Show that $\mathcal{E}(s_1) = \frac{2}{3}$.

(b) Show that $\langle s_1, s_2 \rangle = \frac{4}{3}$.

(c) Show that the set $W = \{v_2, v_3\}$ may be fired legally on s_2, producing the configuration $s_2' = (-1, 0, 1)$.

(d) Check that $\langle s_1, s_2 \rangle - \langle s_1, s_2' \rangle$ is a positive integer.

(e) Show that $\langle s_3, s_1 \rangle$ and $\langle s_3, s_2 \rangle$ are both integers.

(f) Show that the set $W = \{v_2, v_3\}$ may be fired legally on s_3, producing the configuration $s_3' = (0, 0, 0)$.

Remark 4.4.4. An example of a table of values of the induced energy pairing for the critical group of the "diamond graph" is given in Example 4.10.2.

Proposition 4.4.5. *Let L be any generalized inverse of the Laplacian matrix Q (see §2.3 for the definition). Let s_1 and s_2 be degree 0 configurations on Γ. Then*

$$\langle s_1, s_2 \rangle = s_1^t L s_2,$$

i.e., any generalized inverse may be substituted for the Moore–Penrose pseudoinverse in calculating the energy pairing.

Proof. Recall from Corollary 2.3.8 that for any degree 0 configuration s, we have $QQ^+ s = s$. Recall that $Q^t = Q$ and $(Q^+)^t = Q^+$. Thus $Q = QLQ = Q^t L Q$. Recall also that $Q^+ Q Q^+ = Q^+$. We now have

$$\langle s_1, s_2 \rangle = s_1^t Q^+ s_2$$
$$= s_1^t Q^+ Q Q^+ s_2$$
$$= (Q^+ s_1)^t Q (Q^+ s_2)$$
$$= (Q^+ s_1)^t Q^t L Q (Q^+ s_2)$$
$$= (Q Q^+ s_1)^t L (Q Q^+ s_2)$$
$$= s_1^t L s_2.$$

$\square$

Next we show that the energy of a q-legal configuration decreases by a positive integer after a legal set-firing.

Proposition 4.4.6. *Suppose that s is a q-legal configuration and W is a non-empty subset of $V \setminus \{q\}$ such that W can be fired q-legally. Let s' be the configuration after W is fired, i.e., $s' = s - Q x_W$, where x_W is the characteristic vector of W. Then $\mathcal{E}(s) - \mathcal{E}(s')$ is a positive integer.*

Proof. We have $s' = s - Q x_W$, where $x_W(v) = 1$ if $v \in W$ and $x_W(v) = 0$ if $v \notin W$. Expanding $\mathcal{E}(s')$, and using properties of Q^+ from Proposition 2.3.7 and the fact that Q is symmetric, gives

$$\mathcal{E}(s') = (s - Q x_W)^t \, Q^+ \, (s - Q x_W)$$
$$= s^t Q^+ s - s^t Q^+ Q x_W - x_W^t Q^t Q^+ s + x_W^t Q^t Q^+ Q x_W$$
$$= \mathcal{E}(s) - s^t \left(I - \frac{1}{n} J \right) x_W - x_W^t \left(I - \frac{1}{n} J \right) s + x_W^t Q x_W.$$

Note that $Js = \mathbf{0}$, since s has degree 0. Thus, by Lemmas 4.3.12 and 4.3.9,

$$\mathcal{E}(s') = \mathcal{E}(s) - 2\deg_W(s) + \operatorname{outdeg}(W). \tag{4.5}$$

But $\deg_W(s) \geq \operatorname{outdeg}(W)$, because W may be fired. Thus

$$\mathcal{E}(s) - \mathcal{E}(s') \geq \deg_W(s) \geq \operatorname{outdeg}(W),$$

where $\operatorname{outdeg}(W)$ is a positive integer since W is non-empty (and Γ is connected). $\square$

We now use the alternative formula for the Moore–Penrose pseudoinverse from Lemma 2.3.10 to describe the energy of a q-legal configuration. Let $\lambda_1 = 0, \lambda_2, \ldots, \lambda_n$ be the eigenvalues of the Laplacian matrix Q. Since Q is a real $n \times n$ symmetric matrix, we can choose a basis of $\mathbb{R}^n$ consisting of n orthonormal eigenvectors $w_1, w_2, \ldots, w_n$ corresponding to eigenvalues $\lambda_1, \lambda_2, \ldots, \lambda_n$. Let U be the orthogonal matrix with columns $w_1, w_2, \ldots, w_n$ and let Σ^+ be the diagonal matrix with diagonal entries $0, \frac{1}{\lambda_2}, \frac{1}{\lambda_3}, \ldots, \frac{1}{\lambda_n}$.

Recall from Lemma 2.3.10 that the Moore–Penrose pseudoinverse of Q is given by $Q^+ = U\Sigma^+U^t$.

Suppose that s is a q-legal configuration (and thus of degree 0). Note that w_1 spans the kernel of Q, so s must be orthogonal to w_1 and must lie in the span of the remaining eigenvectors $w_2, w_3, \ldots, w_n$. The following lemma gives a description of the energy of a q-legal configuration in terms of its components with respect to the orthonormal vectors $w_2, w_3, \ldots, w_n$.

Lemma 4.4.7. *Let s be a q-legal configuration. Let $0 = a_1, a_2, a_3, \ldots, a_n$ be the components of s with respect to the orthonormal basis $w_1, w_2, w_3, \ldots, w_n$ of $\mathbb{R}^n$, so that*

$$s = \sum_{i=2}^{n} a_i w_i.$$

Then the energy of s is

$$\mathcal{E}(s) = \sum_{i=2}^{n} \frac{1}{\lambda_i} a_i^2.$$

In particular, $\mathcal{E}(s) \geq 0$ and $\mathcal{E}(s) = 0$ if and only if $s = 0$.

Proof. Both statements follow from the decomposition

$$Q^+ = U\Sigma^+U^t$$

and the fact that the coefficients of the vector U^ts are $(0, a_2, a_3, \ldots, a_n)$. $\square$

Example 4.4.8. Consider the complete graph K_4 on four vertices, $q = v_1, v_2, v_3, v_4$. The Laplacian matrix of K_4 is

$$Q = \begin{pmatrix} 3 & -1 & -1 & -1 \\ -1 & 3 & -1 & -1 \\ -1 & -1 & 3 & -1 \\ -1 & -1 & -1 & 3 \end{pmatrix}.$$

The eigenvalues of Q are $0, 4, 4, 4$, corresponding to orthonormal eigenvectors

$$w_1 = \begin{pmatrix} \frac{1}{2} \\ \frac{1}{2} \\ \frac{1}{2} \\ \frac{1}{2} \end{pmatrix}, w_2 = \begin{pmatrix} -\frac{1}{\sqrt{2}} \\ 0 \\ 0 \\ \frac{1}{\sqrt{2}} \end{pmatrix}, w_3 = \begin{pmatrix} -\frac{1}{\sqrt{6}} \\ 0 \\ \sqrt{\frac{2}{3}} \\ -\frac{1}{\sqrt{6}} \end{pmatrix}, \text{ and } w_4 = \begin{pmatrix} -\frac{1}{2\sqrt{3}} \\ \frac{\sqrt{3}}{2} \\ -\frac{1}{2\sqrt{3}} \\ -\frac{1}{2\sqrt{3}} \end{pmatrix}.$$

The Moore–Penrose pseudoinverse of Q is

$$
Q^+ = \begin{pmatrix}
\frac{3}{16} & -\frac{1}{16} & -\frac{1}{16} & -\frac{1}{16} \\
-\frac{1}{16} & \frac{3}{16} & -\frac{1}{16} & -\frac{1}{16} \\
-\frac{1}{16} & -\frac{1}{16} & \frac{3}{16} & -\frac{1}{16} \\
-\frac{1}{16} & -\frac{1}{16} & -\frac{1}{16} & \frac{3}{16}
\end{pmatrix}.
$$

There are 16 q-reduced degree 0 configurations.

Exercise 4.10. Consider the following four degree 0 configurations on K_4: $s_1 = \{-1, 1, 0, 0\}$, $s_2 = \{-2, 1, 1, 0\}$, $s_3 = \{-2, 2, 0, 0\}$, and $s_4 = \{-3, 2, 1, 0\}$. Let q be the first vertex.

(a) Show that s_1, s_2, s_3, and s_4 are q-reduced.

(b) Show that the energies of s_1, s_2, s_3, and s_4 are $\mathcal{E}(s_1) = \frac{1}{2}$, $\mathcal{E}(s_2) = \frac{3}{2}$, $\mathcal{E}(s_3) = 2$, and $\mathcal{E}(s_4) = \frac{7}{2}$.

(c) Find the remaining 12 degree 0 q-reduced configurations on K_4 and their energies.

4.5 Equivalence classes of configurations

In this section, we define an equivalence relation on configurations and show that the equivalence class of each degree 0 configuration contains a unique q-critical configuration and a unique q-reduced configuration. In §4.6 we will show that these equivalence classes form a group, the critical group or Jacobian.

Definition 4.5.1. We will say that configurations s and s' are *equivalent* if they differ by an element of the image of the Laplacian Q, i.e., if there is an integer vector x such that
$$
s' = s - Qx.
$$

It follows from the definition of complementary configurations that s and s' are equivalent configurations if and only if their complements s^* and $(s')^*$ are equivalent configurations.

We will now show that replacing degree 0 configurations by equivalent configurations in the energy pairing changes the value by an integer.

Proposition 4.5.2. *If s_1 and s_2 are degree 0 configurations such that s_1 is equivalent to s_1' and s_2 is equivalent to s_2', then $\langle s_1', s_2' \rangle - \langle s_1, s_2 \rangle$ is an integer.*

Proof. We may assume that $s_1' = s_1 - Qx_1$ and $s_2' = s_2 - Qx_2$, for some integer vectors x_1 and x_2. Recall that Q and Q^+ are symmetric. Recall also, from Corollary 2.3.8, that $QQ^+s_i = s_i$ and $s_i^tQ^+Q = s_i^t$ for $i = 1, 2$. Then

$$
\begin{aligned}
\langle s_1', s_2' \rangle &= (s_1 - Qx_1)^tQ^+(s_2 - Qx_2) \\
&= s_1^tQ^+s_2 - s_1^tQ^+Qx_2 - x_1^tQ^tQ^+s_2 + x_1^tQ^tQ^+Qx_2 \\
&= \langle s_1, s_2 \rangle - s_1^tQ^+Qx_2 - x_1^tQQ^+s_2 + x_1^tQQ^+Qx_2 \\
&= \langle s_1, s_2 \rangle - s_1^tx_2 - x_1^ts_2 + x_1^tQx_2.
\end{aligned}
$$

Since the terms $s_1^tx_2$, $x_1^ts_2$, and $x_1^tQx_2$ are all integers, the result follows. $\square$

The next lemma shows that if s is a q-legal configuration such that $s - Qx$ is also q-legal, for some integer vector x, with $x(v) \geq 0$ for all vertices v, $x(v) > 0$ for at least one vertex v, and $x(q) = 0$, then s cannot be q-reduced.

Lemma 4.5.3. *Let s be a q-legal configuration. Suppose that there is an integer vector x such that $x(v) \geq 0$ for all vertices v, $x(v) > 0$ for at least one vertex v, and $x(q) = 0$, and such that $s - Qx$ is also q-legal. Then there exists a sequence of non-empty subsets of $V \setminus \{q\}$ of the form $W_1 \subseteq W_2 \subseteq \cdots \subseteq W_r$ such that $x = x_{W_1} + x_{W_2} + \cdots + x_{W_r}$ and the sets $W_1, W_2, \ldots, W_r$ may be fired q-legally in that order. In particular, s is not q-reduced.*

Proof. We choose W_1 to be the set of all vertices on which x takes its maximum value, W_2 the set of all vertices on which $x - x_{W_1}$ takes its maximum value, and so on. The final set, W_r will be the support of x, i.e., the set of vertices v such that $x(v) > 0$. By definition of the sets W_i, $x = x_{W_1} + x_{W_2} + \cdots + x_{W_r}$ and $s - Qx = s - Qx_{W_1} - Qx_{W_2} - \cdots - Qx_{W_r}$.
Consider any vertex v in W_1. By Lemma 4.3.9,

$$
s(v) - Qx(v) = s(v) - \text{outdeg}_{W_1}(v) - \text{outdeg}_{W_2}(v) - \cdots - \text{outdeg}_{W_r}(v).
$$

Since $s - Qx$ is q-legal, $s(v) - Qx(v) \geq 0$. It follows that $s(v) \geq \text{outdeg}_{W_1}(v)$ for all vertices v in W_1. Thus, W_1 can be fired q-legally, giving a q-legal configuration $s - Qx_{W_1}$. The result follows by induction. $\square$

We now show that distinct q-reduced configurations cannot be equivalent to each other.

Lemma 4.5.4. *Suppose that s_1 and s_2 are equivalent q-reduced configurations, i.e., such that $s_1 = s_2 - Qx$, for some integer vector x. Then $s_1 = s_2$.*

Proof. We may assume that $x(q) = 0$, since $Q\mathbf{1} = 0$, where $\mathbf{1}$ is the all 1's vector, so that we can add a multiple of $\mathbf{1}$ to x if necessary.
Next decompose x into its positive and negative parts, $x = x_2 - x_1$, where the supports of x_1 and x_2 (the sets of vertices on which they are nonzero)

are disjoint, and, for each i, $x_i(q) = 0$ and $x_i(v) \geq 0$ for $v \neq q$. Let s' be the configuration $s_1 - Qx_1 = s_2 - Qx_2$.

We will now show that s' is q-legal. Consider any vertex $v \neq q$. Since the supports of x_1 and x_2 are disjoint, either $x_1(v) = 0$ or $x_2(v) = 0$. Suppose that $x_1 = 0$. Then $s'(v) = s_1(v)$, which is nonnegative since s_1 is q-legal. Similarly, if $x_2(v) = 0$, then $s'(v) = s_2(v)$ is nonnegative. Thus s' is q-legal. But we assumed that s_1 and s_2 are q-reduced, so by Lemma 4.5.3, $s_1 - Qx_1$ and $s_2 - Qx_2$ cannot be q-legal unless x_1 and x_2 are 0. Therefore $s_1 = s_2$. $\square$

We next prove that every q-legal configuration is equivalent to a q-reduced configuration.

Lemma 4.5.5. *If s is a q-legal configuration, then there is an integer vector x with $x(q) = 0$ and $x(v) \geq 0$ for $v \neq q$, such that $s - Qx$ is q-reduced.*

Proof. Suppose that s is a q-legal configuration which is not q-reduced. Then there is a set $W_1 \subset V \setminus \{q\}$ such that W_1 may be fired q-legally to obtain a q-legal configuration s_1. By Lemma 4.4.7 and Proposition 4.4.6,

$$0 \leq \mathcal{E}(s_1) \leq \mathcal{E}(s) - 1.$$

Since the energy decreases by at least 1 each time a set is fired q-legally, only a finite number of sets $W_1, W_2, ..., W_N$ may be fired q-legally before a q-reduced configuration is reached. Let x be the sum of the characteristic vectors of $W_1, W_2, ..., W_N$. $\square$

Next we show that every degree 0 configuration is equivalent to a q-legal configuration.

Lemma 4.5.6. *For every degree 0 configuration s, there exists an equivalent configuration s' such that s' is q-legal.*

Proof. We may decompose s into the difference of two q-legal configurations, $s = s^+ - s^-$, where

$$s^+(v) = \begin{cases} s(v), & \text{if } s(v) > 0 \text{ and } v \neq q, \\ 0, & \text{if } s(v) \leq 0 \text{ and } v \neq q, \\ -\sum_{v \neq q} s^+(v), & \text{if } v = q, \end{cases}$$

and

$$s^-(v) = \begin{cases} -s(v), & \text{if } s(v) < 0 \text{ and } v \neq q, \\ 0, & \text{if } s(v) \geq 0 \text{ and } v \neq q, \\ -\sum_{v \neq q} s^-(v), & \text{if } v = q. \end{cases}$$

Let M_q be the maximal degree 0 q-stable configuration, given by

$$M_q(v) = \begin{cases} \deg(v) - 1, & \text{if } v \neq q, \\ -\sum_{v \neq q} M_q(v), & \text{if } v = q. \end{cases}$$

Since M_q and s^- are q-legal, so is their sum $M_q + s^-$. It follows from Lemma 4.5.5 that there is an integer vector x with $x(q) = 0$ and $x(v) \geq 0$ for $v \neq q$, such that $M_q + s^- - Qx$ is q-reduced. Therefore

$$0 \leq (M_q + s^- - Qx)(v) \leq M_q(v)$$

for all $v \neq q$. Thus $(-s^- + Qx)(v) \geq 0$ and $s^+ - s^- + Qx)(v) \geq 0$ for all $v \neq q$, i.e., $s' = s + Qx$ is a q-legal configuration which is equivalent to s. $\qquad \square$

We can now show that every equivalence class of degree 0 configurations contains a unique q-reduced configuration.

Proposition 4.5.7. *Let s be a degree 0 configuration. Then the equivalence class of s contains a unique q-reduced configuration s_r. Moreover, if s is also q-legal then s_r may be reached from s by a sequence of q-legal set-firing moves (without firing q).*

For a concrete instance of this, see Example 4.6.1 below.

Proof. By Lemma 4.5.6, there exists a q-legal configuration s' which is equivalent to s. By Lemma 4.5.5, there exists a q-reduced configuration s_r which is equivalent to s'. By Lemma 4.5.4, this q-reduced configuration is unique.

Moreover, by Lemmas 4.5.5 and 4.5.3, if s is also q-legal, then s may be changed to s_r by a sequence of q-legal set-firing moves. $\qquad \square$

It follows from the previous proposition that each equivalence class of degree 0 configurations contains a unique critical configuration.

Proposition 4.5.8. *Let s be a degree 0 configuration. Then the equivalence class of s contains a unique q-critical configuration s_c. Moreover, if s is also q-legal then s_c may be reached from s by a sequence of q-legal chip-firing moves (possibly including firing q, when legal).*

Proof. Let s be a configuration such that $\sum_{v \in V} s(v) = 0$ and let $s^* = M_q - s$ be the complement of s, where M_q is the maximal q-stable configuration. By Lemma 4.5.4, there is a unique q-reduced configuration r which is equivalent to s^*. Let x be an integer vector such that $r = s^* - Qx$. The complement $r^* = M_q - r$ of r is q-critical by Lemma 4.3.33. Then $r^* = M_q - (M_q - s) + Qx$, so that $s = r^* - Qx$. Therefore, s is q-equivalent to the q-critical configuration r^*. Since r is the unique q-reduced configuration in the equivalence class of s^*, its complement r^* must be the only q-critical configuration in the equivalence class of s.

Now suppose that s is also q-legal. We will show that s may be changed to a q-critical configuration c by a sequence of q-legal chip-firing moves. Since there is only one q-critical configuration s_c which is in the equivalence class of s, we must have $c = s_c$.

First we note that, since the energy decreases by a positive integer each time a vertex other than q is fired, by Proposition 4.4.6, only a finite number of vertices other than q may be fired q-legally before we reach a q-stable configuration, in which q is the only vertex which may be fired q-legally, i.e., there is an upper bound on the length of a proper sequence of q-legal chip-firings of s. The vertex q may then be fired until a non-stable configuration is reached (which will occur in a finite number of firings, since each neighbor of q gains at least one chip each time q is fired). We then fire vertices other than q again until a stable configuration is reached. Repeating this process of alternately firing a proper sequence which produces a stable configuration and then firing q until a non-stable configuration is reached, we produced a sequence of q-stable configurations. Since the number of q-stable configurations is finite (see Remark 4.3.16), at least one of them must recur, so this one is critical and can be reached by q-legal chip-firing moves.   □

4.6 Critical group of a graph

The set of all configurations of degree 0 on a graph Γ forms an abelian group under addition. The degree 0 configurations which are in the image of the Laplacian Q, considered as a map $Q : \mathbb{Z}^n \to \mathbb{Z}^n$, form a subgroup, $\text{Im}(Q)$. The quotient of the group of all degree 0 configurations by the subgroup $\text{Im}(Q)$ forms an abelian group, denoted $\mathcal{K}(\Gamma) \simeq Jac(\Gamma)$, whose elements are equivalence classes of configurations. This group is called the critical group, the Jacobian, or the Picard group (depending on the context). Given a fixed vertex q, there is a unique q-critical configuration in each equivalence class. Thus, the critical group is finite, since there are only finitely many q-critical configurations.

Alternatively, the elements of the critical group are sometimes taken to be the set of all q-critical configurations on Γ, for some fixed vertex q. The sum $s_1 \oplus s_2$ of two q-critical configurations s_1 and s_2 is defined to be the unique q-critical configuration which is equivalent to the ordinary sum $s_1 + s_2$. This group is clearly isomorphic to the group of equivalence classes defined above, and is thus independent of the vertex q chosen as the sink (up to isomorphism).

Since a configuration s is q-critical if and only if its complement s^* is q-reduced, by Lemma 4.3.33, it follows that the set of reduced configurations can be given also be given the structure of a group. The sum $s_1 \oplus' s_2$ of two q-reduced configurations s_1 and s_2 is defined to be the unique q-reduced configuration which is equivalent to the ordinary sum $s_1 + s_2$.

It is important to note that the operation of taking the complement with respect to the maximal q-stable configuration is not a homomorphism between the group of q-critical configurations under $\oplus$ and the group of q-reduced configurations under $\oplus'$ (see Remark 4.3.32).

It is well-known that the order of the critical group is the number of spanning trees of the graph Γ. We prove this result in §4.9 using an ordered version of Dhar's burning algorithm due to Cori and Le Borgne (see Corollary 4.9.16).

4.6.1 The Jacobian and the Picard group

The names Jacobian or Picard group are often used for the critical group when making analogies between graphs and algebraic curves. The group of all configurations of degree zero on a graph Γ can be thought of as the group of all *degree zero divisors* on Γ. The subgroup of configurations which are in the image of the Laplacian Q may be considered the *principal divisors* on Σ. The quotient of the divisors of degree zero by the subgroup of principal divisors is called the *Jacobian* $\mathrm{Jac}(\Gamma)$ or *Picard group* $\mathrm{Pic}(\Gamma)$ of the graph Γ. We will use the terms critical group, Jacobian, and Picard group interchangeably when referring to the finite abelian group $\mathcal{K}(\Gamma)$.

4.6.2 Simple examples of critical groups

Example 4.6.1. Consider the graph Γ with configuration $s = (-6, 3, 1, 0, 2)$, depicted in Figure 4.4. Here the vertex labeled 0 is taken to be q.

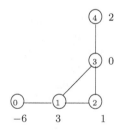

Figure 4.4: A graph having 5 vertices, with configuration $s = (0, 0, 1, 0, 2)$.

Now, we fire the following sequence of vertices: 4, 4, 1, 3, 4, 2. This gives us the configuration $c = (-5, 2, 1, 2, 0)$, depicted in Figure 4.5. Noting that the maximal q-stable configuration M_q on Γ is also $(-5, 2, 1, 2, 0)$, we see that the complement of c is $c^* = M_q - c = (0, 0, 0, 0, 0)$, which is reduced. Therefore, c must be critical.

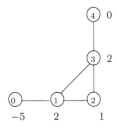

Figure 4.5: A critical configuration $c = (-5, 2, 1, 2, 0)$ which can be reached from s by q-legal chip-firing moves.

It is easy to show that the only reduced configurations on the graph Γ are $(0, 0, 0, 0, 0)$, $(-1, 0, 1, 0, 0)$, and $(-1, 0, 0, 1, 0)$. Therefore, the only critical configurations on Γ are their complements, $(-5, 2, 1, 2, 0)$, $(-4, 2, 0, 2, 0)$, and $(-4, 2, 1, 1, 0)$. Thus, the critical group of Γ is isomorphic to $\mathbb{Z}/3\mathbb{Z}$.

Note that the critical group of a graph does not change (up to isomorphism) if we change the vertex selected as q. Note also that in a reduced configuration, vertices other than q which have degree 1 must have no chips on them. This means that the critical group does not change if we remove a degree 1 vertex and the edge incident to it. Thus, the critical group of Γ is isomorphic to the critical group of C_3, the cyclic group on 3 vertices.

Exercise 4.11. Show that the critical group $\mathcal{K}(K_4)$ of K_4 is isomorphic to $\mathbb{Z}/4\mathbb{Z} \times \mathbb{Z}/4\mathbb{Z}$. Hint: Let q be the first vertex. Check that the equivalence classes of the configurations $s_1 = \{-1, 1, 0, 0\}$ and $s_2 = \{-1, 0, 1, 0\}$ are of order 4 and every q-reduced configuration is equivalent to a configuration of the form $is_1 + js_2$, for $i, j \in \{0, 1, 2, 3\}$.

4.6.3 The Smith normal form and invariant factors

We will show how the Smith normal form of a reduced Laplacian matrix of a graph Γ can be used to calculate the critical group of Γ. The invariant factors $n_1, n_2, \ldots, n_r$ of the Smith normal form give us a decomposition

$$\mathcal{K}(\Gamma) = (\mathbb{Z}/n_1\mathbb{Z}) \times (\mathbb{Z}/n_2\mathbb{Z}) \times \ldots \times (\mathbb{Z}/n_r\mathbb{Z})$$

of the critical group $\mathcal{K}(\Gamma)$, and the numbers $n_1, n_2, \ldots, n_r$ are the invariant factors of the critical group.

Recall that a matrix Q_i obtained by removing the i-th row and i-th column of the Laplacian matrix Q is called a *reduced Laplacian matrix*. Let Q^* be any reduced Laplacian matrix of a graph Γ. Recall also that for a connected graph Γ, a reduced Laplacian matrix Q^* has rank $n - 1$, where n is the number of vertices of Γ. We will show that the critical group of Γ is isomorphic to the lattice quotient $\mathbb{Z}^{n-1}/\mathrm{Col}(Q^*)$, where $\mathrm{Col}(Q^*)$ denotes the $\mathbb{Z}$-span of the columns of Q^* in $\mathbb{Z}^{n-1}$.

Let M be an integral $m \times m$ matrix of full rank. The *Smith normal form*, Σ, of M is a diagonal matrix Σ with nonzero integer diagonal entries $k_1, k_2, \ldots, k_m$ such that $k_1|k_2|\cdots|k_m$, and matrices U_1 and U_2 with integer entries and determinants 1 or -1 such that

$$U_1\Sigma = MU_2. \tag{4.6}$$

The numbers k_i, called *elementary divisors* or *invariant factors*, are unique up to multiplication by -1.

Since all the entries of Q^* are integers and Q^* has (full) rank $n - 1$, the matrix Q^* has a Smith normal form: a diagonal matrix Σ with strictly positive invariant factors $k_1|k_2|\cdots|k_{n-1}$.

Proposition 4.6.2. *Let Γ be a connected graph and let Q^* be a reduced Laplacian matrix of Γ. Then the critical group $\mathcal{K}(\Gamma)$ is isomorphic to the quotient $\mathbb{Z}^{n-1}/\mathrm{Col}(Q^*)$, where $\mathrm{Col}(Q^*)$ denotes the $\mathbb{Z}$-span of the columns of Q^*. Furthermore, if $k_1, k_2, \ldots, k_{n-1}$ are the invariant factors of Q^*, then the critical group of Γ has a decomposition*

$$\mathcal{K}(\Gamma) = (\mathbb{Z}/k_1\mathbb{Z}) \times (\mathbb{Z}/k_2\mathbb{Z}) \times \cdots \times (\mathbb{Z}/k_{n-1}\mathbb{Z}).$$

Proof. Let us choose as our sink vertex q the vertex corresponding to the row and column of Q which were removed to obtain Q^*. For each degree 0 configuration s, the value of $s(q)$ is determined by the values of $s(v)$ for all $v \neq q$. Thus the group of all degree 0 configurations is isomorphic to $\mathbb{Z}^{n-1}$. The subgroup of degree 0 configurations of the form Qx, for x an integer vector, corresponds under this isomorphism to the subgroup of $\mathbb{Z}^{n-1}$ given by $\mathrm{Col}(Q^*)$. Thus, the critical group is isomorphic to $\mathbb{Z}^{n-1}/\mathrm{Col}(Q^*)$.

Let Σ be the Smith normal form of Q^* and let $k_1, k_2, \ldots, k_{n-1}$ be the diagonal entries of Σ. Let U_1 and U_2 be matrices with integer entries and determinants 1 or -1 such that (4.6) holds with $M = Q^*$.

First note that the column span of Q^*U_2 is the same as the column span of Q^*, since U_2 is an automorphism of $\mathbb{Z}^{n-1}$.

Next, let $u_1, u_2, \ldots, u_{n-1}$ be the columns of U_1 and note that these vectors form a basis for $\mathbb{Z}^{n-1}$, since U_1 is an automorphism of $\mathbb{Z}^{n-1}$. The columns of the matrix $U_1\Sigma$ are $k_1u_1, k_2u_2, \ldots, k_{n-1}u_{n-1}$, which must also span the column space of Q^*. Note that $\mathbb{Z}/k_i\mathbb{Z}$ is trivial for $k_i = \pm 1$ and isomorphic to $\mathbb{Z}/|k_i|\mathbb{Z}$ otherwise. Thus, the quotient $\mathbb{Z}^{n-1}/\mathrm{Col}(Q^*)$ must be isomorphic to $(\mathbb{Z}/k_1\mathbb{Z}) \times (\mathbb{Z}/k_2\mathbb{Z}) \times \cdots \times (\mathbb{Z}/k_{n-1}\mathbb{Z})$. □

As a consequence of the previous proposition, we obtain another formula for the number of spanning trees of a graph Γ.

Corollary 4.6.3. *Let Γ be a connected graph with reduced Laplacian matrix Q^* and invariant factors $n_1, n_2, \ldots, n_r$ of Q^*. Then the number of spanning trees of Γ (the spanning tree number) is*

$$\kappa = n_1 n_2 \ldots n_r.$$

Example 4.6.4. Recall that the Laplacian matrix of the complete graph K_4 on 4 vertices is

$$Q = \begin{pmatrix} 3 & -1 & -1 & -1 \\ -1 & 3 & -1 & -1 \\ -1 & -1 & 3 & -1 \\ -1 & -1 & -1 & 3 \end{pmatrix},$$

so that a reduced Laplacian matrix is

$$Q^* = \begin{pmatrix} 3 & -1 & -1 \\ -1 & 3 & -1 \\ -1 & -1 & 3 \end{pmatrix}.$$

A Smith normal form of Q^* is

$$\Sigma = \begin{pmatrix} 1 & 0 & 0 \\ 0 & 4 & 0 \\ 0 & 0 & 4 \end{pmatrix},$$

and we have $U_1\Sigma = Q^*U_2$ where

$$U_1 = \begin{pmatrix} 3 & -1 & -1 \\ -1 & 1 & 0 \\ -1 & 0 & 1 \end{pmatrix} \text{ and } U_2 = \begin{pmatrix} 1 & -1 & -1 \\ 0 & 1 & 0 \\ 0 & 0 & 1 \end{pmatrix}.$$

The columns of U_1 are

$$u_1 = \begin{pmatrix} 3 \\ -1 \\ -1 \end{pmatrix}, \ u_2 = \begin{pmatrix} -1 \\ 1 \\ 0 \end{pmatrix} \text{ and } u_3 = \begin{pmatrix} -1 \\ 0 \\ 1 \end{pmatrix}.$$

It follows that the vectors u_1, $4u_2$, and $4u_3$ span the column space of Q^*, so we can see that the critical group of K_4 must be $\mathcal{K}(K_4) = (\mathbb{Z}/4\mathbb{Z}) \times (\mathbb{Z}/4\mathbb{Z})$.

Exercise 4.12. Suppose that Γ is the union of subgraphs Γ_1 and Γ_2 which are joined at a single vertex. Show that the critical group of Γ is the product of the critical groups of Γ_1 and Γ_2. (Hint: We may order the vertices of Γ such that the reduced Laplacian matrix Q^* of Γ, with respect to the common vertex, is block diagonal of the form

$$\begin{pmatrix} Q_1^* & 0 \\ 0 & Q_2^* \end{pmatrix},$$

where Q_1^* and Q_2^* are the reduced Laplacian matrices of Γ_1 and Γ_2, respectively.)

Example 4.6.5. The graphs Γ_1 and Γ_2 in Figure 4.6 are found in Clancy, Leake, and Payne [CLP15]. They have the same critical group,

$$Jac(\Gamma_1) = Jac(\Gamma_2) = \mathbb{Z}/24\mathbb{Z},$$

and the same Tutte polynomial (see Definition 1.2.2),

$$x^5 + 3x^4 + 3x^3 y + x^2 y^2 + 3x^3 + 5x^2 y + 3xy^2 + y^3 + x^2 + 2xy + y^2,$$

but distinct Duursma zeta functions (see Definition 1.2.21). The Duursma zeta polynomial of Γ_1 is

$$-(64x^7 - 54x^6 + 62x^5 - 7x^4 + 23x^3 - 4x^2 + x - 1)(3x^2 + x + 1)(2x^2 + 1)(x^2 - 1)^2$$

while the Duursma zeta polynomial of Γ_2 is

$$-(32x^7 - 56x^6 + 44x^5 - 20x^4 + 21x^3 - 4x^2 + 2x - 1)(4x^2 + x + 1)(3x^2 + x + 1)(x^2 - 1)^2.$$

Each graph consists of a 3-cycle C_3 joined to the diamond graph Γ at a single vertex. It follows that the Tutte polynomial of each graph is the product of the Tutte polynomial of C_3 and the Tutte polynomial of Γ. Similarly, the critical group of each is a product of the critical group of C_3 and the critical group of the diamond graph Γ.

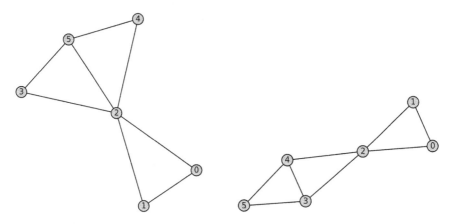

Figure 4.6: Two graphs with the same critical group but distinct Duursma zeta functions.

```
───────────────────────── Sage ─────────────────────────

sage: D1 = {1:{0,2},2:{0,1,3,4,5},5:{3,4}}
sage: Gamma1 = Graph(D1)
sage: Gamma1.show()
sage: D2 = {1:{0,2},2:{0,1,3,4},3:{4,5},4:{5}}
sage: Gamma2 = Graph(D2)
sage: Gamma1.tutte_polynomial()
x^5 + 3*x^4 + 3*x^3*y + x^2*y^2 + 3*x^3 + 5*x^2*y + 3*x*y^2 + y^3 +
x^2 + 2*x*y + y^2
sage: Gamma2.tutte_polynomial()
x^5 + 3*x^4 + 3*x^3*y + x^2*y^2 + 3*x^3 + 5*x^2*y + 3*x*y^2 + y^3 +
x^2 + 2*x*y + y^2
```

4.6.4 Energy pairing on the critical group

We can now define an energy pairing $\langle \cdot, \cdot \rangle$ on the critical group, with values
in $\mathbb{Q}$ mod $\mathbb{Z}$. We will show that this pairing is non-degenerate, i.e., if S is an
element of the critical group, then $\langle S, T \rangle = 0$ for all T in the critical group if
and only if $S = 0$ (i.e., if and only if S is the identity element of the group).

Let Γ be a connected graph and let Q be its Laplacian matrix. Recall that
if s_1 and s_2 are degree 0 configurations on Γ and Q^+ is the Moore–Penrose
pseudoinverse of Q, the energy pairing of s_1 and s_2 is given by

$$\langle s_1, s_2 \rangle = s_1^t Q^+ s_2.$$

Recall also, from Proposition 4.5.2, that if s_1, s_1', s_2, and s_2' are degree 0 configurations on Γ such that s_1 is equivalent to s_1' and s_2 is equivalent to s_1', i.e., such that there exist integer n-vectors x_1 and x_2 such that

$$s_1' = s_1 - Qx_1 \qquad \text{and} \qquad s_2' = s_2 - Qx_2,$$

then

$$\langle s_1', s_2' \rangle - \langle s_1, s_2 \rangle \in \mathbb{Z}.$$

It follows that the energy pairing on degree 0 configurations determines a well-defined pairing on the critical group Γ, by taking values in $\mathbb{Q}$ mod $\mathbb{Z}$ as follows:

Definition 4.6.6. Let S_1 and S_2 be elements of the critical group $\mathcal{K}(\Gamma)$ of Γ with representatives s_1 and s_2. The energy pairing of S_1 and S_2 is given by

$$\langle S_1, S_2 \rangle = \langle s_1, s_2 \rangle \qquad \text{mod } \mathbb{Z}.$$

Exercise 4.13. Show that the energy pairing on $\mathcal{K}(\Gamma)$ is symmetric, i.e., if S_1 and S_2 are elements of the critical group, then $\langle S_1, S_2 \rangle = \langle S_2, S_1 \rangle$.

Exercise 4.14. Show that the energy pairing on $\mathcal{K}(\Gamma)$ is bilinear i.e., if S_1, S_2, and S_3 are elements of the critical group, then $\langle S_1 + S_2, S_3 \rangle = \langle S_1, S_3 \rangle + \langle S_2, S_3 \rangle$ and $\langle S_1, S_2 + S_3 \rangle = \langle S_1, S_2 \rangle + \langle S_1, S_3 \rangle$.

Proposition 4.6.7. *The energy pairing on $\mathcal{K}(\Gamma)$ is non-degenerate, i.e., an element S of the critical group satisfies the condition*

$$\langle S, T \rangle = 0 \qquad mod \ \mathbb{Z} \qquad for \ all \ T \in \mathcal{K}(\Gamma) \tag{4.7}$$

if and only if S is the identity element of $\mathcal{K}(\Gamma)$, i.e., if and only if every representative of S is of the form Qx for some integer vector x.

Proof. Suppose that S is the identity element of $\mathcal{K}(\Gamma)$ and s is a representative of S, so that $s = Qx$ for some integer vector x. Let T be any element of $\mathcal{K}(\Gamma)$ and let t be a degree 0 configuration which is a representative of T. Then

$$\begin{aligned}
\langle s, t \rangle &= (Qx)^t Q^+ t \\
&= x^t Q^t Q^+ t \\
&= x^t Q Q^+ t \\
&= x^t t \\
&\in \mathbb{Z}
\end{aligned}$$

so $\langle S, T \rangle = 0$ mod $\mathbb{Z}$.

Next suppose that S satisfies the condition 4.7. Let s be any degree 0 configuration in S. Let $\mathbf{1} = (1, 1, \ldots, 1)$. We will show that the entries of $u = Q^+s$ all differ by integers, so that if $u_1 = (Q^+s)_1$ is the first entry of u, then the vector $x = u - u_1\mathbf{1}$ is an integer vector. We then show that $Qx = s$.

For $2 \leq i \leq n$, let t_i be the degree 0 configuration given by

$$t_i(v_j) = \begin{cases} -1, & \text{if } j = 1, \\ 1, & \text{if } j = i, \\ 0, & \text{otherwise,} \end{cases}$$

and let $m_i = \langle s, t_i \rangle$. Then m_i is an integer, for $2 \leq i \leq n$, by our assumption. We have

$$\begin{aligned} m_i &= \langle s, t_i \rangle \\ &= s^t Q^+ t_i \\ &= s^t (Q^+)^t t_i \\ &= (Q^+ s)^t t_i \\ &= u^t t_i \\ &= -u_1 + u_i. \end{aligned}$$

Thus the vector $x = u - u_1\mathbf{1} = (0, m_2, m_3, \ldots, m_n)$ is an integer vector. We now note that

$$\begin{aligned} Qx &= Q(u - u_1\mathbf{1}) \\ &= Q(Q^+ s - u_1\mathbf{1}) \\ &= QQ^+ s && \text{since } Q\mathbf{1} = 0 \\ &= s && \text{by Corollary 2.3.8,} \end{aligned}$$

so that S is the identity element of $\mathcal{K}(\Gamma)$. □

The energy pairing of the critical group of the diamond graph is given in Example 4.10.2.

4.7 Examples of critical groups

We find the critical groups of trees, cycle graphs, and complete graphs. We use computer calculations to find the critical groups of some Cayley graphs. We also include some interesting examples due to Biggs and to Clancy, Leake, and Payne.

4.7.1 Trees

We begin with a simple lemma. Recall that a tree is a connected graph having no cycles.

Lemma 4.7.1. *The critical group does not change (up to isomorphism) if an edge with a vertex of degree one is removed from a graph. Consequently, if T is a tree then the critical group of T is trivial.*

Proof. A vertex $v \neq q$ of degree one is stable for a configuration s if and only if $s(v) = 0$. Also note that the critical group does not change (up to isomorphism) if a different vertex is selected as the sink, or government. Thus, if q is a vertex of degree one, we may choose another vertex q' as the sink or government, and then remove the edge adjacent to q. $\square$

Example 4.7.2. If Γ is a path graph on n vertices, then the critical group of Γ is trivial. This follows from Lemma 4.7.1.

4.7.2 Cycle graphs

The cyclic group provides an example which shows that the map taking a critical configuration to its complement (a reduced configuration) is not a homomorphism on the corresponding groups. See also Remark 4.3.32.

Example 4.7.3. The critical of a cycle graph C_n on n vertices, for $n \geq 3$, is a cyclic group of order n, i.e., $\mathcal{K}(C_n)$ is isomorphic to $\mathbb{Z}/n\mathbb{Z}$.

Let $q = v_1, v_2, \ldots, v_n$ be the vertices of C_n, with edges connecting v_i and v_{i+1}, for $1 \leq i \leq n-1$, and v_n and v_1. The configuration $s = (-1, 1, 0, \ldots, 0)$ is q-reduced and its equivalence class is a generator of $\mathcal{K}(C_n)$.

Remark 4.7.4. The example C_4 shows that although the map $s \mapsto s^*$ taking a configuration to its complement is a bijection from the set of q-reduced configuration to the set of q-critical configurations, the induced map on the critical group is not necessarily a homomorphism. The q-reduced configurations on C_4 are

$$\begin{aligned} s_1 &= (0,0,0,0) \\ s_2 &= (-1,1,0,0) \\ s_3 &= (-1,0,1,0) \\ s_4 &= (-1,0,0,1) \end{aligned}$$

and the corresponding q-critical configurations are

$$\begin{aligned} s_1^* &= (-3,1,1,1) \\ s_2^* &= (-2,0,1,1) \\ s_3^* &= (-2,1,0,1) \\ s_4^* &= (-2,1,1,0). \end{aligned}$$

Note that s_1 is a q-reduced representative of the identity element of $\mathcal{K}(C_4)$ and s_3^* is a q-critical representative of the identity element of $\mathcal{K}(C_4)$. Thus, the operation of taking the complement of a configuration does not induce a homomorphism from the critical group to itself.

4.7.3 Complete graphs

There are many ways to derive the structure of the critical group of the complete graph K_n on n vertices. We will use the Smith normal form of the reduced Laplacian matrix. The familiar formula n^{n-2} for the number of spanning trees of the complete graph on n vertices is a consequence of the following proposition.

Proposition 4.7.5. *The critical group of the complete graph K_n on n vertices is isomorphic to the direct product of $n-2$ copies of $\mathbb{Z}/n\mathbb{Z}$.*

Proof. The reduced Laplacian matrix Q^* of K_n is an $n-1$ by $n-1$ matrix with every diagonal entry equal to $n-1$ and every non-diagonal entry equal to -1:

$$
Q^* = \begin{pmatrix}
n-1 & -1 & -1 & \cdots & -1 \\
-1 & n-1 & -1 & \cdots & -1 \\
-1 & -1 & n-1 & \cdots & -1 \\
\vdots & \vdots & \vdots & & \vdots \\
-1 & -1 & -1 & \cdots & n-1
\end{pmatrix}.
$$

It straightforward to check that $U_1 \Sigma = Q^* U_2$, where

$$
U_1 = \begin{pmatrix}
n-1 & -1 & -1 & \cdots & -1 \\
-1 & 1 & 0 & \cdots & 0 \\
-1 & 0 & 1 & \cdots & 0 \\
\vdots & \vdots & \vdots & & \vdots \\
-1 & 0 & 0 & \cdots & 1
\end{pmatrix},
$$

i.e., U_1 is the matrix obtained from the $n \times n$ identity matrix by replacing the $(1,1)$ entry by $n-1$ and replacing all other entries of the first row and column by -1,

$$
\Sigma = \begin{pmatrix}
1 & 0 & 0 & \cdots & 0 \\
0 & n & 0 & \cdots & 0 \\
0 & 0 & n & \cdots & 0 \\
\vdots & \vdots & \vdots & & \vdots \\
0 & 0 & 0 & \cdots & n
\end{pmatrix},
$$

and

$$U_2 = \begin{pmatrix} 1 & -1 & -1 & \cdots & -1 \\ 0 & 1 & 0 & \cdots & 0 \\ 0 & 0 & 1 & \cdots & 0 \\ \vdots & \vdots & \vdots & & \vdots \\ 0 & 0 & 0 & \cdots & 1 \end{pmatrix}.$$

The determinant of U_2 is clearly 1. To find the determinant of U_1, we let U_1' be the matrix obtained from U_1 by replacing the first row of U_1 by the sum of all rows of U_1, i.e.,

$$U_1' = \begin{pmatrix} 1 & 0 & 0 & \cdots & 0 \\ -1 & 1 & 0 & \cdots & 0 \\ -1 & 0 & 1 & \cdots & 0 \\ \vdots & \vdots & \vdots & & \vdots \\ -1 & 0 & 0 & \cdots & 1 \end{pmatrix}.$$

Then $\det(U_1) = \det(U_1') = 1$. It follows that Σ is a Smith normal form of Q^*. Thus, the invariant factors of the critical group $\mathcal{K}(K_n)$ are n repeated $n - 2$ times, so $\mathcal{K}(K_n) = (\mathbb{Z}/n\mathbb{Z})^{n-2}$.   □

4.7.4 Wheel graphs

Recall that the *Fibonacci numbers* f_n are given by $f_0 = 1$, $f_1 = 1$, and $f_n = f_{n-1} + f_{n-2}$ for $n > 1$. The *Lucas numbers* ℓ_n are given by $\ell_0 = 2$, $\ell_1 = 1$, and $\ell_n = \ell_{n-1} + \ell_{n-2}$ for $n > 1$.

Example 4.7.6. Biggs [Bi99] shows that the critical group of the wheel graph W_n on n vertices has order ℓ_{n-1}^2 if n is even and $5f_{n-2}^2$ if n is odd, where ℓ_n is the n-th Lucas number and f_n is the n-th Fibonacci number. The wheel graph will be discussed further in §4.10.

4.7.5 Example of Clancy, Leake, and Payne

Example 4.7.7. The graphs Γ_1, Γ_2 in Figure 4.7 are found in Clancy, Leake, and Payne [CLP15]. They have the same Tutte polynomial but distinct critical groups:

$$\mathcal{K}(\Gamma_1) = \mathbb{Z}/5\mathbb{Z} \times \mathbb{Z}/25\mathbb{Z},$$

$$Jac(\Gamma_2) = \mathbb{Z}/125\mathbb{Z}.$$

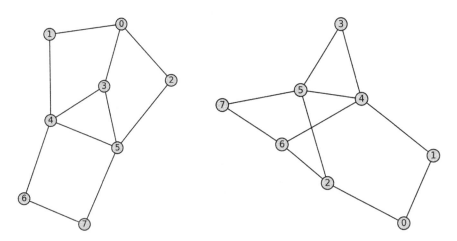

Figure 4.7:   Two graphs with the same Tutte polynomial but distinct critical groups.

```
──────────────────────── Sage ────────────────────────
sage: D3 = {0:{1,2,3},4:{1,3,5,6},5:{2,3,7},6:{7}}
sage: Gamma3 = Graph(D3)
sage: Gamma3.show(dpi=300)
sage: D4 = {0:{1,2},4:{1,3,5,6},5:{2,3,7},6:{2,7}}
sage: Gamma4 = Graph(D4)
sage: Gamma4.show(dpi=300)
sage: Gamma3.tutte_polynomial()
x^7 + 4*x^6 + x^5*y + 9*x^5 + 6*x^4*y + 3*x^3*y^2 + x^2*y^3 + 13*x^4 +
13*x^3*y + 7*x^2*y^2 + 3*x*y^3 + y^4 + 12*x^3 + 15*x^2*y + 9*x*y^2 +
3^y'3 + 7^x'2 + 9^x^y + 4^y'2 + 2^x + 2^y
sage: Gamma4.tutte_polynomial()
x^7 + 4*x^6 + x^5*y + 9*x^5 + 6*x^4*y + 3*x^3*y^2 + x^2*y^3 + 13*x^4 +
13*x^3*y + 7*x^2*y^2 + 3*x*y^3 + y^4 + 12*x^3 + 15*x^2*y + 9*x*y^2 +
3*y^3 + 7*x^2 + 9*x*y + 4*y^2 + 2*x + 2*y
%sage: critical_group(Gamma3)
%Finitely generated module V/W over Integer Ring with invariants (5, 25)
%sage: critical_group(Gamma4)
%Finitely generated module V/W over Integer Ring with invariants (125)
```

4.7.6 Some Cayley graphs

Cayley graphs were introduced in §2.6 and will be discussed more in Chapter 6.

Cayley graphs of groups

The Cayley graph of a group has a lot of symmetries. This does not mean that the critical group is large. In fact, by Lemma 4.7.1, if we add a vertex and an edge to a graph (so the vertex has degree 1), the critical group remains the same, while the automorphism group typically decreases.

Example 4.7.8. Consider the symmetric group on $\{1, 2, 3\}$, G generated by $S = \{(1, 2, 3), (1, 2), (1, 3, 2)\}$. The Cayley graph Γ_2 of (G, S), depicted in Figure 4.8, is 3-regular and has critical group

$$Jac(\Gamma_2) = \mathbb{Z}/5\mathbb{Z} \times \mathbb{Z}/15\mathbb{Z} \cong (\mathbb{Z}/5\mathbb{Z})^2 \times \mathbb{Z}/3\mathbb{Z}.$$

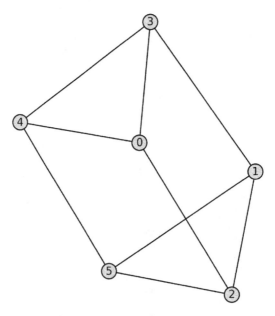

Figure 4.8: The Cayley graph of $(G, S) = (S_3, \{(1, 2, 3), (1, 2), (1, 3, 2)\})$.

The labeling of the vertices in Figure 4.8 is given as follows:

$$0 : 1, 1 : (1, 2), 2 : (1, 2, 3), 3 : (1, 3, 2), 4 : (2, 3), 5 : (1, 3).$$

The subgroup G_0 of the automorphism group of Γ_2 given by $G_0 = \langle (0, 2)(1, 3)(4, 5) \rangle$ has quotient $\Gamma_1 = \Gamma_2/G_0 \cong C_3$, where C_3 denotes the cycle graph on 3 letters. The critical group of Γ_1 is $\mathbb{Z}/3\mathbb{Z}$: $Jac(\Gamma_1) \cong \mathbb{Z}/3\mathbb{Z}$.

The quotient map $\phi : \Gamma_2 \to \Gamma_1$ is harmonic. Propositions 4.8.2 and 4.8.4 below predict that there is an injection $Jac(\Gamma_1) \to Jac(\Gamma_2)$ and a surjection $Jac(\Gamma_2) \to Jac(\Gamma_1)$. Both of these facts are easy to verify directly.

Note that if $s = (-3, 1, 1, 1, 0, 0)$ is a configuration on Γ_2 then the quotient map $\phi : \Gamma_2 \to \Gamma_1$ induces, via the pushforward (4.1), the configuration $\phi(s) = (-2, 2, 0)$. Note that s is stable but $\phi(s)$ is not.

Example 4.7.9. As in Example 4.7.8, we consider the symmetric group on $\{1, 2, 3\}$, G, but with a different set of generators: $S = \{(1, 2), (2, 3), (1, 3)\}$. The Cayley graph Γ_2 of (G, S), depicted in Figure 4.9, is 3-regular and has critical group
$$Jac(\Gamma_2) = (\mathbb{Z}/3\mathbb{Z})^2 \times \mathbb{Z}/9\mathbb{Z}.$$

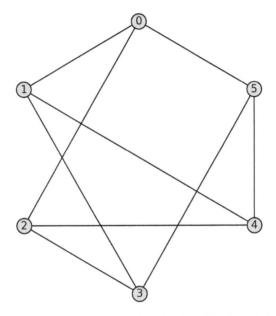

Figure 4.9: The Cayley graph of $(G, S) = (S_3, \{(1, 2), (2, 3), (1, 3)\})$.

The labeling of the vertices in Figure 4.9 is given as follows:

$$0 : 1, 1 : (1, 2), 2 : (1, 2, 3), 3 : (1, 3, 2), 4 : (2, 3), 5 : (1, 3).$$

The subgroup G_0 of the automorphism group of Γ_2 given by $G_0 = \langle (1, 2)(3, 4) \rangle$ has quotient $\Gamma_1 = \Gamma_2/G_0 \cong C_4$, where C_4 denotes the cycle graph on 4 letters. The critical group of Γ_1 is $\mathbb{Z}/4\mathbb{Z}$: $Jac(\Gamma_1) \cong \mathbb{Z}/4\mathbb{Z}$.

The quotient map
$$\phi : \Gamma_2 \to \Gamma_1,$$

$$0 \mapsto 2, \{1, 2\} \mapsto 0, \{3, 4\} \mapsto 3, 5 \mapsto 1,$$

is not harmonic. It is easy to see directly that there is no injection $Jac(\Gamma_1) \to Jac(\Gamma_2)$, nor is there a surjection $Jac(\Gamma_2) \to Jac(\Gamma_1)$.

Cayley graphs of p-ary functions

We consider the Boolean case first.

Example 4.7.10. Consider the function $f : GF(2)^3 \to GF(2)$ defined by $f(x_0, x_1, x_2) = x_0 x_1 + x_2^2$. The Cayley graph Γ_2 of f, depicted in Figure 4.10, is 4-regular and has critical group

$$Jac(\Gamma_2) = \mathbb{Z}/6\mathbb{Z} \times (\mathbb{Z}/24\mathbb{Z})^2 \cong \mathbb{Z}/2\mathbb{Z} \times (\mathbb{Z}/3\mathbb{Z})^3 \times (\mathbb{Z}/8\mathbb{Z})^2.$$

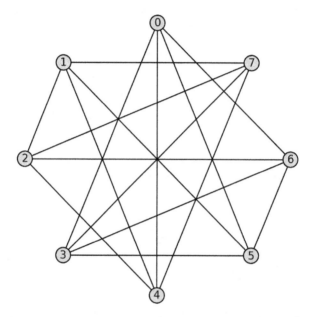

Figure 4.10: The Cayley graph of $f(x_0, x_1, x_2) = x_0 x_1 + x_2^2$.

The labeling of the vertices of Γ_2 in Figure 4.10 is given as follows:

$$0 : (0,0,0), 1 : (1,0,0), 2 : (0,1,0), 3 : (1,1,0),$$

$$4 : (0,0,1), 5 : (1,0,1), 6 : (0,1,1), 7 : (1,1,1).$$

The automorphism group of Γ_2 has order 48 and contains the element $(2,7)(3,6)$. It happens that the quotient graph $\Gamma_1 = \Gamma_2/G_0$, where $G_0 = \langle (2,7)(3,6) \rangle$, under this action is isomorphic to the graph in Figure 4.8. Therefore, Γ_1 has critical group

$$Jac(\Gamma_1) = \mathbb{Z}/5\mathbb{Z} \times \mathbb{Z}/15\mathbb{Z} \cong \mathbb{Z}/3\mathbb{Z} \times (\mathbb{Z}/5\mathbb{Z})^2.$$

The quotient map $\Gamma_2 \to \Gamma_1$ is not harmonic. It is easy to see directly that there is no injection $Jac(\Gamma_1) \to Jac(\Gamma_2)$, nor is there a surjection $Jac(\Gamma_2) \to Jac(\Gamma_1)$.

Example 4.7.11. Here we use the notation $GF(5) = \mathbb{Z}/5\mathbb{Z}$. Consider the following bent functions from $GF(5)^2$ to $GF(5)$ which represent six different equivalence classes of even bent functions f such that $f(0) = 0$ under non-degenerate linear transformations of the coordinates:

$$f_1(x_1, x_2) = x_1^2 + 3x_2^2$$
$$f_2(x_1, x_2) = x_1^4 + 4x_1^2 + 4x_1 x_2$$
$$f_3(x_1, x_2) = x_1^4 + 2x_1^3 x_2 + 4x_1^2$$
$$f_4(x_1, x_2) = x_1^2 + 4x_1 x_2$$
$$f_5(x_1, x_2) = x_1^4 + 3x_1 x_2^3$$
$$f_6(x_1, x_2) = x_1^4 + 4x_1^3 x_2.$$

We let Γ_i be the "unweighted" Cayley graph whose vertices are the 25 elements of $GF(5)^2$ and whose edges are pairs $\{v, w\}$ such that $v, w \in GF(5)^2$ and $f_i(w - v) \neq 0$. By checking that the rank of the Laplacian of each graph is 24, we see that the graphs are all connected graphs.

The graph Γ_1 is a complete graph on 25 vertices, therefore the results of §4.7.3 imply that the critical group $Jac(\Gamma_1)$ is the direct product of $\mathbb{Z}/25\mathbb{Z}$ with itself 23 times.

Computer calculations using the Smith normal form of the Laplacian matrices of the graphs Γ_i were used to find the invariant factors of the remaining five graphs. The invariant factors of Γ_2 and Γ_3 are both

$$\{2, 2, 2, 2, 10, 10, 50, 50, 201300, 201300, 201300, 201300\}.$$

The invariant factors of Γ_4, Γ_5, and Γ_6 are all

$$\{15, 15, 15, 15, 15, 15, 15, 15, 60, 60, 300, 300, 300, 300, 300, 300\}.$$

Example 4.7.12. Here we use the notation $GF(3) = \mathbb{Z}/3\mathbb{Z}$. Consider the following bent functions from $GF(3)^3$ to $GF(3)$ which represent two different equivalence classes of even bent functions f such that $f(0) = 0$ under non-degenerate linear transformations of the coordinates:

$$f_1(x_1, x_2, x_3) = x_1^2 + x_2^2 + x_3^2$$
$$f_2(x_1, x_2, x_3) = x_1 x_3 + 2x_2^2 + 2x_1^2 x_2^2.$$

We let G_i be the "unweighted Cayley graph" whose vertices are the 27 elements of $GF(3)^3$ and whose edges are pairs $\{v, w\}$ such that $v, w \in GF(3)^3$ and $f_i(w - v) \neq 0$. By checking that the rank of the Laplacian of each graph is 26, we see that the graphs are both connected graphs.

Computer calculations using the Smith normal form of the Laplacian matrices of the graphs Γ_i were used to find the invariant factors of the two graphs. The invariant factors of both graphs are

$$\{21, 21, 21, 63, 126, 378, 1890, 1890, 1890, 5670, 5670, 5670\}.$$

4.7.7 Graphs with cyclic critical groups

Biggs has constructed a family of modified wheel graphs with cyclic critical groups.

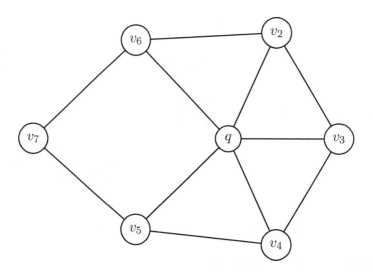

Figure 4.11: A graph with a cyclic critical group of order 176.

Biggs' family of modified wheel graphs with cyclic critical groups will be discussed further in §4.10. That section discusses Shokrieh's cryptanalytic break of Biggs' proposed cryptosystem based on critical groups.

Example 4.7.13. Biggs' construction starts with a wheel graph with an odd number $2n + 1$ of spoke vertices (and an even total number of vertices

$2n + 2$). One vertex is added between any two of the spoke vertices. This new vertex will be of degree two. The resulting graph has a critical group which is cyclic of order $2\ell_{2n+1}f_{2n+2}$, where ℓ_i is the i-th Lucas number and f_i the i-th Fibonacci number.

For example, the graph shown in Figure 4.11 has a cyclic critical group of order 176. Further details may be found in Biggs [Bi07].

4.8 Harmonic morphisms and Jacobians

Suppose that Γ_1 and Γ_2 are connected graphs and Γ_1 has at least one edge. Let $\phi : \Gamma_2 \to \Gamma_1$ be a non-constant harmonic morphism. We will use the terminology common to the literature on harmonic morphisms of graphs and refer to critical groups as Jacobians, and degree 0 configurations as degree 0 divisors. We will show that the map ϕ induces a surjective pushforward homomorphism

$$\phi_* : \mathrm{Jac}(\Gamma_2) \to \mathrm{Jac}(\Gamma_1)$$

and an injective pullback homomorphism

$$\phi^* : \mathrm{Jac}(\Gamma_1) \to \mathrm{Jac}(\Gamma_2)$$

on Jacobians. These results are due to Baker and Norine [BN09], but our proof of injectivity is somewhat different, as it uses the energy pairing on the critical group.

Exercise 4.15. Show if $\phi : \Gamma_2 \to \Gamma_1$ is harmonic then the size of the critical group of Γ_1 divides the size of the critical group of Γ_2.

Before proving the injectivity of the map on Jacobians induced by pullbacks, we prove the following adjoint property of pushforward and pullback maps.

Lemma 4.8.1. *Suppose that Γ_1 and Γ_2 are connected graphs and Γ_1 has at least one edge. Let $\phi : \Gamma_2 \to \Gamma_1$ be a non-constant harmonic morphism. Let D be a degree 0 divisor on Γ_2 and let D' be a degree 0 divisor on Γ_1. Then*

$$\langle \phi^*(D'), D \rangle = \langle D', \phi_*(D) \rangle.$$

Proof. Let a be the coefficient vector of D and let b be the coefficient vector of D'. Then

$$
\begin{aligned}
\langle \phi^*(D'), D \rangle &= (\Phi_{mV} b)^t Q_2^+ a \\
&= b^t \Phi_{mV}^t Q_2^+ a \\
&= b^t Q_1^+ Q_1 \Phi_{mV}^t Q_2^+ a \qquad \text{by Corollary 2.3.8} \\
&= b^t Q_1^+ \Phi_V^t Q_2 Q_2^+ a \qquad \text{by Proposition 3.3.25} \\
&= b^t Q_1^+ \Phi_V^t a \qquad\qquad \text{by Corollary 2.3.8} \\
&= \langle D', \phi_*(D) \rangle.
\end{aligned}
$$

$\square$

We now use this adjoint property to prove injectivity of the map on Jacobians induced by the pullback map of divisors.

Proposition 4.8.2. *Suppose that* Γ_1 *and* Γ_2 *are connected graphs and* Γ_1 *has at least one edge. If* $\phi : \Gamma_2 \to \Gamma_1$ *is a non-constant harmonic morphism, then* ϕ^* *determines an injective map*

$$
\phi^* : Jac(\Gamma_1) \to Jac(\Gamma_2).
$$

Proof. First, we check that if D' is a principal divisor on Γ_1, then $\phi^*(D')$ is a principal divisor on Γ_2. Suppose that the coefficient vector b of D' satisfies $b = Q_1 y$ for some integer n_1-vector y. The coefficient vector of $\phi^*(D')$ is

$$
a = \Phi_{mV} Q_1 y = Q_2 \Phi_V y,
$$

where the second equality follows from Proposition 3.3.25. Thus $\phi^*(D')$ is principal on Γ_2.

Next we wish to show that if $\phi^*(D')$ is principal on Γ_2, then D' is principal on Γ_1. Suppose that D' is a degree 0 divisor on Γ_1 such that $\phi^*(D')$ is principal on Γ_2. Then $\langle \phi^*(D'), D \rangle = 0 \bmod \mathbb{Z}$, for all degree 0 divisors D on Γ_2. By Lemma 4.8.1, it follows that

$$
\langle D', \phi_*(D) \rangle = 0 \bmod \mathbb{Z}
$$

for all degree 0 divisors D on Γ_2. Recall that the map ϕ_* is surjective, so every degree 0 divisor D'' on Γ_1 is of the form $\phi_*(D)$ for some degree 0 divisor D on Γ_2. Thus $\langle D', D'' \rangle = 0 \bmod \mathbb{Z}$, for all degree 0 divisors D'' on Γ_1. Thus D' is principal. It follows that the map ϕ^* is injective. $\square$

Example 4.8.3. In this example, we show that the morphism induced by a quotient, $\Gamma \to \Gamma/G$ where $G \subset Aut(\Gamma)$, need not be harmonic. In this situation, there need not be any relationship between the critical group of Γ and that of Γ/G.

Denote by Γ_2 the Paley graph on 13 vertices, depicted in Figure 5.16 (see §5.17 below). The automorphism group of Γ_2 has order 78 and its 2-Sylow subgroup G acts via the element $(1, 12)(2, 11)(3, 10)(4, 9)(5, 8)(6, 7)$.

The quotient $\Gamma_1 = \Gamma_2/G$ is depicted in Figure 4.12. The 7 vertices on Γ_1 are G-orbits of the 13 vertices on Γ_2, as follows:

$$0 : \{1, 12\}, \quad 1 : \{5, 8\}, \quad 2 : \{3, 10\}, \quad 3 : \{0\}, \quad 4 : \{2, 11\}, \quad 5 : \{6, 7\}, \quad 6 : \{4, 9\}.$$

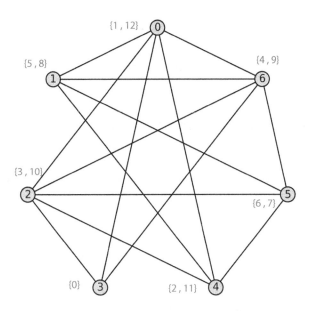

Figure 4.12:   The quotient graph Γ_2/G.

Computations using Sage show that:

- The morphism $\phi : \Gamma_2 \to \Gamma_1$ is not harmonic.

- The critical group of Γ_1 is $\mathbb{Z}/29\mathbb{Z} \times \mathbb{Z}/58\mathbb{Z}$.

- The critical group of Γ_2 is $\mathbb{Z}/3\mathbb{Z} \times (\mathbb{Z}/39\mathbb{Z})^5$.

Exercise 4.16. Suppose that $\phi : C_n \to C_m$ is a non-constant map of cycle graphs. Show that if v is a non-special divisor on C_m, then $\phi^*(v)$ is a non-special divisor on C_n.

Recall the pushforward map from Definition 3.6.1. Next, we extend the surjectivity results of Lemma 3.6.3 and Lemma 3.3.30 to Jacobians. The result below asserts the surjectivity of the induced pushforward map on Jacobians.

Proposition 4.8.4. *Suppose that Γ_1 and Γ_2 are connected graphs and Γ_1 has at least one edge. If $\phi : \Gamma_2 \to \Gamma_1$ is a non-constant harmonic morphism, then ϕ_* determines a surjective homomorphism*

$$\phi_* : Jac(\Gamma_2) \to Jac(\Gamma_1).$$

Proof. To prove that ϕ_* determines a well-defined map on equivalence classes, we must prove that if D is a principal divisor on Γ_2, then $\phi_*(D)$ is a principal divisor on Γ_1.

Recall that D is principal if there exists an n_2-vector x with integer entries such that the coefficient vector of D is $Q_2 x$. In this case, the coefficient vector of $\phi_*(D)$ is

$$\Phi_V^t Q_2 x = Q_1 \Phi_{mV}^t x$$

which is a principal divisor on Γ_1.

The surjectivity follows from the surjectivity of ϕ_* on divisors.   □

4.8.1 Example: morphism from cube to K_4

Let K_4 be the complete graph on 4 vertices and let Γ be the cube graph.

Let τ be the morphism taking opposite vertices in Γ to vertices in K_4. We will label the vertices of K_4 as v_1, v_2, v_3, and v_4. We will label the vertices of Γ as w_i, for $i = 1, 2, \ldots, 8$, where τ maps w_i to $v_{i \pmod 4}$, i.e., $\tau(w_1) = \tau(w_5) = v_1$, $\tau(w_2) = \tau(w_6) = v_2$, etc.

We will often denote the divisor

$$D = \sum_{i=1}^{4} a_i v_i$$

on K_4 as (a_1, a_2, a_3, a_4) and use a similar convention for divisors on Γ.

Let Q_{K_4} be the Laplacian matrix of K_4 with respect to v_1, v_2, v_3, and v_4.

There is an ordering of the vertices of the cube satisfying the condition above, such that the Laplacian matrix of Γ is

$$Q_\Gamma = \begin{pmatrix} 3 & -1 & 0 & -1 & 0 & 0 & -1 & 0 \\ -1 & 3 & -1 & 0 & 0 & 0 & 0 & -1 \\ 0 & -1 & 3 & -1 & -1 & 0 & 0 & 0 \\ -1 & 0 & -1 & 3 & 0 & -1 & 0 & 0 \\ 0 & 0 & -1 & 0 & 3 & -1 & 0 & -1 \\ 0 & 0 & 0 & -1 & -1 & 3 & -1 & 0 \\ -1 & 0 & 0 & 0 & 0 & -1 & 3 & -1 \\ 0 & -1 & 0 & 0 & -1 & 0 & -1 & 3 \end{pmatrix}.$$

We will use v_1 as the sink vertex for K_4 and w_1 for the sink vertex for Γ.

We calculate the Smith normal form of the Laplacian matrices and use them to find the critical groups.

There are integer matrices Σ_{K_4}, U_1, and U_2 such that

$$U_1 \Sigma_{K_4} = Q_{K_4} U_2, \tag{4.8}$$

where Σ_{K_4} is the Smith normal form of Q_{K_4} and the matrices U_1 and U_2 are invertible. Computer calculations show that

$$\Sigma_{K_4} = \begin{pmatrix} 1 & 0 & 0 & 0 \\ 0 & 4 & 0 & 0 \\ 0 & 0 & 4 & 0 \\ 0 & 0 & 0 & 0 \end{pmatrix},$$

and that the following matrices have determinant 1 and satisfy Equation (4.8):

$$U_1 = \begin{pmatrix} -1 & 0 & 0 & 0 \\ 3 & -1 & 0 & 0 \\ -1 & 1 & 1 & 0 \\ -1 & 0 & -1 & 1 \end{pmatrix} \quad \text{and} \quad U_2 = \begin{pmatrix} 0 & 0 & 1 & 1 \\ 1 & -1 & 1 & 1 \\ 0 & 1 & 2 & 1 \\ 0 & 0 & 0 & 1 \end{pmatrix}.$$

From the diagonal elements of the Smith normal form of Q_{K_4}, we can see that the critical group $\mathcal{K}(K_4)$ of K_4 has order 16 and is isomorphic to

$$\mathbb{Z}/(4\mathbb{Z}) \times \mathbb{Z}/(4\mathbb{Z}).$$

Furthermore, the two factor groups are generated by the equivalence classes of the second and third columns of U_1 (corresponding to the invariant factors 4 and 4 of the Smith normal form). These columns are $(0, -1, 1, 0)$ and $(0, 0, 1, -1)$. The reduced divisors of these equivalence classes are

$$u_1 = (-3, 0, 2, 1) \quad \text{and} \quad u_2 = (-3, 1, 2, 0).$$

Thus, we can describe any element of the critical group by a pair (i, j), with $0 \leq i \leq 3$ and $0 \leq j \leq 3$, corresponding to the equivalence class of the divisor $iu_1 + ju_2$. After finding generators of the critical group of Γ, we will find a different pair of generators of the critical group of K_4, with respect to which the mapping of Jacobians takes a simple form.

Similarly, there are integer matrices Σ_Γ, V_1, and V_2 such that

$$V_1 \Sigma_\Gamma = Q_\Gamma V_2, \tag{4.9}$$

where Σ_Γ is the Smith normal form of Q_Γ and the matrices V_1 and V_2 are invertible. Computer calculations show that

$$\Sigma_\Gamma = \begin{pmatrix} 1 & 0 & 0 & 0 & 0 & 0 & 0 & 0 \\ 0 & 1 & 0 & 0 & 0 & 0 & 0 & 0 \\ 0 & 0 & 1 & 0 & 0 & 0 & 0 & 0 \\ 0 & 0 & 0 & 1 & 0 & 0 & 0 & 0 \\ 0 & 0 & 0 & 0 & 2 & 0 & 0 & 0 \\ 0 & 0 & 0 & 0 & 0 & 8 & 0 & 0 \\ 0 & 0 & 0 & 0 & 0 & 0 & 24 & 0 \\ 0 & 0 & 0 & 0 & 0 & 0 & 0 & 0 \end{pmatrix},$$

and that the following matrices have determinant 1 and satisfy Equation (4.9):

$$V_1 = \begin{pmatrix} -1 & 0 & 0 & 0 & 0 & 0 & 0 & 0 \\ 3 & -1 & 0 & 0 & 0 & 0 & 0 & 0 \\ -1 & 3 & -1 & 0 & 0 & 0 & 0 & 0 \\ 0 & -1 & 0 & -1 & 0 & 0 & 0 & 0 \\ 0 & -1 & 3 & -1 & -5 & 0 & -1 & 0 \\ 0 & 0 & -1 & 3 & 3 & 1 & 0 & 0 \\ 0 & 0 & 0 & -1 & -1 & 0 & 0 & 0 \\ -1 & 0 & -1 & 0 & 3 & -1 & 1 & 1 \end{pmatrix}$$

and

$$V_2 = \begin{pmatrix} 0 & 0 & 0 & 0 & 0 & 0 & 0 & 1 \\ 1 & 0 & 0 & 0 & 0 & -1 & 1 & 1 \\ 0 & 1 & 0 & 0 & -1 & 0 & -3 & 1 \\ 0 & 0 & 0 & 0 & 0 & 1 & -2 & 1 \\ 0 & 0 & 1 & 0 & -3 & 0 & -8 & 1 \\ 0 & 0 & 0 & 1 & 1 & 3 & -3 & 1 \\ 0 & 0 & 0 & 0 & 0 & 0 & 1 & 1 \\ 0 & 0 & 0 & 0 & 1 & -3 & 6 & 1 \end{pmatrix}.$$

From the diagonal elements of the Smith normal form of Q_Γ, we can see that the critical group $\mathcal{K}(\Gamma)$ of Γ has order 384 and is isomorphic to

$$\mathbb{Z}/(2\mathbb{Z}) \times \mathbb{Z}/(8\mathbb{Z}) \times \mathbb{Z}/(24\mathbb{Z}).$$

Furthermore, the three factor groups are generated by the equivalence classes of the fifth, sixth, and seventh columns of V_1 (corresponding to the invariant factors 2, 8, and 24 of the Smith normal form). These columns are $(0,0,0,0,-5,3,-1,3)$, $(0,0,0,0,0,1,0,-1)$, and $(0,0,0,0,-1,0,0,1)$. The reduced divisors of these equivalence classes are

$$y_1 = (-3,0,1,0,0,1,0,1),$$
$$y_2 = (-4,2,1,0,1,0,0,0), \text{ and}$$
$$y_3 = (4,0,1,2,0,1,0,0),$$

respectively. Thus, we can describe any element of the critical group by a triple (i, j, k), with $0 \le i \le 1$, $0 \le j \le 7$, and $0 \le k \le 23$, corresponding to the equivalence class of the divisor $iy_1 + jy_2 + ky_3$. We will find it convenient to use a different set of generators of the critical group of K_4, with respect to which the pullbacks of the generators of the critical group of K_4 take a particularly simple form. We will calculate these pullbacks and find a new set of generators for K_4.

For each divisor D on K_4, there is a pullback divisor $\tau^*(D)$ on Γ. If

$$D = a_1 v_1 + a_2 v_2 + a_3 v_3 + a_4 v_4,$$

then

$$\tau^*(D) = a_1 w_1 + a_2 w_2 + a_3 w_3 + a_4 w_4 + a_1 w_5 + a_2 w_6 + a_3 w_7 + a_4 w_8.$$

Let T be the matrix given by

$$T = \begin{pmatrix} 1 & 0 & 0 & 0 \\ 0 & 1 & 0 & 0 \\ 0 & 0 & 1 & 0 \\ 0 & 0 & 0 & 1 \\ 1 & 0 & 0 & 0 \\ 0 & 1 & 0 & 0 \\ 0 & 0 & 1 & 0 \\ 0 & 0 & 0 & 1 \end{pmatrix}.$$

Note that if the coefficients of a divisor D on K_4 are $a = (a_1, a_2, a_3, a_4)$, then the coefficients of $\tau^*(D)$ are Ta. Note also that $Q_\Gamma T = T Q_{K_4}$.

We now show that the pullback map extends to a map of equivalence classes. We wish to show that if a divisor $D = \sum a_i v_i$ on K_4 is principal, i.e., if $a = Q_{K_4} x$ for some vector x, then $\tau^*(D)$ is principal on Γ, i.e., $Ta = Q_\Gamma y$ for some vector y. Note that

$$Ta = T Q_{K_4} x = Q_\Gamma T x,$$

so let $y = Tx$. Therefore τ^* determines a map on equivalence classes (which we will denote by the same symbol, by abuse of notation)

$$\tau^* : \mathcal{K}(K_4) \to \mathcal{K}(\Gamma).$$

Next let us check that this map is a homomorphism. If $D_1 = \sum a_i v_i$ and $D_2 = \sum b_i v_i$ then

$$\tau^*(D_1 + D_2) = \sum_{i=1}^{4}(a_i + b_i)w_i + \sum_{i=5}^{8}(a_{i-4} + b_{i-4})w_i$$

$$= \sum_{i=1}^{4} a_i w_i + \sum_{i=5}^{8} a_{i-4}w_i + \sum_{i=1}^{4} b_i w_i + \sum_{i=5}^{8} b_{i-4}w_i$$

$$= \tau^*(D_1) + \tau^*(D_2).$$

The pullbacks to Γ of u_1 and u_2 (reduced representatives of a pair of generators of the critical group of K_4), are

$$\tau^*(u_1) = (-3, 0, 2, 1, -3, 0, 2, 1) \qquad \text{and}$$
$$\tau^*(u_2) = (-3, 1, 2, 0, -3, 1, 2, 0).$$

The corresponding reduced divisors on Γ are

$$(-5, 1, 1, 2, 0, 0, 0, 1) \sim 4w_2 + 6w_3$$

and

$$(-5, 2, 1, 1, 0, 1, 0, 0) \sim 2w_2 + 6w_3.$$

Note that u_1 and u_2 are both of order 4. It follows (and can be checked directly, using the fact that w_2 has order 8 and w_3 has order 24) that the divisors $4w_2 + 6w_3$ and $2w_2 + 6w_3$ are also both of order 4. In fact, the pullback map $\tau^* : \mathcal{K}(K_4) \to \mathcal{K}(\Gamma)$ is injective. We can see this directly by noting that $\tau^*(u_1) - \tau^*(u_2) \sim 2w_2$ and $-\tau^*(u_1) + 2\tau^*(u_2) \sim 6w_3$.

Let x_1 be the reduced representative of the equivalence class of $u_1 - u_2$ and let x_2 be the reduced representative of the equivalence class of $-u_1 + 2u_2$. We have

$$x_1 = (-3, 0, 1, 2) \qquad \text{and} \qquad x_2 = (-1, 0, 0, 1)$$

and

$$\tau^*(x_1) \sim 2w_2 \qquad \text{and} \qquad \tau^*(x_2) \sim 6w_3.$$

Note that x_1 and x_2 are also generators for $\mathcal{K}(K_4)$, since $u_1 = 2x_1 + x_2$ and $u_2 = x_1 + x_2$. Thus, we can describe any element of the critical group $\mathcal{K}(K_4)$ by a pair (i, j), with $0 \le i \le 3$ and $0 \le j \le 3$, corresponding to the equivalence class of the divisor $ix_1 + jx_2$. We can describe the map τ^* on equivalence classes in terms of ordered pairs on K_4 and ordered triples on Γ (with respect to the generators x_1 and x_2 for $\mathcal{K}(K_4)$ and y_1, y_2, and y_3 for $\mathcal{K}(\Gamma)$) as

$$\tau^*(i, j) \sim (0, 2i, 6j).$$

4.9 Dhar's burning algorithm

Dhar's algorithm gives an efficient procedure for checking whether or not a configuration (or divisor) on a graph is q-reduced. An ordered version of Dhar's algorithm due to Cori and Le Borgne, using a fixed ordering of the edges of a graph, gives a bijection between spanning trees and q-reduced degree 0 divisors (and hence elements of the critical group). A careful look at this bijection reveals that it can be used to prove a result of Merino [Me97] relating the Tutte polynomial to a partition of the collection of q-reduced degree 0 divisors by weight (total degree off q). Baker and Shokrieh [BS13] is a good source for additional applications of Dhar's algorithm. Compare also to Lemma 4.3.33.

Suppose that we start with a q-legal configuration s. We will construct a finite sequence of sets of marked or "burned" vertices, $V_0 = \{q\} \subset V_1 \subset \cdots \subset V_l$ and a corresponding sequence of sets of marked or "burned" edges $E_0 = \phi \subset E_1 \subset \cdots \subset E_l$. We will let $W_i = V_i^c$ be the complement of V_i, i.e., W_i is the i-th set of unmarked or unburned vertices. Each burning move consists of two steps, the first of which marks or burns edges, and the second of which marks or burns vertices. The first step of the first burning move consists of adding all edges adjacent to $V_0 = \{q\}$ to E_0 to form E_1. The second step consists of identifying all vertices $v \neq q$ for which $\text{outdeg}_{W_0}(v) > s(v)$ and adding them to V_0 to form V_1. Note that, for $v \in W_0$, $\text{outdeg}_{W_0}(v)$ is the number of edges in E_1 adjacent to v. We can think of the vertices added in the first step as the obstacles to being able to perform a set-firing move on W_0.

Similarly, the first step of the i-th burning move consists of adding all edges adjacent to V_{i-1} to E_{i-1} to form E_i. Note that for $v \in W_{i-1}$, $\text{outdeg}_{W_i}(v)$ is the number of edges in E_i adjacent to v. The second step consists of identifying all vertices $v \in W_{i-1}$ for which $\text{outdeg}_{W_{i-1}}(v) > s(v)$ and adding them to V_{i-1} to form V_i. In terms of the burning metaphor, a vertex is burned in the i-th step if the number of burned edges adjacent to it in the i-th step is larger than $s(v)$. The process stops when it is no longer possible to add vertices to V_i, which will be in at most $n-1$ moves.

Lemma 4.9.1. *Let s be a q-legal configuration and let V_f be the final set of vertices produced by the burning moves described above. Then s is q-reduced if and only if $V_f = V$, i.e., if and only if all vertices of the graph are eventually burned.*

Proof. Suppose that $V_f \neq V$ and let W be the complement of V_f. Then $\text{outdeg}_W(v) \leq s(v)$, for all v in W. This means that the set W can be fired q-legally, so s is not q-reduced.

Conversely, suppose that s is not q-reduced. Then there exists a set $W \subset V \setminus \{q\}$ that can be fired q-legally, i.e., such that $\text{outdeg}_W(v) \leq s(v)$, for all v in W. Consider the sequence of sets of vertices $V_0 = \{q\} \subset V_1 \subset \cdots \subset V_l = V_f$

constructed by the burning moves described above. If V_f contains a vertex of W, let k be the least integer such that V_k contains a vertex of W and let v be such a vertex. This means that v was burned at the k-th step, so that $\operatorname{outdeg}_{W_{k-1}}(v) > s(v)$. Since V_{k-1} did not contain any vertex of W, we have $W \subset W_{k-1}$. But outdegree cannot increase as we make a set larger, so $\operatorname{outdeg}_W(v) \geq \operatorname{outdeg}_{W_{k-1}}(v)$. Thus $\operatorname{outdeg}_W(v) > s(v)$, which contradicts our assumption that W can be fired q-legally. Thus V_f can't contain any vertex of W, so $V_f \neq V$.   □

Example 4.9.2. Let Γ be the diamond graph shown in Figure 4.13. Consider the configurations $s_1 = (-2, 1, 0, 1)$ and $s_2 = (-2, 1, 1, 0)$. We will use Dhar's algorithm to show that s_1 is q-reduced but s_2 is not.

At each step, V_i is the set of burned vertices, E_i is the set of burned edges, and W_i is the set of unburned vertices. We start with

$$V_0 = \{q\}, \quad E_0 = \phi, \quad \text{and} \quad W_0 = \{v_2, v_3, v_4\}.$$

For the next step, we add the edges e_1 and e_2 adjacent to q to the set of burned edges. We note that

$$\operatorname{outdeg}_{W_0}(v_2) = 1, \quad \operatorname{outdeg}_{W_0}(v_3) = 1, \quad \text{and} \quad \operatorname{outdeg}_{W_0}(v_4) = 0.$$

Since $\operatorname{outdeg}_{W_0}(v_3) > s_1(v_3) = 0$ (the number of burned edges adjacent to v_3 exceeds the number of firefighters at v_3) we add v_3 to the set of burned vertices to obtain

$$V_1 = \{q, v_3\}, \quad E_1 = \{e_1, e_2\}, \quad \text{and} \quad W_1 = \{v_2, v_4\}.$$

Now
$$\operatorname{outdeg}_{W_1}(v_2) = 2 \quad \text{and} \quad \operatorname{outdeg}_{W_1}(v_4) = 1.$$

Since $\operatorname{outdeg}_{W_1}(v_2) > s_1(v_2) = 1$, we burn v_2 to obtain

$$V_2 = \{q, v_2, v_3\}, \quad E_2 = \{e_1, e_2, e_3, e_5\}, \quad \text{and} \quad W_2 = \{v_4\}.$$

Finally, $\operatorname{outdeg}_{W_2}(v_4) = 2 > s_1(v_4) = 1$, so we add v_4 to the set of burned vertices to obtain $V_3 = \{q, v_2, v_3, v_4\}$. Since all vertices are burned by Dhar's algorithm, s_1 is q-reduced.

Now consider $s_2 = (-2, 1, 1, 0)$. We start with the same sets V_0, E_0, and W_0. The outdegrees of the vertices of W_0 are the same as before. Since there is no vertex v_i in W_0 whose outdegree exceeds $s(v_i)$, no vertex in W_0 can be burned. Therefore, Dhar's algorithm applied to s_2 terminates before all vertices are burned, so s_2 is not reduced.

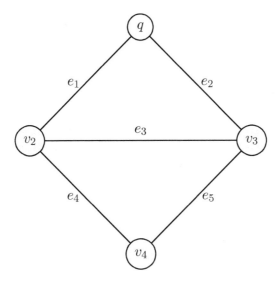

Figure 4.13:   A diamond graph with an edge ordering.

4.9.1 Finding reduced configurations

Here we give a brief description of an algorithm for calculating the reduced
representatives of the critical group (Jacobian) of a connected graph Γ. As
usual, we assume that Γ may have multi-edges but does not have loops.
Suppose that $v_1, v_2, \ldots, v_n$ is an ordering of the vertices of Γ. We let A be
the adjacency matrix of Γ with respect to this ordering, so that A_{ij} is the
number of edges between v_i and v_j. We let Q be the Laplacian matrix of Γ
with respect to the same ordering. We will suppose that $q = v_1$ is the sink
vertex.

Let s be a q-legal configuration on Γ. First we recall Dhar's algorithm (in
a slightly different form) for determining whether there is a non-empty set of
vertices in $V \setminus \{q\}$ that can be fired q-legally. If s is q-reduced, the algorithm
below will return the empty set. If s is not q-reduced, the algorithm will
return a set of vertices in $V \setminus \{q\}$ that can be fired q-legally. We then fire this
set to obtain a new q-legal configuration s' and repeat the algorithm. We have
shown previously that if s is q-legal, we will obtain a q-reduced configuration
after firing a finite number of sets in this manner.

Initialize V_{old} as the empty set and $V_{\text{new}} = \{q\}$. **While** $V_{\text{new}} \neq V$ and
$V_{\text{old}} \neq V_{\text{new}}$, do the following.

1. Set $V_{\text{old}} = V_{\text{new}}$.

2. Let W be the complement of V_{old} and let x_W be the characteristic
   vector of W, i.e.,

$$x_W(i) = \begin{cases} 0, & \text{if } v_i \in V_{\text{old}}, \\ 1, & \text{if } v_i \notin V_{\text{old}}. \end{cases}$$

3. Let V_{new} be the union of V_{old} and all vertices $v_i \in W$ such that $Qx_W(i) > s(i)$.

Return the complement of V_{new} in V.

If the set returned by Dhar's algorithm is empty, then s is q-reduced. If not, the set returned may be fired q-legally. We fire this set and repeat the process until we obtain a reduced configuration.

Example 4.9.3. Consider the diamond graph of Example 4.9.2 and let $s = (-4, 2, 1, 1)$. We will use Dhar's algorithm to find a reduced configuration equivalent to s. We start with $W_0 = \{v_2, v_3, v_4\}$ and note that $s(v_i) \geq \text{outdeg}_{W_0}(v_i)$ for all i. Thus we can fire the set W_0 q-legally. After firing, the new configuration is $s' = (-2, 1, 0, 1)$, which is q-reduced, by the calculations of Example 4.9.2.

Exercise 4.17. Consider the configuration $s = (-6, 2, 1, 3)$ on the diamond graph of Example 4.9.2. Use Dhar's algorithm to find a sequence of q-legal set-firings that take s to an equivalent reduced configuration.

4.9.2 Finding a q-legal configuration equivalent to s

Suppose now that s is any configuration (equivalently, s is a degree 0 divisor). We wish to find an equivalent q-legal configuration s'. First note that we can write s in the form $s = s^+ - s^-$, where s^+ and s^- are q-legal configurations. Let M be the maximal stable configuration on Γ. Since $M + s^-$ is q-legal, we may find (by the algorithm above) a q-reduced divisor t which is equivalent to $M + s^-$. Since t is q-reduced, t is also q-stable, so $M - t$ is q-legal. Note that $-s^-$ is equivalent to $M - t$. Thus $s^+ - s^-$ is equivalent to $s^+ + (M - t)$, which is q-legal.

Now we have a method of finding a q-reduced configuration equivalent to a given configuration s, even when s is not q-legal.

Example 4.9.4. Consider the configuration $s = (2, -1, 1, -2)$ on the diamond graph Γ of Example 4.9.2. We wish to find a q-legal configuration which is equivalent to s.

We first decompose s into the difference of q-legal configurations,

$$s = s^+ - s^-,$$

where

$$s^+ = (-1, 0, 1, 0) \quad \text{and} \quad s^- = (-3, 1, 0, 2).$$

The maximal q-stable degree 0 configuration on Γ is $M = (-5, 2, 2, 1)$. The configuration $M + s^- = (-8, 3, 2, 3)$ is q-legal. We use Dhar's algorithm to find that the sequence of sets $\{v_2, v_3, v_4\}$, $\{v_2, v_3, v_4\}$, $\{v_4\}$, $\{v_2, v_3, v_4\}$ may be fired q-legally to obtain the equivalent q-reduced configuration $t = (-2, 1, 0, 1)$.

The configuration $M - t = (-3, 1, 2, 0)$ is equivalent to $-s^-$. Thus the original configuration $s = s^+ - s^-$ is equivalent to $s^+ + (M - t) = (-4, 1, 3, 0)$, which is q-legal.

Note that further application of Dhar's algorithm to $(-4, 1, 3, 0)$ gives the q-reduced configuration $(-2, 0, 2, 0)$ which is equivalent to $(-4, 1, 3, 0)$ and thus to s.

Exercise 4.18. Recall that $s = (-2, 1, 0, 1)$ is a q-reduced configuration on the diamond graph Γ of Example 4.9.2. Use Dhar's algorithm and the method of Example 4.9.4 to find the q-reduced configuration equivalent to $-s$.

4.9.3 Ordered Dhar's algorithm

An ordered version of Dhar's algorithm (due to Cori and Le Borgne), with respect to a fixed ordering on the edges of a graph, will be used to prove a result of Merino relating the Tutte polynomial to a partition of the collection of q-reduced divisors of degree 0 by weight. The result is stated by Merino in the dual terminology of q-critical divisors rather than q-reduced divisors. We approach it using a variation of Dhar's burning algorithm, as in Baker and Shokrieh [BS13].

Suppose that $\Gamma = (V, E)$ is a connected graph with a distinguished vertex q.

Definition 4.9.5. Let $D = \sum_{v \in V} a_v v$ be a divisor on Γ. The *weight* of D is defined to be

$$w(D) = \sum_{v \neq q} a_v.$$

Recall that the maximum possible weight of a reduced divisor is the genus g of Γ.

In the terminology of chip-firing games, a divisor is a configuration of chips and the weight is the number of chips on the non-sink vertices. In terms of Dhar's algorithm, the chips are firefighters, each of which can put out a fire approaching on one adjacent edge.

Algorithm 4.9.6. *(Ordered Dhar's algorithm)* Suppose that we have a fixed ordering on the edges of Γ. The following ordered version of Dhar's algorithm, due to Cori and Le Borgne, assigns to each q-reduced degree 0 divisor $D = \sum_{v \in V} a_v v$ on Γ a spanning tree T_D.

We think of a_v as the number of firefighters at vertex v, for $v \neq q$. The weight of the divisor D is thus the total number of firefighters. Each firefighter can prevent a fire from spreading to v from one adjacent edge. When fires approach v from more than a_v adjacent edges, we burn v. We think of the ordering of edges as giving us their flammability, so that higher order edges are more flammable than lower order. We start by burning the sink vertex q. At each step, we pick the highest order unburned edge e which is adjacent to a burned vertex and burn it. If the burned edge e is adjacent to an unburned vertex v, and if there are now more than a_v burned edges adjacent to v, we immediately burn v and mark e as one of the tree edges. Since D is reduced, we will eventually burn through all vertices of Γ, by Dhar's criterion for reduced divisors. When all vertices have been burned, we stop. We obtain a set T_D of $|V| - 1$ marked edges, which form a connected spanning subgraph T_D of Γ. Thus T_D must be a spanning tree.

Example 4.9.7. Consider K_4 with the edge ordering as shown in Figure 4.14.

Let $q = v_1$ be the sink vertex. We will list the sequence in which vertices and edges are burned for one v_1-reduced degree 0 divisor, in order to illustrate the process of finding the corresponding spanning tree by the ordered Dhar's algorithm. We then, in the Appendix, give a table showing all 16 v_1-reduced divisors of degree 0 and the corresponding spanning trees for the given ordering of edges.

Consider the divisor

$$D = -2v_1 + 2v_2.$$

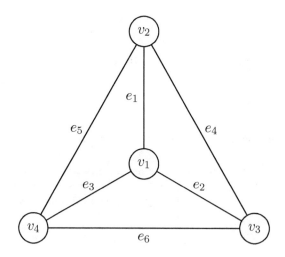

Figure 4.14: K_4 with an edge ordering.

The burning sequence is

$$v_1, e_3, v_4, e_6, v_3, e_5, e_4, e_2, e_1, v_2.$$

Thus the spanning tree produced, given in terms of its three edges, is

$$T_D = \{e_3, e_6, e_1\}.$$

The edge e_2 is externally active. The edges e_5 and e_4 are externally passive (external but not active).

Exercise 4.19. Use the method of Example 4.9.7 to find the spanning trees corresponding to the remaining 15 v_1-reduced divisors on K_4 under the given ordering of the edges. Specify each tree as a list of three edges.

It is useful to classify the edges of Γ into three types: the edges in T, the externally active edges, and the externally passive edges (all edges in $E \setminus T$ which are not externally active).

Lemma 4.9.8. *For every spanning tree T of a connected graph $\Gamma = (V, E)$ there are g external edges, where $g = |E| - |V| + 1$ is the genus of the graph.*

Proof. The $|V| - 1$ edges in T are its internal edges. The remaining $|E| - (|V| - 1) = g$ edges of Γ are the external edges. □

Lemma 4.9.9. *Let D be a q-reduced divisor and let $T = T_D$ be the spanning tree of Γ produced by the ordered Dhar's burning process. If an externally active edge for T was burned in the ordered Dhar's burning process, it was burned at some step after both its adjacent vertices were burned.*

Proof. Consider an externally active edge e for T and let C be the unique cycle in $T \cup \{e\}$. Then, by definition, e is the lowest order edge of C. If e was burned during the burning process, then at least one of its adjacent vertices, call it v, must have been burned before e.

Let P be the longest connected path in C containing v and such that each edge in P was burned before e. Let t be an edge in P. Since t is in T, one endpoint of t was burned before t and the other endpoint was burned immediately after t. Thus all vertices in P were burned before e.

Let w be the terminal vertex of P (the vertex in P farthest from v) and let f be the edge in C which is adjacent to w and not in P. Suppose that $f \neq e$. Since f has higher order than e, f must have been burned before e. This is a contradiction to our assumption that f is not in P. Thus P contains all edges in C except e, so both vertices of e must have been burned before e. □

Example 4.9.10. In Example 4.9.7, the edge e_2 is the only externally active edge. The vertices adjacent to e_2 are v_1 and v_3, both of which are burned before e_2 is burned.

The following lemma tells us that in the ordered Dhar's burning process there is one firefighter (i.e., one chip on a non-sink vertex) for each externally passive edge for the resulting spanning tree.

Lemma 4.9.11. *Let D be a q-reduced divisor and let $T = T_D$ be the spanning tree of Γ produced by the ordered Dhar's burning process. The edges which are externally passive for T (i.e., the edges in $E \setminus T$ which are not externally active) are the edges that get burned and put out by firefighters in the burning process, and consequently do not burn through new vertices.*

Let $w(D)$ be the weight of D, i.e., the number of chips (or firefighters) on vertices other than q. The number of externally passive edges for T is equal to the $w(D)$.

Proof. Consider an edge e which is not in T and which is not externally active for T. Suppose that e does not get burned during the ordered Dhar's burning process. Let C be the unique cycle in $T \cup \{e\}$. At least one edge of C must be of lower order than e, since e is not externally active. Every edge in C other than e is in T, so every edge in C other than e must be burned. For each such edge t, the fire burns through a vertex v_t adjacent to t immediately after burning through t. The one remaining vertex in C, call it w, must have been the first vertex in the cycle to be burned.

Suppose that w is an endpoint of e. Let P be the shortest path in C from w to an edge t of lower order than e. The edges e and t are both adjacent to P and e is of higher order, so e must be burned before t, which is a contradiction of our assumption.

Suppose now that w is not an endpoint of e. Then w partitions the edges of C, other than e, into two paths from w to e. Let t be the edge of minimum order in C, and let P be the path which does not contain t. Then all edges of P must be burned before t. Since e is adjacent to P, e must be burned before t, which again is a contradiction.

Thus e must be burned before the burning process terminates. Since e is not in T, the fire in e is put out by a firefighter.

Now suppose that e' is an edge that is burned and put out by a firefighter. Then e' cannot be in T, by the definition of T. Furthermore, e' cannot be externally active, by Lemma 4.9.9. Thus e' must be externally passive. □

Example 4.9.12. In Example 4.9.7, the weight of the divisor $D = -2v_1 + 2v_2$ is $w(D) = 2$, corresponding to the two externally passive edges e_4 and e_5 for T_D. The edges e_4 and e_5 get burned and put out by firefighters at v_2, so the fires on these edges do not burn through any new vertices.

The following corollaries are immediate consequences of Lemma 4.9.11.

Corollary 4.9.13. *Let D be a q-reduced divisor and let $T = T_D$ be the spanning tree of Γ produced by the ordered Dhar's burning process. Any edge which is not burned must be externally active for T.*

Corollary 4.9.14. *Let D be a q-reduced divisor and let $T = T_D$ be the spanning tree of Γ produced by the ordered Dhar's burning process. Let $w(D)$ be the weight of D, i.e., the number of chips (or firefighters) on vertices other than q. Then the number of externally active edges for T is $g - w(D)$.*

We can now restate our classification of the edges of Γ as follows: The edges of T are those which are burned, are not put out by firefighters, and burn through a vertex. The externally active edges are those which either never burn or burn only after both endpoints have already been burned. The externally passive edges are those which are burned and put out by firefighters, so do not burn through a vertex.

Proposition 4.9.15. *The map from q-reduced divisors of degree 0 to spanning trees of Γ given by the ordered Dhar's burning algorithm is a bijection.*

The well-known spanning tree theorem is an immediate consequence.

Corollary 4.9.16 (Spanning tree theorem). *The order of the critical group of a connected graph is equal to the number of spanning trees.*

To prove the proposition, we give the inverse map from spanning trees to reduced divisors of degree 0.

Algorithm 4.9.17. *(Reverse ordered Dhar's algorithm)* Let T be a spanning tree of Γ with j externally active edges. We will construct a q-reduced divisor $D_T = \sum a_v v$ of degree 0 and weight $g - j$ and such that this map from spanning trees to q-reduced divisors of degree 0 is the inverse of the map given by the ordered Dhar's burning algorithm.

We start a burning process at q, always burning the highest order edge adjacent to a burned vertex. The fire spreads from an edge to an unburned adjacent vertex if and only if the edge is in T. We terminate the process when all edges of T have been burned (i.e., when all vertices of the graph have been burned).

When we burn through a vertex $v \neq q$ from a tree edge t, we assign the number of firefighters a_v at v to be one fewer than the number of burned edges adjacent to v at that time (so t is the a_v-th burned edge adjacent to v). The value of a_q is chosen so that the degree of D is 0.

Showing that the weight of the divisor $D = \sum a_v v$ determined by this process is $g - j$ amounts to reversing the arguments of Lemmas 4.9.11 and 4.9.9 above: we show that externally passive edges are all burned and put out by firefighters, and externally active edges are either unburned, or burned only after both adjacent vertices have been burned.

Similarly, it is straightforward to check that the map determined by this algorithm is the inverse of the map determined by the ordered Dhar's algorithm.

4.9.4 Merino's theorem

We now have the tools to prove the following result of Merino [Me97], stated in terms of q-reduced divisors, rather than q-critical divisors.

Theorem 4.9.18. *(C. Merino) Let $\Gamma = (V, E)$ be a connected graph and let $g = |E| - |V| + 1$ be the genus of Γ. Suppose that an ordering has been fixed on the edge set E. Then the number of spanning trees of Γ with j externally active edges (with respect to the given ordering) is equal to the number of q-reduced degree 0 divisors with weight $g - j$. In terms of the Tutte polynomial*

$$T_\Gamma(x, y) = \sum_{i,j} t_{i,j} x^i y^j,$$

this means that the coefficient of y in $T_\Gamma(1, y)$ is the number of q-reduced divisors of weight $g - j$.

Proof. By Proposition 4.9.15, there is a bijection between spanning trees of Γ and q-reduced degree 0 divisors. Furthermore, by Corollary 4.9.14, a q-reduced degree 0 divisor D of weight $w(D)$ corresponds to spanning tree with $j = g - w(D)$ externally active edges. Equivalently, a spanning tree with j externally active edges corresponds to a q-reduced degree 0 divisor of weight $g - j$.   □

Example 4.9.19. We showed in Example 1.2.10 that, with respect to the edge ordering in that example, there are 6 spanning trees of K_4 with $j = 0$ externally active edges, 6 spanning trees with $j = 1$, 3 spanning trees with $j = 2$, and 1 spanning tree with $j = 3$. These correspond to the 6 q-reduced degree 0 divisors with weight 3, 6 q-reduced divisors with weight 2, 3 q-reduced divisors with weight 1, and 1 q-reduced divisor with weight 0. See also Example 4.9.7 and Exercise 4.19 and its solution.

Recall that a divisor D is called non-special if it has degree $g - 1$ and the dimension of its linear system $|D|$ is $r(D) = -1$.

Corollary 4.9.20. *The number of q-reduced non-special divisors on Γ (which is equal to the number of q-reduced degree 0 divisors of weight g) is given by $T_\Gamma(1, 0)$.*

Example 4.9.21. Using the formula for the Tutte polynomial of K_4 given in Example 1.2.4, we obtain the number of q-reduced non-special divisors on K_4 as

$$T_{K_4}(1, 0) = 6.$$

Example 4.9.22. Using the formula for the Tutte polynomial of the cube graph given in Example 1.2.5, we obtain the number of q-reduced non-special divisors on the cube graph as

$$T_{\mathrm{cube}}(1,0) = 133.$$

Corollary 4.9.20 can be restated as follows: on a connected graph Γ, the number of q-reduced divisors D of degree $g-1$ and with $r(D) = -1$ is given by $T_\Gamma(1,0)$. In light of this result, it is natural to ask whether the number of q-reduced divisors of degree d and $r(D) = s$, for any d and s, can be given by some evaluation of the Tutte polynomial. The answer is no, as shown by an example due to Clancy, Leake, and Payne [CLP15] and described in the following exercise.

Exercise 4.20. Show that one of the graphs of Example 4.6.5 has a divisor D with degree 2 and $r(D) = 1$ and the other doesn't.

Remark 4.9.23. An orientation of the edges of Γ is said to be an *acyclic orientation* if it does not contain any directed cycles. It can be shown that for a connected graph Γ, and for any vertex q of Γ, $T_\Gamma(1,0)$ is the number of acyclic orientations on Γ with a unique source q. Thus the number of acyclic orientations with unique source q is the number of non-special divisors on Γ. See Benson, Chakrabarty, and Tetali [BCT10] for more details.

4.9.5 Computing the critical group of a graph

Dhar's algorithm gives a way to find reduced representatives of the elements of the critical group of a small graph without using the Smith normal form, as follows:

First create a list of all q-stable configurations. Each vertex v other than q can have between 0 and $deg(v) - 1$ chips on it. Thus, the number of stable configurations is

$$\prod_{i=2}^{n} \deg(v_i).$$

Next, find the reduced form of each stable configuration using Dhar's algorithm. Delete all duplicates, and the remaining list is the required set of reduced representatives of elements of the critical group.

Recall that the reduced representatives of the elements of the critical group may be found using the Smith normal form as follows: Suppose that U_1 and U_2 are integer matrices with determinant 1 and Σ is a diagonal matrix with diagonal entries $1, 1, \ldots, 1, n_k, n_{k+1}, \ldots, n_r, 0$, where each n_i is a positive integer and $n_i | n_{i-1}$ for $i = k, k+1, \ldots, r-1$, and such that $U_1 \Sigma = QU_2$. Let x_i be the i-th column of U_i, for $k \leq i \leq r$. Then $n_i x_i$ is in the image of Q, for $k \leq i \leq r$. Thus the equivalence class of x_i has order n_i in the critical group, and the configurations $x_k, x_{k+1}, \ldots, x_r$ generate the critical group. We can then compute the q-reduced form of each of these generators. We then compute an element in the equivalence class of each member of the critical group, and the q-reduced form of each such element.

4.10 Application: Biggs' cryptosystem

In this section we will discuss a cryptosystem proposed by Biggs, based on critical groups.

In [Bi99], Biggs showed that the critical group of the n-vertex wheel graph W_n is given by

$$Jac(W_n) \cong \mathbb{Z}/\ell_{n-1}\mathbb{Z} \oplus \mathbb{Z}/\ell_{n-1}\mathbb{Z},$$

when n is even, and

$$Jac(W_n) \cong \mathbb{Z}/5\mathbb{Z} \oplus \mathbb{Z}/f_{n-2}\mathbb{Z} \oplus \mathbb{Z}/f_{n-2}\mathbb{Z},$$

when n is odd and not a multiple of 5. Here, f_n is the n-th Fibonacci number and ℓ_n is the n-th Lucas number:

$$f_{n+1} = f_n + f_{n-1}, \quad f_0 = 1, \quad f_1 = 1,$$

$$\ell_{n+1} = \ell_n + \ell_{n-1}, \quad \ell_0 = 2, \quad \ell_1 = 1,$$

so $\ell_n = f_n + f_{n-2}$. Biggs [Bi07], modifies these wheel graphs slightly to obtain a family of graphs whose critical groups are cyclic. Using these cyclic critical groups, Biggs proposes a new cryptosystem based on (the hoped difficulty of) the discrete logarithm on this group.

What is surprising is that the discrete log problem (DLP) is actually not computationally difficult to solve on critical groups. Indeed, Shokrieh [Sh09] showed that all such critical groups K have a pairing (the energy pairing described in §4.4) which is efficiently computable. This pairing can be used to solve the discrete log problem.

Algorithm 4.10.1. *(DLP on $Jac(\Gamma)$)*
Input: $D, D' \in Div_0(\Gamma)$ such that $\overline{D'} = x \cdot \overline{D}$ in $Jac(\Gamma)$.
Output: x (mod $ord(D)$), the order of $\overline{D}$ in $Jac(\Gamma)$.

(1) *Compute $\langle D, g \rangle = r + \mathbb{Z}$ and $\langle D', g \rangle = r' + \mathbb{Z}$ using formula for the pairing in terms of the Moore–Penrose inverse.*

(2) *Solve the Diophantine equation $r' = rx + y$ (for variables $x, y \in \mathbb{Z}$) by clearing the denominators of r and r' and using the extended Euclidean algorithm, to get x (mod $ord(D)$).*

Indeed, the following example shows how to carry this out in a simple case.

Example 4.10.2. Consider the graph Γ with adjacency matrix

$$A = \begin{pmatrix} 0 & 1 & 0 & 1 \\ 1 & 0 & 1 & 1 \\ 0 & 1 & 0 & 1 \\ 1 & 1 & 1 & 0 \end{pmatrix}$$

and Laplacian

$$Q = \begin{pmatrix} 2 & -1 & 0 & -1 \\ -1 & 3 & -1 & -1 \\ 0 & -1 & 2 & -1 \\ -1 & -1 & -1 & 3 \end{pmatrix}$$

depicted in Figure 4.15.

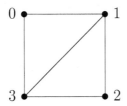

Figure 4.15: A graph having 4 vertices, with critical group $\mathbb{Z}/8\mathbb{Z}$.

The critical group $K = Jac(\Gamma)$ is a cyclic group of order 8. The following list enumerates the critical configurations:

$$c_0 = (-5, 2, 1, 2), c_1 = (-4, 2, 1, 1), c_2 = (-4, 1, 1, 2), c_3 = (-4, 2, 0, 2),$$

$$c_4 = (-3, 2, 0, 1), c_5 = (-3, 1, 0, 2), c_6 = (-3, 0, 1, 2), c_7 = (-3, 2, 1, 0).$$

The corresponding reduced configurations are

$$r_0 = (0, 0, 0, 0), r_1 = (-1, 0, 0, 1), r_2 = (-1, 1, 0, 0), r_3 = (-1, 0, 1, 0),$$

$$r_4 = (-2, 0, 1, 1), r_5 = (-2, 1, 1, 0), r_6 = (-2, 2, 0, 0), r_7 = (-2, 0, 0, 2).$$

The group table of K, in terms of the reduced configurations, is as follows:

$*$	r_0	r_1	r_2	r_3	r_4	r_5	r_6	r_7
r_0	r_0	r_1	r_2	r_3	r_4	r_5	r_6	r_7
r_1	r_1	r_7	r_0	r_4	r_6	r_3	r_2	r_5
r_2	r_2	r_0	r_6	r_5	r_3	r_7	r_4	r_1
r_3	r_3	r_4	r_5	r_0	r_1	r_2	r_7	r_6
r_4	r_4	r_6	r_3	r_1	r_7	r_0	r_5	r_2
r_5	r_5	r_3	r_7	r_2	r_0	r_6	r_1	r_4
r_6	r_6	r_2	r_4	r_7	r_5	r_1	r_3	r_0
r_7	r_7	r_5	r_1	r_6	r_2	r_4	r_0	r_3

The group table of K, in terms of the critical configurations, is as follows:

$*$	c_0	c_1	c_2	c_3	c_4	c_5	c_6	c_7
c_0	c_3	c_4	c_5	c_0	c_1	c_2	c_7	c_6
c_1	c_4	c_6	c_3	c_1	c_7	c_0	c_5	c_2
c_2	c_5	c_3	c_7	c_2	c_0	c_6	c_1	c_4
c_3	c_0	c_1	c_2	c_3	c_4	c_5	c_6	c_7
c_4	c_1	c_7	c_0	c_4	c_6	c_3	c_2	c_5
c_5	c_2	c_0	c_6	c_5	c_3	c_7	c_4	c_1
c_6	c_7	c_5	c_1	c_6	c_2	c_4	c_0	c_3
c_7	c_6	c_2	c_4	c_7	c_5	c_1	c_3	c_0

Next we compute the values of the energy pairing, $\langle c_i, c_j \rangle$.

$\langle , \rangle$	c_0	c_1	c_2	c_3	c_4	c_5	c_6	c_7
c_0	0	$\frac{1}{2}$	$\frac{1}{2}$	0	$\frac{1}{2}$	$\frac{1}{2}$	0	0
c_1	$\frac{1}{2}$	$\frac{5}{8}$	$\frac{3}{8}$	0	$\frac{1}{8}$	$\frac{7}{8}$	$\frac{1}{4}$	$\frac{3}{4}$
c_2	$\frac{1}{2}$	$\frac{3}{8}$	$\frac{5}{8}$	0	$\frac{7}{8}$	$\frac{1}{8}$	$\frac{3}{4}$	$\frac{1}{4}$
c_3	0	0	0	0	0	0	0	0
c_4	$\frac{1}{2}$	$\frac{1}{8}$	$\frac{7}{8}$	0	$\frac{5}{8}$	$\frac{3}{8}$	$\frac{1}{4}$	$\frac{3}{4}$
c_5	$\frac{1}{2}$	$\frac{7}{8}$	$\frac{1}{8}$	0	$\frac{3}{8}$	$\frac{5}{8}$	$\frac{3}{4}$	$\frac{1}{4}$
c_6	0	$\frac{1}{4}$	$\frac{3}{4}$	0	$\frac{1}{4}$	$\frac{3}{4}$	$\frac{1}{2}$	$\frac{1}{2}$
c_7	0	$\frac{3}{4}$	$\frac{1}{4}$	0	$\frac{3}{4}$	$\frac{1}{4}$	$\frac{1}{2}$	$\frac{1}{2}$

A discrete log problem example. Let $a = c_1$ and $b = c_5$. Compute x such that $b = a^x$. We compute the pairing using the table above:

$$\langle a, b \rangle = 7/8, \quad \langle a, a \rangle = 5/8.$$

Note $3 \cdot 5 \equiv 7 \pmod 8$, so $3 \cdot \frac{5}{8} \equiv \frac{7}{8} \pmod 1$, that is, $3 \cdot \frac{5}{8} = \frac{7}{8}$ in $\mathbb{Q}/\mathbb{Z}$. Because $7/8 = \langle a, b \rangle = x \langle a, a \rangle = x \cdot 5/8$, this implies $x = 3$. Indeed, it is easy to check, using the multiplication table above, $b = a^3$.

For further reading on the critical group of a graph and other topics in this chapter, see also Bacher and de la Harpe [BdlH97]; Baker and Norine [BN07] and [BN09]; Baker and Shokrieh [BS13]; Biggs [Bi97], [Bi99], and [Bi99b], and [Bi07]; Blackburn [Bl08]; Bosch and Lorenzini [BL02]; Cori and Le Borgne [CLB03]; Dhar [Dh90]; Menezes, Oorschot, and Vanstone [MvOV96]; Merino [Me97], [Me99], and [Me04]; and Shokrieh [Sh09].

Chapter 5
Interesting Graphs

This chapter collects some examples of notable graphs, along with related Sage computations and exercises. First, we recall some definitions.

For any graph $\Gamma = (V, E)$, let dist: $V \times V \to \mathbb{Z} \cup \{\infty\}$ denote the distance function. In other words, for any $v_1, v_2 \in V$, $\text{dist}(v_1, v_2)$ is the length of the shortest path from v_1 to v_2 (if it exists) and ∞ (if it does not). The diameter of Γ, denoted $\text{diam}(\Gamma)$, is the maximum value (possibly ∞) of this distance function.

Let $\Gamma = (V, E)$ be a graph, let dist: $V \times V \to \mathbb{Z} \cup \{\infty\}$ denote the distance function, and let $G = \text{Aut}(\Gamma)$ denote the automorphism group of Γ. For any $v \in V$, and any $k \geq 0$, let

$$\Gamma_k(v) - \{u \in V \mid \text{dist}(u, v) = k\}.$$

Thus $\Gamma_1(v) = N(v)$. For any subset $S \subset V$ and any $u \in V$, let $N_u(S)$ denote the set of all $s \in S$ which are neighbors of u, i.e., let

$$N_u(S) = S \cap \Gamma_1(u).$$

We say a graph is *distance-transitive* if, for any $k \geq 0$, and any $(u_1, v_1) \in V \times V$ and $(u_2, v_2) \in V \times V$ with $\text{dist}(u_i, v_i) = k$ (for $i = 1, 2$), there is a $g \in G$ such that $g(u_i) = v_i$ (for $i = 1, 2$).

We say a graph is *distance-regular* if for any $k \geq 0$ and any $(v_1, v_2) \in V \times V$ with $\text{dist}(v_1, v_2) = k$, the numbers

$$a_k = |N_{v_1}(\Gamma_k(v_2))|,$$

$$b_k = |N_{v_1}(\Gamma_{k+1}(v_2))|,$$

© Springer International Publishing AG 2017
W.D. Joyner and C.G. Melles, *Adventures in Graph Theory*,
Applied and Numerical Harmonic Analysis,
https://doi.org/10.1007/978-3-319-68383-6_5

$$c_k = |N_{v_1}(\Gamma_{k-1}(v_2)|,$$

are independent of v_1 and v_2.

An *edge-transitive* graph is a graph Γ such that, given any two edges e_1 and e_2 of Γ, there is an automorphism of Γ that maps e_1 to e_2. Analogously, a *vertex-transitive* graph is a graph Γ such that, given any two vertices v_1 and v_2 of Γ, there is an automorphism of Γ that maps v_1 to v_2.

5.1 Biggs–Smith graph

This is a 3-regular graph with 102 vertices and 153 edges, depicted[1] in Figure 5.1. It is the largest 3-regular distance-transitive graph.

It has chromatic number 3, chromatic index 3, radius 7, diameter 7, and girth 9. It is also a 3-vertex-connected graph and a 3-edge-connected graph.

The characteristic polynomial of the Biggs–Smith graph is

$$x^{17}(x-2)^{18}(x-3)^1(x^2-x-4)^9(x^3+3x^2-3)^{16}.$$

Its automorphism group is isomorphic to $PSL(2,17)$.

Computer calculations show that the critical group of the Biggs–Smith graph has invariant factors

$$(9, 153, 153, 153, 153, 153, 153, 153, 306, 306, 306, 306, 306, 306, 306, 306).$$

```
────────────────────── Sage ──────────────────────
sage: Gamma = graphs.BiggsSmithGraph()
sage: Gamma.is_regular()
True
sage: Gamma.is_regular(3)
True
sage: Gamma.diameter()
7
sage: Gamma.is_distance_regular()
True
sage: Gamma.is_edge_transitive()
True
sage: Gamma.vertex_connectivity()
3
sage: len(Gamma.edges())
153
sage: len(Gamma.vertices())
102
```

---

[1] This plot contributed by Stolee found on the Wikipedia page for the Biggs–Smith Graph is in the public domain.

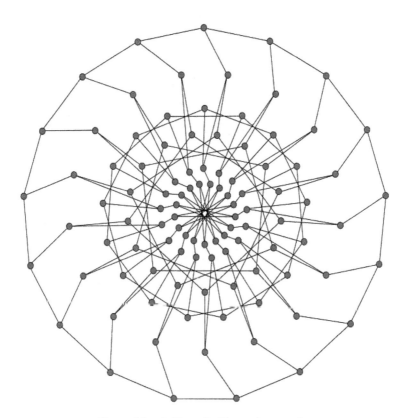

Figure 5.1: A Biggs–Smith graph example.

Exercise 5.1. Use Sage to verify that the girth of the Biggs–Smith graph is 9.

5.2 Brinkmann graph

This graph, depicted in Figure 5.2, a 4-regular graph with 21 vertices and 42 edges discovered by Gunnar Brinkmann.

It has chromatic number 4, chromatic index 5, radius 3, diameter 3, and girth 5. It is also a 3-vertex-connected graph and a 3-edge-connected graph. It's characteristic polynomial is given by

$$(x-4)\cdot(x-2)\cdot(x+2)\cdot(x^3-x^2-2x+1)^2\cdot(x^6+3x^5-8x^4-21x^3+27x^2+38x-41)^2.$$

Computer calculations show that the critical group of the Brinkmann graph has invariant factors $(50594, 354158)$.

```
─────────────────────────── Sage ───────────────────────────
sage: Gamma = graphs.BrinkmannGraph()
sage: Gamma.is_regular(4)
True
sage: Gamma.diameter()
3
sage: Gamma.is_distance_regular()
False
sage: Gamma.girth()
5
sage: Gamma.edge_connectivity()
4
sage: Gamma.is_edge_transitive()
False
sage: Gamma.is_vertex_transitive()
False
sage: Gamma.vertex_connectivity()
4
sage: len(Gamma.edges())
42
sage: len(Gamma.vertices())
21
sage: Gamma.automorphism_group().order()
14
```

Exercise 5.2. Use Sage to verify that the radius of the Brinkmann graph is 3.

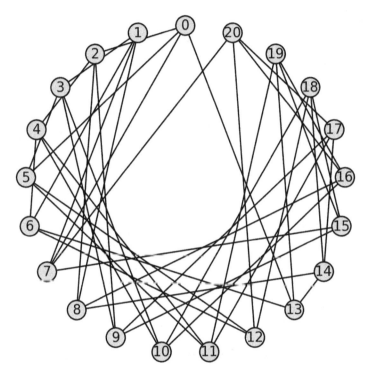

Figure 5.2: A Brinkmann graph example.

5.3 Chvátal graph

The Chvátal graph, depicted[2] in Figure 5.3, is an undirected graph with 12
vertices and 24 edges, with girth 4. It is 4-regular and has chromatic number
4.

Computer calculations show that the critical group of the Chvátal graph
has invariant factors $(6, 42, 1680)$.

─────────────────────── Sage ───────────────────────

```
sage: Gamma = graphs.ChvatalGraph()
sage: Gamma.edge_connectivity()
4
sage: Gamma.girth()
4
sage: len(Gamma.vertices())
12
sage: len(Gamma.edges())
24
sage: Gamma.vertex_connectivity()
4
sage: Gamma.is_vertex_transitive()
False
sage: Gamma.is_edge_transitive()
False
sage: Gamma.diameter()
2
sage: Gamma.is_regular(4)
True
sage: A = Gamma.adjacency_matrix(); A
[0 1 0 0 1 0 1 0 0 1 0 0]
[1 0 1 0 0 1 0 1 0 0 0 0]
[0 1 0 1 0 0 1 0 1 0 0 0]
[0 0 1 0 1 0 0 1 0 1 0 0]
[1 0 0 1 0 1 0 0 1 0 0 0]
[0 1 0 0 1 0 0 0 0 0 1 1]
[1 0 1 0 0 0 0 0 0 0 1 1]
[0 1 0 1 0 0 0 0 1 0 0 1]
[0 0 1 0 1 0 0 1 0 0 1 0]
[1 0 0 1 0 0 0 0 0 0 1 1]
[0 0 0 0 0 1 1 0 1 1 0 0]
[0 0 0 0 0 1 1 1 0 1 0 0]
sage: A.characteristic_polynomial().factor()
(x - 4) * (x + 1) * x^2 * (x + 3)^2 * (x - 1)^4 * (x^2 + x - 4)
sage: Gamma.automorphism_group().order()
8
```

Exercise 5.3. Use Sage to verify that the chromatic number of the Chvátal
graph is 4.

──────────────────────────────────

[2] This plot by David Eppstein found on the Wikipedia page for the Chvátal Graph is in
the public domain.

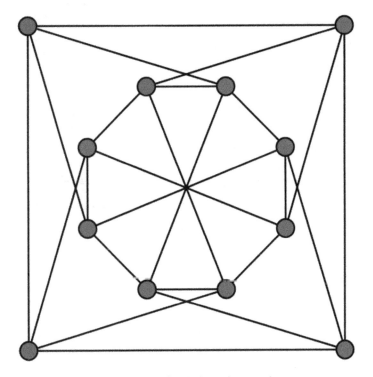

Figure 5.3: A Chvátal graph example.

5.4 Coxeter graph

The Coxeter graph, depicted in Figure 5.4, is a 3-regular graph with 28 vertices and 42 edges. It has chromatic number 3, chromatic index 3, radius 4, diameter 4, and girth 7. It is also distance-regular, 3-vertex-connected, and 3-edge-connected.

Computer calculations show that the critical group of the Coxeter graph has invariant factors $(8, 56, 56, 56, 56, 56)$ and has characteristic polynomial

$$(x - 3) \cdot (x + 1)^7 \cdot (x - 2)^8 \cdot (x^2 + 2x - 1)^6.$$

–––––––––––––––––––––––––––– Sage ––––––––––––––––––––––––––––

```
sage: Gamma = graphs.CoxeterGraph()
sage: G = Gamma.automorphism_group()
sage: G.cardinality()
336
sage: Gamma.girth()
7
sage: Gamma.is_distance_regular()
True
sage: Gamma.is_edge_transitive()
True
sage: Gamma.is_vertex_transitive()
True
```

Exercise 5.4. Use Sage to verify that the diameter of the Coxeter graph is 4.

5.5 Desargues graph

This graph, depicted[3] in Figure 5.5, is a distance-transitive 3-regular graph with 20 vertices and 30 edges. Its automorphism group acts regularly and is order 240.

The characteristic polynomial of the Desargues graph is

––

[3] This plot by David Eppstein found on the Wikipedia page for the Desargues graph is in the public domain.

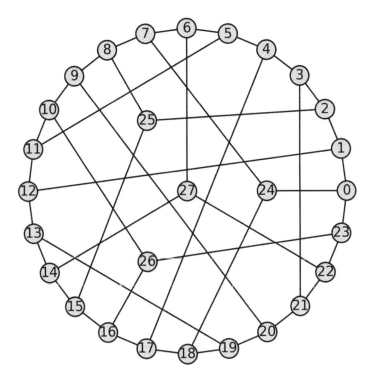

Figure 5.4: Coxeter graph.

$$(x-3)(x-2)^4(x-1)^5(x+1)^5(x+2)^4(x+3).$$

Computer calculations show that the critical group of the Desargues graph has invariant factors $(4, 8, 40, 40, 120)$.

```
──────────────────────────────── Sage ────────────────────────────────
sage: Gamma = graphs.DesarguesGraph()
sage: len(Gamma.edges())
30
sage: len(Gamma.vertices())
20
sage: Gamma.is_regular(3)
True
sage: Gamma.is_distance_regular()
True
sage: Gamma.is_edge_transitive()
True
sage: Gamma.is_vertex_transitive()
True
sage: Gamma.show(layout="circular", dpi = 300)
sage: Gamma.automorphism_group().order()
240
```

Exercise 5.5. Use Sage to verify that the diameter of the Desargues graph is 5.

5.6 Dürer graph

Depicted in Figure 5.6 with 12 vertices and 18 edges, it's named after Albrecht Dürer, whose 1514 engraving Melencolia I includes a depiction of a solid convex polyhedron having the Dürer graph as its skeleton.

It is a 3-vertex-connected simple planar graph. The automorphism group of the Dürer graph is isomorphic to the dihedral group of order 12, D_{12}, and the characteristic polynomial is

$$(x-3) \cdot (x-1) \cdot x^2 \cdot (x+2)^2 \cdot (x^2-5) \cdot (x^2-2)^2.$$

Computer calculations show that the critical group of the Dürer graph has invariant factors $(35, 210)$.

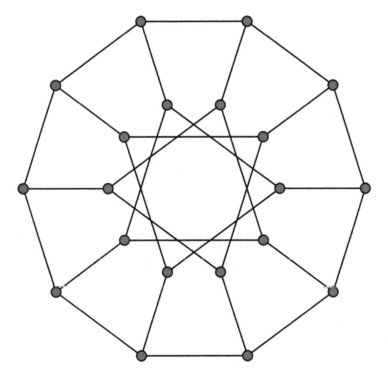

Figure 5.5: A Desargues graph example.

```
─────────────────────────────── Sage ───────────────────────────────
sage: Gamma = graphs.DurerGraph()
sage: Gamma.is_edge_transitive()
False
sage: Gamma.is_regular(4)
False
sage: Gamma.is_regular(3)
True
sage: Gamma.is_distance_regular()
False
sage: Gamma.is_vertex_transitive()
False
sage: Gamma.is_perfect()
False
sage: Gamma.automorphism_group().order()
12
```

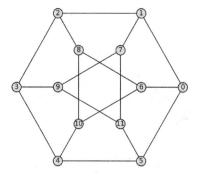

Figure 5.6: A Dürer graph example.

Exercise 5.6. Use Sage to verify that the diameter of the Dürer graph is 4.

5.7 Dyck graph

Depicted in Figure 5.7, this is a 3-regular graph Γ with 32 vertices and 48 edges, named after Walther von Dyck.

It is Hamiltonian with chromatic number 2, chromatic index 3, radius 5, diameter 5, and girth 6. It is also a 3-vertex-connected and a 3-edge-connected graph.

Computer calculations show that the critical group of the Dyck graph has invariant factors $(2, 8, 8, 16, 16, 32, 32, 32, 96)$, so

$$K(\Gamma) = \mathbb{Z}/2\mathbb{Z} \times (\mathbb{Z}/8\mathbb{Z})^2 \times (\mathbb{Z}/16\mathbb{Z})^2 \times (\mathbb{Z}/32\mathbb{Z})^4 \times \mathbb{Z}/3\mathbb{Z}.$$

Moreover, Γ has characteristic polynomial

$$(x - 3) \cdot (x + 3) \cdot (x - 1)^9 \cdot (x + 1)^9 \cdot (x^2 - 5)^6.$$

```
─────────────────────── Sage ───────────────────────
sage: Gamma = graphs.DyckGraph(); Gamma
sage: Gamma.is_bipartite()
True
sage: Gamma.is_vertex_transitive()
True
```

The automorphism group of Γ is order $192 = 3 \cdot 64$. The 3-Sylow subgroup G of this automorphism group has orbits (with respect to the vertex labels in Figure 5.7)

$$\{20\}, \{0\}, \{9, 6, 17\}, \{3, 22, 31\}, \{8, 1, 7\}, \{11, 28, 13\}, \{26, 4, 30\},$$

$$\{2, 23, 15\}, \{16, 10, 14\}, \{25, 18, 5\}, \{24, 19, 21\}, \{27, 12, 29\}.$$

The quotient graph Γ/G has critical group

$$K(\Gamma/G) = \mathbb{Z}/2\mathbb{Z} \times \mathbb{Z}/9\mathbb{Z} \times \mathbb{Z}/16\mathbb{Z}.$$

The element

$$(0, 1, 2, 3, 4, 5, 6, 7)(8, 9, 10, 11, 12, 13, 14, 15) \times$$

$$\times (16, 17, 18, 19, 20, 21, 22, 23)(24, 25, 26, 27, 28, 29, 30, 31) \in Aut(\Gamma)$$

acts by rotating the graph Γ counterclockwise in each shell depicted in Figure 5.7. This element generates a subgroup of order 8, $H \subset Aut(\Gamma)$. The quotient Γ/H is the path graph on 4 vertices (whose critical group is trivial, as it is a tree, by Lemma 4.7.1).

Exercise 5.7. Use Sage to verify that the diameter of the Dyck graph is 5.

Exercise 5.8. Assuming the critical group computations above, verify that the quotient map $\Gamma \to \Gamma/G$ is not harmonic.

Exercise 5.9. Verify that the quotient map $\Gamma \to \Gamma/H$ is not harmonic.

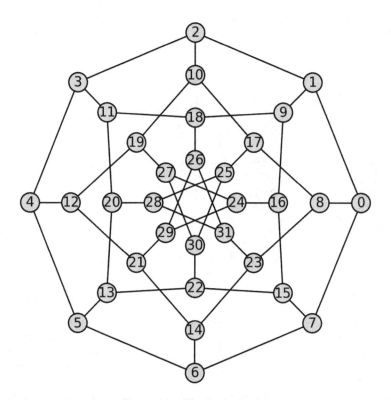

Figure 5.7:   The Dyck graph.

5.8 Errera graph

This is a graph with 17 vertices and 45 edges discovered by Alfred Errera. Depicted in Figure 5.4, it provides an example of how Kempe's 1879 "proof" of the four color theorem cannot work.

The Errera graph Γ is planar and has chromatic number 4, chromatic index 6, radius 3, diameter 4, and girth 3. It is a 5-vertex-connected graph and a 5-edge-connected graph. The graph Γ is not a vertex-transitive graph and its full automorphism group is isomorphic to the dihedral group of order 20, D_{20}.

The quotient Γ/D_{20} is the tree with three vertices, $\bar{0}, \bar{1}, \bar{2}$, associated with the following D_{20}-orbits:

$$\bar{0} = \{0, 4\}, \bar{1} = \{1, 3, 5, 7, 10, 11, 12, 14, 15, 16\}, \bar{2} = \{8, 9, 2, 13, 6\}.$$

Computer calculations show that the critical group of the Errera graph has invariant factors $(52838, 264190)$ and the characteristic polynomial is

$$(x^2 - 2x - 5) \cdot (x^2 + x - 1)^2 \cdot (x^3 - 4x^2 - 9x + 10) \times$$

$$\times (x^4 + 2x^3 - 7x^2 - 18x - 9)^2.$$

```
                              —— Sage ——
sage: Gamma = graphs.ErreraGraph()
sage: G = Gamma.automorphism_group()
sage: G.cardinality()
20
sage: G.is_isomorphic(DihedralGroup(10))
True
sage: Gamma.is_vertex_transitive()
False
sage: Gamma.is_distance_regular()
False
sage: Gamma.girth()
3
```

Exercise 5.10. Use Sage to verify that the radius of the Errera graph is 3.

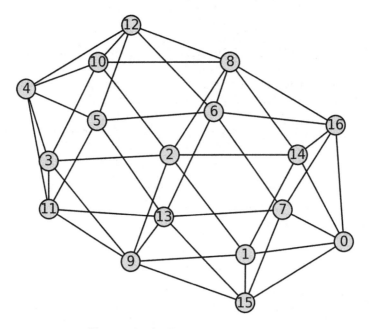

Figure 5.8:   An Errera graph example.

5.9 Foster graph

Foster graph is a 3-regular graph with 90 vertices and 135 edges.

The Foster graph is Hamiltonian, with chromatic number 2, chromatic index 3, radius 8, diameter 8 and girth 10. It is also bipartite, 3-vertex-connected and 3-edge-connected graph.

Computer calculations show that the critical group of the Foster graph has invariant factors

$$(8, 8, 8, 8, 8, 8, 8, 24, 24, 72, 360, 360, 360, 360, 360, 360, 360, 1080)$$

and characteristic polynomial

$$(x - 3) \cdot (x + 3) \cdot (x - 2)^9 \cdot (x + 2)^9 \cdot x^{10} \times$$

$$\times (x - 1)^{18} \cdot (x + 1)^{18} \cdot (x^2 - 6)^{12}.$$

```
                              Sage
sage: Gamma = graphs.FosterGraph()
sage: Gamma.is_vertex_transitive()
True
sage: Gamma.is_hamiltonian()
True
sage: Gamma.girth()
10
sage: Gamma.chromatic_number()
2
sage: Gamma.is_bipartite()
True
sage: Gamma.automorphism_group().order()
4320
sage: Gamma.is_regular()
True
sage: Gamma.is_edge_transitive()
True
```

Exercise 5.11. Use Sage to verify that the diameter of the Foster graph is 8.

5.10 Franklin graph

This graph, depicted in Figure 5.10, is a 3-regular graph with 12 vertices and 18 edges named after Philip Franklin. It is Hamiltonian and has chromatic number 2, chromatic index 3, radius 3, diameter 3, and girth 4. It is also a 3-vertex-connected and 3-edge-connected perfect graph. Its characteristic polynomial is

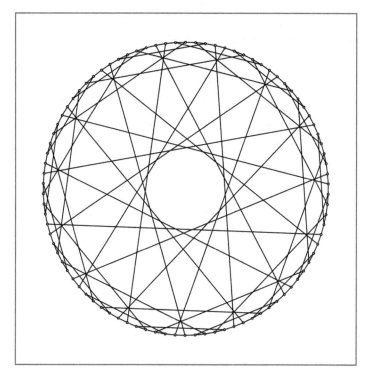

Figure 5.9: A Foster graph example.

$$(x-3) \cdot (x+3) \cdot (x-1)^3 \cdot (x+1)^3 \cdot (x^2-3)^2.$$

Computer calculations show that the critical group of the Franklin graph has invariant factors $(4, 16, 144)$.

```
─────────────────────────────── Sage ───────────────────────────────

sage: G = graphs.FranklinGraph()
sage: G.automorphism_group().order()
48
sage: G.is_bipartite()
True
sage: G.chromatic_number()
2
sage: G.girth()
4
sage: G.is_hamiltonian()
True
sage: G.is_vertex_transitive()
True
sage: G.is_edge_transitive()
False
sage: G.is_planar()
False
sage: G.is_regular()
True
```

It has an automorphism group G of order 48 which acts transitively on the vertices but not on the edges of Γ. The quotient Γ/G is the empty graph having one vertex.

Exercise 5.12. Use Sage to verify that the diameter of the Franklin graph is 3.

5.11 Gray graph

This is an undirected bipartite graph with 54 vertices and 81 edges. It is a 3-regular graph discovered by Marion C. Gray in 1932.

The Gray graph has chromatic number 2, chromatic index 3, radius 6, and diameter 6. It is also a 3-vertex-connected and 3-edge-connected non-planar graph.

The Gray graph is an example of a graph which is edge-transitive but not vertex-transitive.

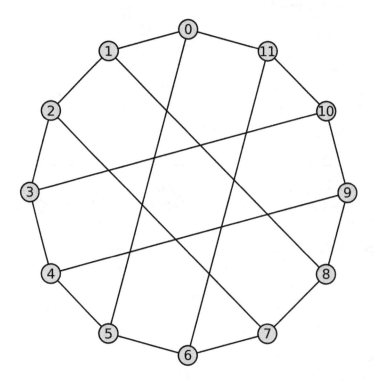

Figure 5.10: A Franklin graph example.

Computer calculations show that the critical group of the Gray graph has invariant factors $(3, 3, 3, 6, 18, 18, 18, 18, 18, 18, 54, 54, 54, 54, 16)$ and has characteristic polynomial

$$(x - 3) \cdot (x + 3) \cdot x^{16} \cdot (x^2 - 6)^6 \cdot (x^2 - 3)^{12}.$$

```
——————————————————————— Sage ———————————————————————
sage: Gamma = graphs.GrayGraph()
sage: Gamma.is_bipartite()
True
sage: Gamma.is_edge_transitive()
True
sage: Gamma.is_vertex_transitive()
False
sage: Gamma.is_hamiltonian()
True
sage: Gamma.diameter()
6
```

Exercise 5.13. Use Sage to verify that the radius of the Gray graph is 6.

5.12 Grötzsch graph

This is a triangle-free graph with 11 vertices, 20 edges, and chromatic number 4, named after German mathematician Herbert Grötzsch. Depicted in Figure 5.12, it has chromatic index 5, radius 2, girth 4, and diameter 2. It is also a 3-vertex-connected and 3-edge-connected graph.

Computer calculations show that the critical group of the Grötzsch graph has invariant factors $(139, 278)$ and has characteristic polynomial

$$(x - 1)^5 \cdot (x^2 - x - 10) \cdot (x^2 + 3x + 1)^2.$$

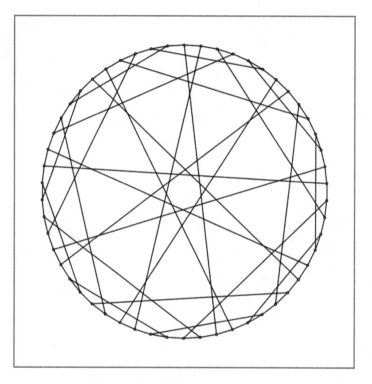

Figure 5.11:   A Gray graph example.

```
                          ── Sage ──
sage: Gamma = graphs.GrotzschGraph()
sage: G = Gamma.automorphism_group()
sage: G.cardinality()
10
sage: G.is_isomorphic(DihedralGroup(5))
True
sage: Gamma.girth()
4
sage: Gamma.diameter()
2
sage: Gamma.is_distance_regular()
False
sage: Gamma.is_vertex_transitive()
False
sage: Gamma.is_edge_transitive()
False
```

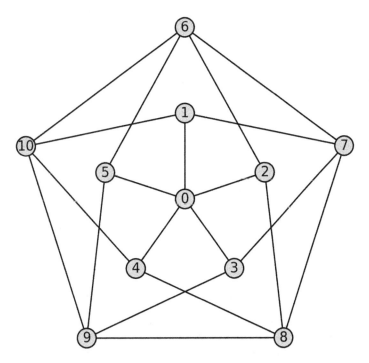

Figure 5.12:   A Grötzsch graph example.

Exercise 5.14. Use Sage to verify that the radius of the Grötzsch graph is 2.

5.13 Heawood graph

This graph, depicted in Figure 5.13, is an undirected graph with 14 vertices and 21 edges, named after Percy John Heawood. It is a 3-regular, distance-transitive, distance-regular graph. It has chromatic number 2 and chromatic index 3.

The automorphism group of the Heawood graph is isomorphic to the projective linear group $PGL2(7)$, a group of order 336. It acts transitively on the vertices and on the edges of the graph. Therefore, this graph is vertex-transitive.

Computer calculations show that the critical group of the Heawood graph has invariant factors $(7, 7, 7, 7, 21)$ and has characteristic polynomial

$$(x - 3) \cdot (x + 3) \cdot (x^2 - 2)^6.$$

```
────────────────────────── Sage ──────────────────────────
sage: Gamma = graphs.HeawoodGraph()
sage: Gamma.radius()
3
sage: Gamma.is_edge_transitive()
True
sage: Gamma.is_vertex_transitive()
True
sage: Gamma.chromatic_number()
2
sage: Gamma.diameter()
3
sage: Gamma.line_graph().chromatic_number()   # chromatic index
3
sage: Gamma.automorphism_group().order()
336
```

Exercise 5.15. Use Sage to verify that the girth of the Heawood graph is 6.

5.14 Hoffman graph

The Hoffman graph is a 4-regular graph with 16 vertices and 32 edges. It is has the same spectrum as the hypercube graph of $GF(2)^4$.

Computer calculations show that the critical group of the Hoffman graph has invariant factors $(2, 2, 8, 24, 24, 24, 96)$ and has characteristic polynomial

$$(x - 4) \cdot (x + 4) \cdot (x - 2)^4 \cdot (x + 2)^4 \cdot x^6.$$

This graph has automorphism group G of size 48 with a center Z of order 2 for which $G/Z \cong S_4$, the symmetric group on 4 letters.

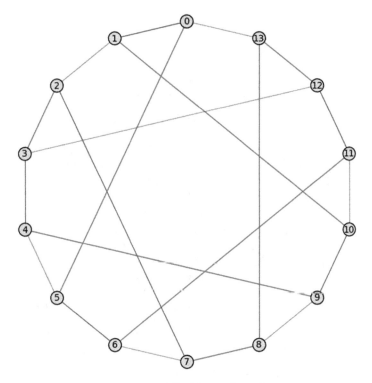

Figure 5.13:   The Heawood graph.

──────────────── Sage ────────────────

```
sage: Gamma = graphs.HoffmanGraph()
sage: Gamma.is_distance_regular()
False
sage: Gamma.is_vertex_transitive()
False
sage: Gamma.is_strongly_regular()
False
sage: Gamma.is_regular()
True
sage: Gamma.is_edge_transitive()
False
sage: Gamma.is_hamiltonian()
True
```

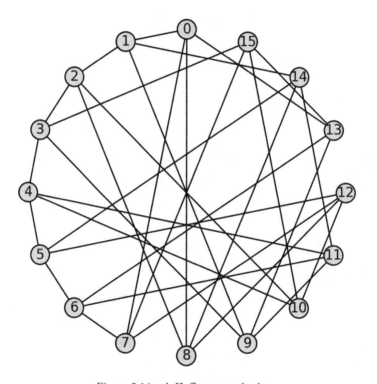

Figure 5.14: A Hoffman graph plot.

Exercise 5.16. Use Sage to verify that the girth of the Hoffman graph is 4.

5.15 Hoffman–Singleton graph

The Hoffman–Singleton graph is a 7-regular undirected graph with 50 vertices and 175 edges. It is the unique strongly regular graph with parameters
$(50, 7, 0, 1)$. Its automorphism group G has the simple group $PSU(3, 5)$ as a
subgroup of index 3.

Calculations using Sage show that the critical group of the Hoffman–
Singleton graph has invariant factors 5 (8 times), 10, and 50 (19 times).

```
────────────────────────────  Sage  ────────────────────────────
sage: Gamma = graphs.HoffmanSingletonGraph()
sage: Gamma.is_hamiltonian()
True
sage: Gamma.is_edge_transitive()
True
sage: Gamma.is_strongly_regular()
True
sage: Gamma.is_vertex_transitive()
True
sage: Gamma.is_distance_regular()
True
sage: G = Gamma.automorphism_group()
sage: H = G.normal_subgroups()[1]
sage: H.structure_description()
'PSU(3,5)'
sage: G.quotient(H)
Permutation Group with generators [(), (1,2)]
```

The characteristic polynomial of the Hoffman–Singleton graph is equal to

$$(x - 7)(x - 2)^{28}(x + 3)^{21}.$$

Exercise 5.17. Use Sage to verify that the girth of the Hoffman–Singleton
graph is 5.

5.16 Nauru graph

Nauru graph, depicted[4] in Figure 5.15, is a symmetric bipartite cubic graph
with 24 vertices and 36 edges, chromatic number 2, chromatic index 3, diameter 4, radius 4, and girth 6. It is also a 3-vertex-connected and 3-edge-
connected graph.

Computer calculations show that the critical group of the Nauru graph
has invariant factors $(5, 5, 5, 30, 120, 360)$ and characteristic polynomial

$$(x - 3) \cdot (x + 3) \cdot (x - 1)^3 \cdot (x + 1)^3 \cdot x^4 \cdot (x - 2)^6 \cdot (x + 2)^6.$$

---

[4] This plot by David Eppstein found on the Wikipedia page for the Nauru graph is in the
public domain.

─────────────── Sage ───────────────

```
sage: Gamma = graphs.NauruGraph()
sage: Gamma.automorphism_group().order()
144
sage: Gamma.is_vertex_transitive()
True
sage: Gamma.is_edge_transitive()
True
sage: Gamma.is_distance_regular()
False
sage: Gamma.is_regular(3)
True
sage: Gamma.is_regular(4)
False
sage: Gamma.diameter()
4
sage: Gamma.girth()
6
```

Exercise 5.18. Use Sage to verify that the girth of the Hoffman–Singleton graph is 5.

5.17 Paley graphs

These, one of which is depicted in Figure 5.16, were named after Raymond Paley.

Let $q = p^r$ be either an arbitrary power of a prime p congruent to 1 (mod 4), or an even power of any odd prime. Note that this implies that $GF(q)$ has a square root of -1. Let $V = GF(q)$ and let

$$E = \{(u, v) \in GF(q) \times GF(q) \mid u - v \in (GF(q)^{\times})^2\}.$$

These graphs are strongly regular graphs. They give rise to a family of expander graphs. When q is prime, the Paley graph is a Hamiltonian circulant graph.

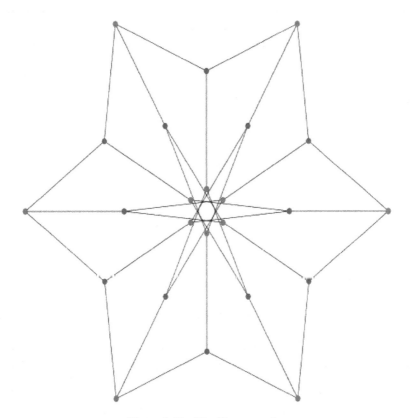

Figure 5.15: The Nauru graph.

Exercise 5.19. Show that for each nonzero square $a \in GF(q)^2 - \{0\}$ and each $b \in GF(q)$, the each map

$$t_{a,b} : x \mapsto ax + b$$

is an automorphism of the Paley graph.

Exercise 5.20. Show that the Frobenius map $x \mapsto x^p$ is an automorphism of $GF(q)$ with order r which induces an automorphism of the Paley graph.

Combining these exercises, we get a group of automorphisms of order $rq(q-1)/2$. Proving that this is the entire automorphism group is more difficult, but details can be found in Kim and Praeger [KP09].

```
─────────────────────────────── Sage ───────────────────────────────
sage: Gamma = graphs.PaleyGraph(13)
sage: G = Gamma.automorphism_group()
sage: G.cardinality()
78
sage: Gamma.is_distance_regular()
True
sage: Gamma.is_vertex_transitive()
True
sage: Gamma.is_edge_transitive()
True
sage: Gamma.diameter()
2
sage: Gamma.girth()
3
```

5.18 Pappus graph

The Pappus graph, depicted in Figure 5.17, is 3-regular, symmetric, and distance-regular.

The Pappus graph is a 3-regular undirected graph with 18 vertices and 27 edges, named after Pappus of Alexandria. The Pappus graph has girth 6, diameter 4, radius 4, chromatic number 2, chromatic index 3, and is both 3-vertex-connected and 3-edge-connected.

Computer calculations show that the critical group of the Pappus graph has invariant factors $(2, 6, 6, 18, 18, 54)$. It has characteristic polynomial

$$(x - 3) \cdot (x + 3) \cdot x^4 \cdot (x^2 - 3)^6.$$

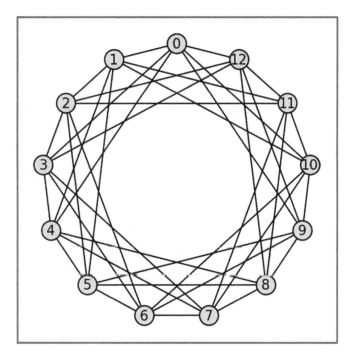

Figure 5.16:   A Paley graph example.

―――――――――――――― Sage ――――――――――――――

```
sage: Gamma = graphs.PappusGraph()
sage: Gamma.is_regular()
True
sage: Gamma.is_planar()
False
sage: Gamma.is_vertex_transitive()
True
sage: Gamma.is_hamiltonian()
True
sage: Gamma.girth()
6
sage: Gamma.is_bipartite()
True
sage: Gamma.line_graph().chromatic_number()
3
```

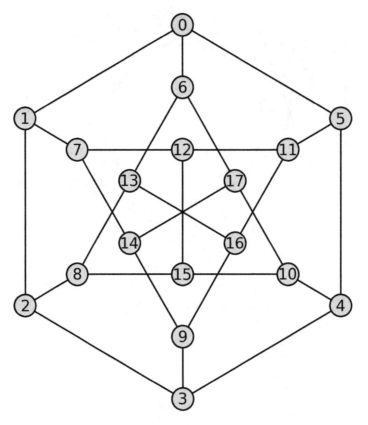

Figure 5.17:   The Pappus graph.

Exercise 5.21. Use Sage to verify that the diameter of the Pappas graph is 4.

5.19 Petersen graph

The Petersen graph is depicted in Figure 5.18. It is an undirected cubic, strongly regular, distance-transitive graph with 10 vertices and 15 edges. It has radius 2, diameter 2, girth 5, and chromatic number 3.

Computer calculations show that the critical group of the Petersen graph has invariant factors $(2, 10, 10, 10)$. It has automorphism group isomorphic to S_5 and characteristic polynomial

$$(x - 3) \cdot (x + 2)^4 \cdot (x - 1)^5.$$

```
—————————————————— Sage ——————————————————

sage: Gamma = graphs.PetersenGraph()
sage: Gamma.is_distance_regular()
True
sage: Gamma.diameter()
2
sage: Gamma.automorphism_group().cardinality()
120
sage: G = Gamma.automorphism_group()
sage: G.is_isomorphic(SymmetricGroup(5))
True
```

Exercise 5.22. Use Sage to verify that the girth of the Petersen graph is 5.

5.20 Shrikhande graph

The Shrikhande graph, depicted in Figure 5.19, is a strongly regular graph with 16 vertices and 48 edges. Each vertex has a degree of 6. The Shrikhande graph is not a distance-transitive graph. It is the smallest distance-regular graph that is not distance-transitive.

The Shrikhande graph is isomorphic to the Cayley graph of $\mathbb{Z}/4\mathbb{Z} \times \mathbb{Z}/4\mathbb{Z}$, where two vertices are adjacent if and only if the difference is in $S = \{\pm(0, 1), \pm(1, 0), \pm(1, 1)\}$.

The characteristic polynomial of the Shrikhande graph is

$$(x - 6)(x - 2)^6(x + 2)^9.$$

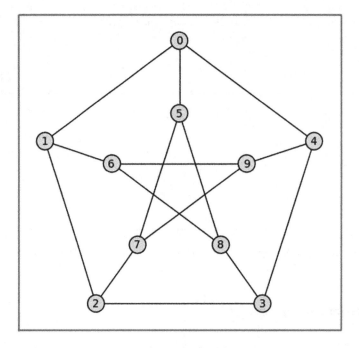

Figure 5.18:   A Petersen graph example.

Computer calculations show that the critical group of the Shrikhande graph has invariant factors $(2, 8, 8, 16, 16, 32, 32, 32, 32)$.

```
──────────── Sage ────────────
sage: Gamma = graphs.ShrikhandeGraph()
sage: Gamma.is_strongly_regular()
True
sage: Gamma.is_vertex_transitive()
True
sage: Gamma.is_distance_regular()
True
sage: Gamma.is_regular()
True
sage: Gamma.is_edge_transitive()
True
```

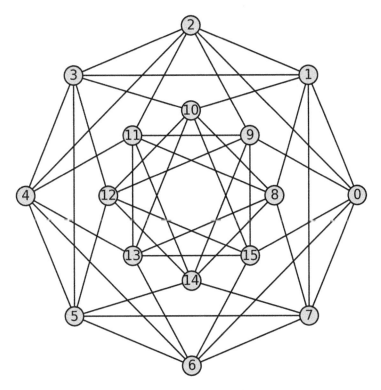

Figure 5.19: The Shrikhande graph.

Exercise 5.23. Use Sage to verify that the girth of the Shrikhande graph is 3.

Chapter 6
Cayley Graphs of Bent Functions and Codes

This chapter introduces some more advanced problems which arise at the intersection of combinatorics, graph theory, error-correcting codes, and cryptography.

6.1 Motivation

Attached to an even function $f : GF(p)^n \to GF(p)$, there is a Cayley graph Γ_f (defined in §6.6). Some natural questions about such graphs arise.

Question 6.1. *For f even, are there necessary and sufficient conditions for Γ_f to be strongly regular?*

Thanks to the work of several mathematicians, the answer to this question is yes when $p = 2$. What about for $p > 2$?

Question 6.2. *Find necessary and sufficient conditions for Γ_f to be connected (and more generally find a formula for the number of connected components of Γ_f).*

It turns out this is easy to answer. We aren't sure who discovered this fact first, but please see Lemma 6.11.1 for more details. See also Lemma 12 in [CJMPW15].

Question 6.3. *Classify the spectrum of Γ_f in terms of the values of the Walsh–Hadamard transform of f.*

The answer is known when $p = 2$ and first appeared in the PhD thesis of A. Bernasconi [B98]. What about when $p > 2$? The answer to the analogous question for the Fourier transform is known for all primes $p \geq 2$ (see Proposition 6 in [CJMPW15]).

© Springer International Publishing AG 2017
W.D. Joyner and C.G. Melles, *Adventures in Graph Theory*,
Applied and Numerical Harmonic Analysis,
https://doi.org/10.1007/978-3-319-68383-6_6

Question 6.4. *Which graph-theoretic properties of Γ_f can be tied to function-theoretic properties of f?*

For further details, see §6.16.

6.2 Introduction

Let p be a prime, and let $GF(p)$ be the finite field of order p. A p-ary function is a function $f : GF(p)^n \to GF(p)$, where n is an integer. We are particularly interested in bent p-ary functions, which are p-ary functions that are in some sense the farthest from linear. We associate to each p-ary function f a combinatorial structure on the level sets $D_k = f^{-1}(k)$, for $k \in GF(p) \setminus \{0\}$ by counting the number of times each element of D_k is represented in the form $d_i - d_j$, for $d_i \in D_i$ and $d_j \in D_j$, where $i, j \in GF(p) \setminus \{0\}$. We also associate to each p-ary function f an edge-weighted Cayley graph Γ_f, with vertex set $GF(p)^n$, and with an edge (u, v) of weight $f(u - v)$ for each pair (u, v) such that $f(u - v) \neq 0$. If $f(0) = 0$, the graph Γ_f has no loops, and if $f(x) = f(-x)$ for all x (i.e., if f is even), then we may regard Γ_f as an undirected graph.

We investigate how the properties of an even, bent p-ary function f with $f(0) = 0$ are related to the combinatorial properties of the level sets D_i and to the graph-theoretic properties of the Cayley graph Γ_f. In the Boolean case ($p = 2$), the theorems of Dillon and Bernasconi–Codenotti–VanderKam characterize bent functions in terms of the combinatorial properties of their level sets and in terms of their Cayley graphs. The p-ary case is more complicated.

We also describe Cayley graphs of binary linear codes.

Much of the material in this chapter was cowritten with Charles Celerier, David Phillips, and Steven Walsh and appeared in [CJMPW15].

6.3 Bent functions

Fix a prime p and an integer $n \geq 1$. Let $GF(p)$ be the finite field of order p (also denoted $\mathbb{Z}/p\mathbb{Z}$). Let $\zeta = e^{2\pi i/p}$. A p-ary function $f : GF(p)^n \to GF(p)$ determines a well-defined function $\zeta^f : GF(p)^n \to \mathbb{C}$.

Let $V = GF(p)^n$.

Definition 6.3.1. The *Walsh* or *Walsh–Hadamard transform*[1] of a p-ary function $f : GF(p)^n \to GF(p)$ is the function $W_f : GF(p)^n \to \mathbb{C}$ given by

---

[1]named for Joseph Walsh and Jacques Hadamard.

$$W_f(y) = \sum_{x \in V} \zeta^{f(x)-\langle y,x \rangle}, \tag{6.1}$$

where $\langle \, , \, \rangle$ is the usual inner product.

The Walsh transform has the following properties:

1. The Walsh coefficients $W_f(y)$ satisfy *Parseval's equation*

$$\sum_{y \in V} |W_f(y)|^2 = p^{2n}.$$

2. If $\sigma : \mathbb{Q}(\zeta) \to \mathbb{Q}(\zeta)$ is defined by $\sigma(\zeta) = \zeta^k$ (where $\mathbb{Q}(\zeta)$ is the field extension of the rational numbers generated by adjoining ζ), and if

$$W_f(y)^\sigma = \sum_{x \in V} (\zeta^k)^{f(x)-\langle y,x \rangle},$$

then

$$W_f(y)^\sigma = W_{kf}(ky).$$

Definition 6.3.2. A p-ary function $f : GF(p)^n \to GF(p)$ is *bent* if

$$|W_f(y)| = p^{n/2},$$

for all $y \in GF(p)^n$.

We will see that bent functions are "maximally nonlinear" in some sense (see Proposition 6.3.9 (c)). They can be used to generate pseudorandom sequences rather easily. Bent functions may also be characterized in terms of an associated matrix (see Proposition 6.3.9 (b)).

Definition 6.3.3. If f is any function on $GF(p)^n$, we say that f is *even* if $f(-x) = f(x)$ for all $x \in GF(p)^n$.

Example 6.3.4. The following table shows the values of an even function $f : GF(3)^2 \to GF(3)$ and its Walsh transform (where $z = \frac{3}{2}(1 + \sqrt{3}i)$).

$GF(3)^2$	(0, 0)	(1, 0)	(2, 0)	(0, 1)	(1, 1)	(2, 1)	(0, 2)	(1, 2)	(2, 2)
f	0	1	1	1	2	2	1	2	2
W_f	-3	z	z	z	$\overline{z}$	$\overline{z}$	z	$\overline{z}$	$\overline{z}$

It can be checked that the values of the Walsh transform (6.1) of f all have absolute value 3. Therefore f is bent.

Definition 6.3.5. Given a complex-valued function $g : GF(p)^n \to \mathbb{C}$, we define the *Fourier transform* of g to be the function $\hat{g} : GF(p)^n \to \mathbb{C}$ (also written $g^{\wedge}$) given by

$$\hat{g}(y) = \sum_{x \in V} g(x)\zeta^{-\langle x,y\rangle}, \tag{6.2}$$

where $\zeta = e^{2\pi i/p}$ and $V = GF(p)^n$.

Note that

$$\hat{g}(0) = \sum_{x \in V} g(x).$$

Note also that if f is a p-ary function, then

$$W_f(y) = (\zeta^f)^{\wedge}(y).$$

It is not hard to see that if g is even and real-valued, then the Fourier transform of g is real valued. (However, this is not necessarily true of the Walsh transform.)

Exercise 6.1. The inverse Fourier transform of a complex-valued function $h : GF(p)^n \to \mathbb{C}$ is the function $h^{\vee} : GF(p)^n \to \mathbb{C}$ given by

$$h^{\vee}(y) = \frac{1}{p^n} \sum_{z \in V} h(z)\zeta^{\langle z,y\rangle}.$$

Show that for any function $g : GF(p)^n \to \mathbb{C}$,

$$\hat{g}^{\vee}(y) = g(y).$$

Hint: Notice that if $a \in GF(p)^n \setminus \{0\}$, then

$$\sum_{z \in V} \zeta^{\langle z,a\rangle} = 0.$$

Exercise 6.2. We say that a function $g : GF(p)^n \to \mathbb{C}$ is *supported at 0* if $g(y) = 0$ for all $y \in GF(p)^n$ such that $y \neq 0$. Show that a function $g : GF(p)^n \to \mathbb{C}$ is supported at 0 if and only if its Fourier transform $\hat{g}$ is constant.

Definition 6.3.6. We call a function $g : GF(p)^n \to GF(p)$ *balanced* if the cardinalities $|g^{-1}(x)|$ (for $x \in GF(p)$) do not depend on x, i.e., $|g^{-1}(x)| = p^{n-1}$ for all $x \in GF(p)$.

Lemma 6.3.7. *Consider a function $g : GF(p)^n \to GF(p)$, where we identify $GF(p)$ with the set $\{0, 1, 2, \ldots, p-1\}$. The following conditions are equivalent:*

(a) *The map g is balanced.*

(b) *The Fourier transform of ζ^g satisfies $\widehat{\zeta^g}(0) = 0$.*

Proof. Let

$$S_i = \{x \mid g(x) = i\} \tag{6.3}$$

for $i \in GF(p)$. Note that

$$\widehat{\zeta^g}(0) = |S_0| + |S_1|\zeta + |S_2|\zeta^2 + \cdots + |S_{p-1}|\zeta^{p-1},$$

which we can regard as an identity in the $(p-1)$-dimensional $\mathbb{Q}$-vector space $\mathbb{Q}(\zeta)$. The identity

$$1 + \zeta + \zeta^2 + \cdots + \zeta^{p-1} = 0,$$

gives

$$\widehat{\zeta^g}(0) - |S_0| + |S_1|$$
$$= (|S_2| - |S_1|)\,\zeta^2 + (|S_3| - |S_1|)\,\zeta^3 + \cdots + (|S_{p-1}| - |S_1|)\,\zeta^{p-1}. \tag{6.4}$$

If $\widehat{\zeta^g}(0)$ is rational (in particular, if $\widehat{\zeta^g}(0) = 0$), then the expressions on both sides of Equation (6.4) must be rational. This is possible only if

$$|S_2| - |S_1| = |S_3| - |S_1| = \cdots = |S_{p-1}| - |S_1| = 0.$$

Furthermore, if $\widehat{\zeta^g}(0) = 0$, then $|S_0| = |S_1|$. Therefore, if $\widehat{\zeta^g}(0) = 0$, then g is balanced.

Conversely, if g is balanced, Equation (6.4) implies that $\widehat{\zeta^g}(0) = 0$.   □

It is convenient to identify $\mathbb{Z}/p^n\mathbb{Z}$ with the set $\{0, 1, \ldots, p^n - 1\}$. We may regard each element $x \in \mathbb{Z}/p^n\mathbb{Z}$ as a polynomial in p of degree at most $n - 1$, and let $\eta(x)$ be the list of coefficients, arranged in order of decreasing degree. The map

$$\eta : \mathbb{Z}/p^n\mathbb{Z} \to GF(p)^n \tag{6.5}$$

is called the *p*-ary representation map and is a bijection. (For our purposes, any bijection will do, but the *p*-ary representation map is the most natural one.)

Definition 6.3.8. We call an $N \times N$ $\{1, -1\}$-matrix M a *Hadamard matrix* if

$$MM^t = NI_N,$$

where I_N is the $N \times N$ identity matrix. More generally, we call an $N \times N$ complex matrix M a *Butson matrix* if

$$M\overline{M}^t = NI_N.$$

The following equivalences are known (see for example Tokareva [T10] and Carlet and Ding [CD04]), but proofs are included for the convenience of the reader.

Proposition 6.3.9. *Let f be a p-ary function $f : GF(p)^n \to GF(p)$. The following are equivalent:*

(a) *The function f is bent.*

(b) *The matrix*

$$\zeta^F = (\zeta^{f(\eta(i) - \eta(j))})_{0 \le i, j \le p^n - 1}$$

is Butson, where η is the p-ary representation map of (6.5).

(c) *The derivative map $D_b f : GF(p)^n \to GF(p)$ given by*

$$D_b f(x) = f(x + b) - f(x) \tag{6.6}$$

is balanced, for each $b \in GF(p)^n$ such that $b \ne 0$.

Proof. Let $h : GF(p)^n \to \mathbb{C}$ be defined by

$$h(b) = (\zeta^{D_b f})^\wedge(0) = \sum_{x \in V} \zeta^{f(x+b) - f(x)}$$

for $b \in GF(p)^n$.

To prove that (a) implies (c), we note that the Fourier transform $\hat{h}$ satisfies

$$
\begin{aligned}
\hat{h}(y) &= \sum_{b \in V} \sum_{x \in V} \zeta^{f(x+b) - f(x)} \zeta^{-\langle y, b \rangle} \\
&= \sum_{b \in V} \sum_{x \in V} \zeta^{f(x+b) - f(x) - \langle y, b \rangle - \langle y, x \rangle + \langle y, x \rangle} \\
&= \sum_{x \in V} \zeta^{-f(x) + \langle y, x \rangle} \sum_{b \in V} \zeta^{f(x+b) - \langle y, x+b \rangle} \\
&= \widehat{\zeta^f}(y) \overline{\widehat{\zeta^f}}(y) \\
&= |\widehat{\zeta^f}(y)|^2 \\
&= |W_f(y)|^2.
\end{aligned}
$$

Therefore, if f is bent then h is a constant, which means that h is supported at 0, by Exercise 6.2. Therefore, $(\zeta^{D_b f})^\wedge(0) = 0$ for $b \ne 0$. By Lemma 6.3.7, $D_b f$ is balanced.

To prove that (c) implies (a), we reverse the argument above. Suppose $D_b f$ is balanced for all $b \in GF(p)^n$ such that $b \neq 0$. By Lemma 6.3.7, h is supported at 0, so $\hat{h}$ is a constant, by Exercise 6.2. Substituting $y = 0$ and using the fact that $D_b f$ is balanced, we see that the constant value of $\hat{h}$ must be $\hat{h}(0) = |V| = p^n$. Thus $|W_f(y)| = p^{n/2}$.

To prove that (c) implies (b), note that

$$(\zeta^F \overline{\zeta^F}^t)_{ik} = \sum_{j=0}^{p^n - 1} \zeta^{f(\eta(i) - \eta(j)) - f(\eta(k) - \eta(j))}$$

$$= \sum_{x \in V} \zeta^{f(x+b) - f(x)}$$

$$= (\zeta^{D_b f})^\wedge(0),$$

where $b = \eta(i) - \eta(k)$. If $D_b f$ is balanced, then by Lemma 6.3.7, $(\zeta^{D_b f})^\wedge(0) = 0$ for all $b \neq 0$. These are the off-diagonal terms in the product $\zeta^F \overline{\zeta^F}^t$. Those terms when $b = 0$ are the diagonal terms. These terms have the value $|V| = p^n$. This implies ζ^F is Butson.

The proof that (b) implies (c) follows by reversing the above argument. The details are omitted.   □

Remark 6.3.10. If $f : GF(2)^n \to GF(2)$ is bent, then n must be even. This is because the Walsh transform of f has magnitude $2^{\frac{n}{2}}$, which must be an integer.

Lemma 6.3.11. *If* $f : GF(2)^n \to GF(2)$ *is bent, then*

$$|f^{-1}(1)| = 2^{n-1} \pm 2^{\frac{n}{2} - 1}.$$

Proof. By Remark 6.3.10, n is even. Since f is bent, $|W_f(0)| = 2^{\frac{n}{2}}$. Let $k = |f^{-1}(1)|$. Then $\zeta^f = (-1)^f$ takes the value -1 k times and the value 1 $2^n - k$ times. Therefore,

$$W_f(0) = k(-1) + (2^n - k)$$
$$= 2^n - 2k$$
$$= \pm 2^{\frac{n}{2}}.$$

Solving for k gives
$$k = 2^{n-1} \pm 2^{\frac{n}{2} - 1}.$$

□

Exercise 6.3. Show that there are 8 functions $f : GF(2)^2 \to GF(2)$ with $f(0) = 0$ (all Boolean functions are even), of which 4 are bent. Express these

4 functions in terms of polynomials in two variables x_0 and x_1 over $GF(2)$, such that each variable occurs with power at most 1 in each term.

For small p and n, the bent functions $f : GF(p)^n \to GF(p)$ can be found by computer programs such as Sage.

Example 6.3.12. There are $3^4 = 81$ even functions $f : GF(3)^2 \to GF(3)$ with $f(0) = 0$, of which exactly 18 are bent. These functions are discussed in more detail in §6.13.1.

Example 6.3.13. There are $3^{13} = 1594323$ even functions $f : GF(3)^3 \to GF(3)$ with $f(0) = 0$, of which exactly 2340 are bent. These functions are discussed in more detail in §6.13.2.

Example 6.3.14. There are $5^{12} = 244140625$ even functions $f : GF(5)^2 \to GF(5)$ with $f(0) = 0$, of which exactly 1420 are bent. These functions are discussed in more detail in §6.13.3.

Remark 6.3.15. Potapov [Pot16] has shown that the minimal Hamming distance between distinct p-ary bent functions on $GF(p)^n$, when n is even, is $p^{\frac{n}{2}}$, where the Hamming distance between p-ary functions f_1 and f_2 on $GF(p)^n$ is

$$d(f_1, f_2) = |\{v \in GF(p)^n \mid f_1(v) \neq f_2(v)\}|.$$

Exercise 6.4. Consider the 4 bent functions $f : GF(2)^2 \to GF(2)$ with $f(0) = 0$ of Exercise 6.3. Show that the Hamming distance between any two of these functions is 2.

6.4 Duals and regularity of bent functions

Let $f : GF(p)^n \to GF(p)$ be a p-ary function.

Definition 6.4.1. Suppose f is bent. We say f is *regular* if and only if $W_f(u)/p^{n/2}$ is a p-th root of unity for all $u \in V = GF(p)$.

If f is regular, then there is a function $f^* : GF(p)^n \to GF(p)$, called the *dual* (or *regular dual*) of f, such that $W_f(u) = \zeta^{f^*(u)} p^{n/2}$, for all $u \in V$. We call f *weakly regular*[2], if there is a function $f^* : GF(p)^n \to GF(p)$, called the *dual* (or μ-*regular dual*) of f, such that $W_f(u) = \mu \zeta^{f^*(u)} p^{n/2}$, for some constant $\mu \in \mathbb{C}$ with absolute value 1.

Proposition 6.4.2. *(Kumar, Scholtz, Welch) If f is bent, then there are functions $f_* : GF(p)^n \to \mathbb{Z}$ and $f^* : GF(p)^n \to GF(p)$ such that*

$$W_f(u)p^{-n/2} = \begin{cases} (-1)^{f_*(u)} \zeta^{f^*(u)}, & \text{for } n \text{ even, or } n \text{ odd and } p \equiv 1 \pmod 4, \\ i^{f_*(u)} \zeta^{f^*(u)}, & \text{for } n \text{ odd and } p \equiv 3 \pmod 4. \end{cases}$$

---

[2] If μ is fixed and we want to be more precise, we call this μ-*regular*.

The result of Proposition 6.4.2 is known (thanks to Kumar, Scholtz, and Welch [KSW85]) but the form above is due to Helleseth and Kholosha [HK06] (although we made a minor correction to their statement). Also, note [KSW85] Property 8 established a more general fact than the statement above.

Remark 6.4.3. Moreover, in the latter case of Proposition 6.4.2, it can be shown that $f_*(u) \notin 2\mathbb{Z}$.

Corollary 6.4.4. *If f is bent and $W_f(0)$ is rational (i.e., belongs to $\mathbb{Q}$), then n must be even.*

The condition $W_f(0) \in \mathbb{Q}$ arises in Lemma 6.10.11.

Suppose $f : GF(p)^n \to GF(p)$ is bent. In this case, for each $u \in GF(p)^n$, the quotient $W_f(u)/p^{n/2}$ is an element of the cyclotomic field $\mathbb{Q}(\zeta)$ having absolute value 1.

Below, we give simple necessary and sufficient conditions for f to be regular. The next three lemmas are well known, but included for the reader's convenience.

Lemma 6.4.5. *Suppose $f : GF(p)^n \to GF(p)$ is bent. The following are equivalent.*

- *f is weakly regular.*

- *$W_f(u)/W_f(0)$ is a p-th root of unity for all $u \in GF(p)^n$.*

Proof. If f is weakly regular with μ-regular dual f^*, then $W_f(u)/W_f(0) = \zeta^{f^*(u) - f^*(0)}$, for each $u \in GF(p)^n$.

Conversely, if $W_f(u)/W_f(0)$ is of the form ζ^{i_u}, for some integer i_u (for $u \in GF(p)^n$), then let $f^*(u)$ be $i_u \pmod{p}$ and let $\mu = W_f(0)/(p^{n/2})$. Then $f^*(u)$ is a μ-regular dual of f. $\square$

Lemma 6.4.6. *Suppose $f : GF(p)^n \to GF(p)$ is bent and weakly regular. The following are equivalent.*

- *f is regular.*

- *$W_f(0)/p^{n/2}$ is a p-th root of unity.*

Proof. One direction is clear. Suppose that f is a weakly regular bent function with μ-regular dual f^* and suppose that $W_f(0)/(p^{n/2}) = \zeta^i$. Note that $W_f(0) = \mu \zeta^{f^*(0)} p^{n/2} = \zeta^i p^{n/2}$ so that $\mu = \zeta^{i - f^*(0)}$. Let $g(u) = f^*(u) - f^*(0) + i$ (where we are treating i as an element of $GF(p)$). Then

$$W_f(u) = \mu \zeta^{f^*(u)} p^{n/2} = \zeta^{i - f^*(0)} \zeta^{g(u) + f^*(0) - i} p^n/2 = \zeta^{g(u)} p^{n/2},$$

so f is regular. $\square$

Lemma 6.4.7. *Suppose that f is bent and weakly regular, with μ-regular dual f^*. Then f^* is bent and weakly regular, with μ^{-1}-regular dual f^{**} given by $f^{**}(x) = f(-x)$. If f is also even, then f^* is even and $f^{**} = f$.*

Proof. Suppose that f is bent and weakly regular with μ-regular dual f^*. Then

$$W_f(u) = \mu \zeta^{f^*(u)} p^{\frac{n}{2}}$$

for all u in $GF(p)^n$. Let $V = GF(p)^n$. The Walsh transform of f^* is given by

$$\begin{aligned}
W_{f^*}(u) &= \sum_{y \in V} \zeta^{f^*(y)} \zeta^{-\langle u,y \rangle} \\
&= \sum_{y \in V} \mu^{-1} p^{-\frac{n}{2}} W_f(y) \zeta^{-\langle u,y \rangle} \\
&= \mu^{-1} p^{-\frac{n}{2}} \sum_{y \in V} \sum_{x \in V} \zeta^{f(x)} \zeta^{-\langle y,x \rangle} \zeta^{-\langle u,y \rangle} \qquad (6.7) \\
&= \mu^{-1} p^{-\frac{n}{2}} \sum_{y \in V} \sum_{x \in V} \zeta^{f(x)} \zeta^{-\langle y,x+u \rangle} \\
&= \mu^{-1} p^{-\frac{n}{2}} \sum_{w \in V} \zeta^{f(w-u)} \sum_{y \in V} \zeta^{-\langle y,w \rangle}.
\end{aligned}$$

Next, we note that

$$\sum_{y \in V} \zeta^{-\langle y,w \rangle} = \begin{cases} p^n, & \text{if } w = 0, \\ 0, & \text{if } w \neq 0, \end{cases}$$

since, if $y = (y_1, y_2, \ldots, y_n)$ and $w = (w_1, w_2, \ldots, w_n)$, we have

$$\begin{aligned}
\sum_{y \in V} \zeta^{-\langle y,w \rangle} &= \sum_{y_1 \in GF(p)} \sum_{y_2 \in GF(p)} \cdots \sum_{y_n \in GF(p)} \zeta^{-y_1 w_1} \zeta^{-y_2 w_2} \ldots \zeta^{-y_n w_n} \\
&= \prod_{i=1}^{n} \left(\sum_{y \in GF(p)} \zeta^{-y w_i} \right)
\end{aligned}$$

and, if $w_i \neq 0$,

$$\sum_{y \in GF(p)} \zeta^{-y w_i} = \zeta^0 + \zeta^1 + \cdots + \zeta^{p-1} = 0.$$

Therefore Equation (6.7) reduces to

$$W_{f^*}(u) = \mu^{-1} p^{-\frac{n}{2}} \zeta^{f(-u)} p^n$$
$$= \mu^{-1} \zeta^{f(-u)} p^{\frac{n}{2}}.$$

It follows that f^* is bent with μ^{-1}-regular dual f^{**} given by $f^{**}(x) = f(-x)$ and that if f is even, $f^{**} = f$.

Furthermore, if f is even,

$$\zeta^{f^*(-u)} = \mu^{-1} p^{-\frac{n}{2}} W_f(-u)$$
$$= \mu^{-1} p^{-\frac{n}{2}} \sum_{x \in V} \zeta^{f(x)} \zeta^{-\langle -u, x \rangle}$$
$$= \mu^{-1} p^{-\frac{n}{2}} \sum_{w \in V} \zeta^{f(-w)} \zeta^{-\langle u, w \rangle}$$
$$= \mu^{-1} p^{-\frac{n}{2}} \sum_{w \in V} \zeta^{f(w)} \zeta^{-\langle u, w \rangle} \qquad \text{since } f \text{ is even}$$
$$= \mu^{-1} p^{-\frac{n}{2}} W_f(u)$$
$$= \zeta^{f^*(u)}.$$

Since f^* takes values in $GF(p)$, it follows that $f^*(-u) = f^*(u)$ for all u in V, so f^* is even. □

6.5 Partial difference sets

A number of combinatorial structures originally arose from statisticians who wanted to design experiments. For example, if a farmer has a number of fertilizers for his soy plants and wants to determine which fertilizer is the most useful, how does one design a statistical experiment to accurately predict the best one? The Indian mathematician Raj Chandra Bose was a leading figure in this development.

A difference set is a combinatorial structure arising naturally from certain groups. These structures arise in the theory of error-correcting codes, as well as signal analysis.

Definition 6.5.1. Let G be a finite abelian multiplicative group of order v, and let D be a subset of G of order k. The set D is a (v, k, λ)-*difference set* (DS) if the list of differences $d_1 d_2^{-1}$, where $d_1, d_2 \in D$, represents every nonidentity element in G exactly λ times. We also use the notation (G, D) to when referring to a difference set $D \subset G$.

A difference set is called *elementary* if G is an elementary abelian 2-group (i.e., isomorphic to $(\mathbb{Z}/2\mathbb{Z})^n$ for some n).

A *Hadamard difference set* D is one whose parameters are of the form $(4m^2, 2m^2 - m, m^2 - m)$ or $(4m^2, 2m^2 + m, m^2 + m)$, for some $m > 1$.

Exercise 6.5. Let $G = \mathbb{Z}/7\mathbb{Z}$ and $D = \{1, 2, 4\}$. Does D form a difference set? If so, what are its parameters?

Exercise 6.6. Let $G = \mathbb{Z}/11\mathbb{Z}$ and $D = \{1, 3, 4, 5, 9\}$. Does D form a difference set? If so, what are its parameters?

Exercise 6.7. Let $G = \mathbb{Z}/2\mathbb{Z} \times \mathbb{Z}/8\mathbb{Z}$ and

$$D = \{(0,0), (0,1), (0,5), (1,0), (1,6)\}.$$

Does D form a difference set? If so, what are its parameters? Is it a Hadamard difference set?

Definition 6.5.2. Let G be a finite abelian multiplicative group of order v, and let D be a subset of G of order k. The set D is a (v, k, λ, μ)-*partial difference set* (PDS) if the list of differences $d_1 d_2^{-1}$, where $d_1, d_2 \in D$, represents every nonidentity element in D exactly λ times and every nonidentity element in the complement $G \setminus D$ exactly μ times.

Let $D^{-1} = \{d^{-1} \mid d \in D\}$.

Exercise 6.8. Let G be a finite abelian multiplicative group of order v, and let D be a subset of G of order k such that $D = D^{-1}$. Show that if (G, D) is an (v, k, λ, λ)-partial difference set, then it is also a (v, k, λ)-difference set.

Example 6.5.3. Consider the finite field

$$GF(9) = GF(3)[x]/(x^2 + 1) = \{0, 1, 2, x, x+1, x+2, 2x, 2x+1, 2x+2\},$$

written additively. The set of nonzero quadratic residues is given by

$$D = \{1, 2, x, 2x\}.$$

One can show that D is a partial difference set with parameters

$$v = 9, \quad k = 4, \quad \lambda = 1, \quad \mu = 2.$$

We shall return to this example (with more details) below, in Example 6.9.7.

Definition 6.5.4. Let (G, D) be a (v, k, λ, μ)-partial difference set. We say it is of *Latin square type* (respectively, *negative Latin square type*) if there exist integers $N > 0$ and $R > 0$ (respectively, $N < 0$ and $R < 0$) such that

$$(v, k, \lambda, \mu) = (N^2, R(N-1), N + R^2 - 3R, R^2 - R).$$

Example 6.5.3 above is of Latin square type ($N = 3$ and $R = 2$) and of negative Latin square type ($N = -3$ and $R = -1$).

6.5.1 Dillon's correspondence

Consider functions

$$f : GF(p)^n \to GF(p),$$

where p is a prime and $n > 1$ is an integer. Dillon's thesis [D74] was one of the first publications to discuss the relationship between bent functions and combinatorial structures, such as difference sets. His work concentrated on the Boolean case. Dillon proved that the support of f, i.e., the level set $f^{-1}(1)$, is a difference set in $GF(2)^n$.

Two (naive) analogs of Dillon's correspondence are formalized in Analog 6.16.1 and Analog 6.16.2.

Theorem 6.5.5. *(Dillon Correspondence, [D74] Theorem 6.2.10 p. 78) The function $f : GF(2)^n \to GF(2)$ is bent if and only if $f^{-1}(1)$ is an elementary Hadamard difference set of $GF(2)^n$.*

Proof. Suppose that f is bent. Then n is even, so $2^n = 4m^2$ for some integer m. Let $V = GF(p)^n$ and let $k = |f^{-1}(1)|$. Since f is bent, $D_b f$ is balanced, so there are $2m^2$ values $x \in V$ such that $D_b f(x) = 1$, It follows that there are m^2 pairs $x, x + b$ such that $f(x) = 1$ and $f(x + b) = 0$. This accounts for m^2 elements in the support $f^{-1}(1)$ of f. The remaining $k - m^2$ elements x in the support of f must satisfy $f(x) = f(x + b) = 1$, so $D_b f(x) = 0$ for these values of x. The differences $(x + b) - x$ represent b a total of $\lambda = k - m^2$ times as differences of elements of the support of f. Note that by Lemma 6.3.11, $k = 2^{n-1} \pm 2^{\frac{n}{2}-1} = 2m^2 \pm m$, so $\lambda = m^2 \pm m$. Thus $(GF(2)^n, f^{-1}(1))$ is a difference set with parameters $(4m^2, 2m^2 \pm m, m^2 \pm m)$.

Conversely, suppose that $(GF(2)^n, f^{-1}(1))$ is a Hadamard difference set. Then there are $\lambda = m^2 \pm m$ elements of $f^{-1}(1)$ in pairs $x, x + b$ (i.e., such that both x and $x + b$ are in $f^{-1}(1)$), and the remaining m^2 elements x of $f^{-1}(1)$ satisfy $f(x + b) = 0$. For these x, $D_b f(x) = 1$ and $D_b f(x + b) = 1$. Thus, there are exactly $2m^2$ elements of V at which $D_b f$ is nonzero. Thus $D_b f$ is balanced, so f is bent.   □

6.6 Cayley graphs

In this section, we define Cayley graphs associated to partial difference sets and to p-ary functions. In the Boolean case ($p = 2$), the Bernasconi–Codenotti–VanderKam correspondence (Theorem 6.6.10) characterizes bent functions in terms of their Cayley graphs.

Let G be a finite multiplicative group, and let D be a nonempty subset of G such that D does not contain the identity element 1. We define a Cayley graph associated to the pair (G, D). We are particularly interested in the case in which (G, D) is a partial difference set.

Definition 6.6.1. The *Cayley graph* $\Gamma = \Gamma(G, D)$ *associated with the pair* (G, D) *is the graph whose vertices are the elements of G and such that a directed edge connects two vertices g_2 and g_2 if $g_2 = dg_1$ for some $d \in D$.*

If $D = D^{-1} = \{d^{-1} | d \in D\}$, the Cayley graph $\Gamma(G, D)$ is an undirected graph, since if $g_2 = dg_1$ for some $d \in D$, we have $g_1 = d^{-1}g_2$, where $d^{-1} \in D$.

Note, that if D is a partial difference set such that $\lambda \neq \mu$, then $D = D^{-1}$ (Proposition 1 in Pohill [Po03]).

Example 6.6.2. Let G be the abelian group $GF(3)^2$ and let D be the subset $\{(0, 1), (0, 2), (1, 0), (2, 0)\}$. Then (G, D) is a $(9, 4, 1, 2)$-partial difference set such that $D = D^{-1}$. **Sage** can be used to plot the Cayley graph of (G, D) (Figure 6.1). Note that the partial difference set (G, D) is isomorphic to the partial difference set of Example 6.5.3.

———— Sage ————

```
sage: V = range(9); V
[0, 1, 2, 3, 4, 5, 6, 7, 8]
sage: E = [(0,1),(0,2),(0,3),(0,6),(1,2),(1,4),(1,7),(2,5),(2,8),(3,4),
(3,5),(3,6),(4,5),(4,7),(5,8),(6,7),(6,8),(7,8)]
sage: Gamma = Graph([V,E])
sage: Gamma.is_regular(4)
True
sage: Gamma.is_strongly_regular()
True
```

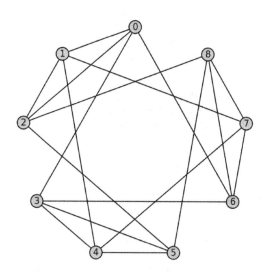

Figure 6.1: The Cayley graph for the PDS of Example 6.6.2.

Remark 6.6.3. We will see in §6.13.1 that the partial difference set D of Example 6.6.2 is the support of the even bent functions b_2 and $b_3 = -b_2$ of Proposition 6.13.1.

6.6.1 Strongly regular graphs

The notion of a strongly regular graph is closely related to the notion of a partial difference set.

Recall that the neighborhood $N(u)$ of a vertex u in a graph is the set of all vertices adjacent to u.

Definition 6.6.4. A connected simple graph Γ (without edge weights) is called *strongly regular* with parameters (v, k, λ, μ) if it has v vertices, each of degree k, and distinct vertices u_1 and u_2 have λ common neighbors if u_1 and u_2 are neighbors, and μ common neighbors if u_1 and u_2 are not neighbors. More concisely,

$$|N(u_1) \cap N(u_2)| = \begin{cases} k, & \text{if } u_1 = u_2, \\ \lambda, & \text{if } u_1 \in N(u_2), \\ \mu, & \text{if } u_1 \notin N(u_2) \text{ and } u_1 \neq u_2. \end{cases}$$

In the usual terminology/notation, such a graph is called an $SRG(\nu, k, \lambda, \mu)$.

It is well known that strongly regular graphs may be characterized by their spectra, i.e., by the eigenvalues of their adjacency matrices (see, e.g., [Bo98]). If Γ is a connected regular graph which is not a complete graph, then Γ is strongly regular if and only if it has exactly three distinct eigenvalues. One of these eigenvalues is k, with corresponding eigenvector the all 1's vector.

The following result relating partial difference sets and strongly regular graphs is well known, but the proof is included for convenience.

Theorem 6.6.5. *Let G be a finite abelian multiplicative group and let $D \subseteq G$ be a subset such that $1 \notin D$. Then D is a (v, k, λ, μ)-partial difference set such that $D = D^{-1}$ if and only if the associated Cayley graph $\Gamma(G, D)$ is a (v, k, λ, μ)-strongly regular graph.*

Proof. Suppose D is a (v, k, λ, μ)-partial difference set such that $D = D^{-1}$. Then, $\Gamma(G, D)$ has v vertices. The set D has k elements, and each vertex g of $\Gamma(G, D)$ has neighbors dg, where $d \in D$. Therefore, $\Gamma(G, D)$ is regular of degree k. Let g_1 and g_2 be distinct vertices in $\Gamma(G, D)$. Let x be a vertex that is a common neighbor of g_1 and g_2, i.e., $x \in N(g_1) \cap N(g_2)$. Then $x = d_1 g_1 = d_2 g_2$ for some $d_1, d_2 \in D$, which implies that $d_1 d_2^{-1} = g_1^{-1} g_2$. If $g_1^{-1} g_2 \in D$, then there are exactly λ ordered pairs (d_1, d_2) that satisfy the previous equation (by Definition 6.5.2). If $g_1^{-1} g_2 \notin D$, then $g_1^{-1} g_2 \in G \setminus (D \cup \{1\})$, so there are exactly μ ordered pairs (d_1, d_2) that satisfy the equation. If $g_1^{-1} g_2 \in D$, then $g_2 = dg_1$ for some $d \in D$, so g_1 and g_2 are adjacent. By a similar argument, if $g_1^{-1} g_2 \in G \setminus (D \cup \{1\})$, then g_1 and g_2 are not adjacent. So $\Gamma(G, D)$ is a (v, k, λ, μ)-strongly regular graph.

Conversely, suppose $\Gamma(G, D)$ is a (v, k, λ, μ)-strongly regular graph. If $\Gamma(G, D)$ is undirected, then for distinct vertices g_1 and g_2 there is an edge from g_1 to g_2 if and only if there is an edge from g_2 to g_1. By definition, g_1 and g_2 are connected by an edge if and only if $g_1 = dg_2$, for some $d \in D$. Thus $g_1 = d_1 g_2$ if and only if $g_2 = d_2 g_1$, for some $d_1, d_2 \in D$. This implies that $d_2 = d_1^{-1}$, so $D = D^{-1}$. Since $\Gamma(G, D)$ is a (v, k, λ, μ)-strongly regular, it is k-regular, so the order of D is k. Let x be a vertex in $\Gamma(G, D)$ such that $x \in N(g_1) \cap N(g_2)$. Then $x = d_1 g_1 = d_2 g_2$ for some $d_1, d_2 \in D$, which implies that $d_1 d_2^{-1} = g_1^{-1} g_2$. If g_1 and g_2 are adjacent, then $g_1^{-1} g_2 \in D$, so there are exactly λ ordered pairs (d_1, d_2) that satisfy the previous equation. If g_1 and g_2 are not adjacent, then $g_1^{-1} g_2 \in G \setminus (D \cup \{1\})$, so there are exactly μ ordered pairs (d_1, d_2) that satisfy the equation. Therefore, D is a (v, k, λ, μ)-partial difference set and $D = D^{-1}$. □

The notion of a partial difference set can be characterized algebraically in terms of the group ring

$$\mathbb{C}[G] = \left\{ \sum_{g \in G} c_g \cdot g \mid c_g \in \mathbb{C} \right\},$$

where addition is componentwise and multiplication is induced by the multiplicative structure of G. We identify a subset S of G with the corresponding formal sum $\sum_{g \in S} g$ in $\mathbb{C}[G]$.

The following lemma uses the group ring notation to identify D and D^{-1} as elements of $\mathbb{C}[G]$. We denote the identity element of G here by 1.

Lemma 6.6.6. *Let G be a finite abelian multiplicative group of order v, and let D be a subset of order k such that $1 \notin D$. The pair (G, D) is a (v, k, λ, μ)-partial difference set if and only if the following identity holds in $\mathbb{C}[G]$:*

$$D \cdot D^{-1} = (k - \mu) \cdot 1 + (\lambda - \mu) \cdot D + \mu \cdot G.$$

The well-known proof is omitted.

If furthermore $D^{-1} = D$ and $G \setminus (D \cup \{1\})$ is non-empty, we obtain from Lemma 6.6.6 the well-known identity

$$k^2 - k = k\lambda + (v - k - 1)\mu. \tag{6.8}$$

Remark 6.6.7. If A is the adjacency matrix of a simple strongly regular graph (without edge weights) having parameters (v, k, λ, μ), then

$$A^2 = kI + \lambda A + \mu(J - I - A), \tag{6.9}$$

where J is the all 1's matrix and I is the identity matrix. This identity is relatively easy to verify by simply computing $(A^2)_{ij}$ in the three separate cases (a) $i = j$, (b) $i \neq j$ and i, j adjacent, and (c) $i \neq j$ and i, j nonadjacent[3].

Remark 6.6.8. Suppose that $f : GF(p)^n \to GF(p)$ is an even function such that $f(0) = 0$, and let

$$V = GF(p)^n, \quad D = \mathrm{supp}(f) = \{v \in V \mid f(v) \neq 0\}, \quad \text{and} \quad D' = V \backslash (D \cup \{0\}).$$

Let $v = |V|$ and $k = |D|$. Because f is even, $D^{-1} = D$ and $(D')^{-1} = D'$.
   For each $d \in D$, let

$$\lambda_d = |\{(g, h) \in D \times D \mid g - h = d\}|,$$

and, for each $d' \in D'$, let

$$\mu_{d'} = |\{(g, h) \in D \times D \mid g - h = d'\}|.$$

If D is a partial difference set then (a) λ_d does not depend on $d \in D$ (we denote the common value by λ), and (b) $\mu_{d'}$ does not depend on $d' \in D'$ (we denote the common value by μ).

   Since $g - h \in D$ if and only if $f(g - h) \neq 0$ (for distinct $g, h \in V$), there are $k\lambda$ pairs $(g, h) \in D \times D$ such that $g - h = d$ for some $d \in D$. Likewise, since $g - h \in D'$ if and only if $f(g - h) = 0$ (for distinct $g, h \in V$), there are $(v - k - 1)\mu$ pairs $(g, h) \in D \times D$ such that $g - h = d'$ for some $d' \in D'$. Therefore, since there are $k^2 - k$ pairs $(g, h) \in D \times D$ such that $g \neq h$, we once again obtain the identity of Equation (6.8):

$$k^2 - k - k\lambda + (v - k - 1)\mu.$$

Exercise 6.9. Find the incidence matrix and girth of the Cayley graph of the partial difference set in Example 6.5.3.

Exercise 6.10. Find the adjacency matrix and diameter of the Cayley graph of the partial difference set in Example 6.5.3.

6.6.2 Cayley graphs of bent functions

Definition 6.6.9. Let f be a p-ary function $f : GF(p)^n \to GF(p)$. The *Cayley graph of f* is the edge-weighted directed graph $\Gamma_f = (V, E)$ whose vertex set is $V = GF(p)^n$ and whose edge set is

---

[3]It can also be proven by character-theoretic methods, but this method seems harder to generalize to the edge-weighted case.

$$E = \{(u, v) \mid u, v \in GF(p)^n,\ f(u - v) \neq 0\},$$

where the edge $(u, v) \in E$ has weight $f(u - v)$. We routinely identify $GF(p)$ with the set $\{0, 1, \ldots, p-1\}$ when referring to the edge weights of Γ_f. If f is even, then we can (and do) regard Γ_f as a weighted undirected graph.

For example, the Cayley graph of the even bent function in Example 6.3.4 is given in Figure 6.2.

For the case of Boolean functions, we have the following Bernasconi-Codenotti-VanderKam correspondence (see [B98], [BC99], and [BCV01]) characterizing bent functions in terms of their Cayley graphs.

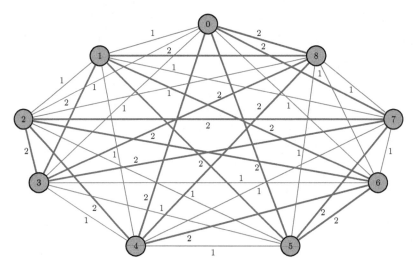

Figure 6.2: The undirected Cayley graph of an even $GF(3)$-valued bent function of two variables from Example 6.3.4. (The vertices are ordered as in the example.)

Theorem 6.6.10. *(Bernasconi-Codenotti-VanderKam correspondence) Let* $f : GF(2)^n \to GF(2)$ *and let* $k = |\mathrm{supp}(f)|$. *The function* f *is bent if and only if the Cayley graph of* f *is a strongly regular graph having parameters* $(2^n, k, \lambda, \lambda)$ *for some* λ.

Remark 6.6.11. The following "conjecture" is false: If $f : GF(p)^n \to GF(p)$ is any even bent function then the (unweighted) Cayley graph of f is distance-transitive. In fact, this fails when $p = 2$ for any bent function of 4 variables having support of size 6. Indeed, in this case the Cayley graph of f is isomorphic to the Shrikhande graph (with strongly regular parameters $(16, 6, 2, 2)$), which is not a distance-transitive (see Brouwer, Cohen, and Neumaier [BrCN89], pp. 104–105, 130).

Analogously, if $f : GF(p)^n \to GF(p)$ is any even bent function then the (unweighted) Cayley graph of f is not in general distance-regular.

6.7 Association schemes

The following definition is standard, but we give Pott, Tan, Feng, and Ling [PTFL11] as a reference.

Definition 6.7.1. Let S be a finite set and let $R_0, R_1, \ldots, R_s$ denote binary relations on S (subsets of $S \times S$). The *dual* of a relation R is the set

$$R^* = \{(x, y) \in S \times S \mid (y, x) \in R\}.$$

Assume $R_0 = \Delta_S = \{(x, x) \in S \times S \mid x \in S\}$. We say $(S, R_0, R_1, \ldots, R_s)$ is an *s-class association scheme on S* if the following properties hold.

- We have a disjoint union

$$S \times S = R_0 \cup R_1 \cup \cdots \cup R_s,$$

  with $R_i \cap R_j = \emptyset$ for all $i \neq j$.
- For each i there is a j such that $R_i^* = R_j$ (and if $R_i^* = R_i$ for all i, then we say the association scheme is *symmetric*).
- For all i, j and all $(x, y) \in S \times S$, define

$$\rho_{ij}(x, y) = |\{z \in S \mid (x, z) \in R_i, (z, y) \in R_j\}|.$$

  For each k, and for all $x, y \in R_k$, the integer $\rho_{ij}(x, y)$ is a constant, denoted ρ_{ij}^k.

These constants ρ_{ij}^k are called the *intersection numbers* or *parameters* or *structure constants* of the association scheme.

6.7.1 Adjacency rings (Bose–Mesner algebras)

Next, we recall (see Herman [He11]) the matrix-theoretic version of this definition.

Let $Mat_{m \times m}(\mathbb{Z})$ denote the non-abelian ring of $m \times m$ matrices with integer coefficients.

Definition 6.7.2. Let S be a finite abelian multiplicative group of order m. Let $(S, R_0, \ldots, R_s)$ denote a tuple consisting of S with relations R_i for which we have a disjoint union

$$S \times S = R_0 \cup R_1 \cup \cdots \cup R_s,$$

with $R_i \cap R_j = \emptyset$ for all $i \neq j$. Let $A_i \in Mat_{m \times m}(\mathbb{Z})$ denote the adjacency matrix of R_i, $i = 0, 1, \ldots, s$.

We say that the subring of $\mathbb{Z}[Mat_{m \times m}(\mathbb{Z})]$ generated by the set $\{A_i\}$ of adjacency matrices is an *adjacency ring* (also called a Bose–Mesner algebra) provided the set of adjacency matrices satisfies the following five properties:

- for each integer $i \in \{0, 1, 2, \ldots, s\}$, A_i is a $(0, 1)$-matrix;

- $\sum_{i=0}^{s} A_i = J$ (where J is the all 1's matrix);

- for each integer $i \in \{0, 1, 2, \ldots, s\}$, $A_i^t = A_j$, for some integer $j \in \{0, 1, 2, \ldots, s\}$;

- there is a subset $K \subset \{0, 1, 2, \ldots, s\}$ such that $\sum_{j \in K} A_j = I$; and

- there is a set of nonnegative integers $\{\rho_{ij}^k \mid i, j, k \in \{0, 1, 2, \ldots, s\}\}$ such that

$$A_i A_j = \sum_{k=0}^{s} \rho_{ij}^k A_k$$

for all such i, j.

It is well known that a partial difference set (G, D) is naturally associated to a 2-class association scheme, namely (G, R_0, R_1, R_2) where R_0 is the diagonal on $G \times G$,

$$R_0 = \Delta_G,$$

$$R_1 = \{(g, h) \mid gh^{-1} \in D\},$$

$$R_2 = \{(g, h) \mid gh^{-1} \notin D, \ g \neq h\}.$$

To verify this, let consider the "Schur ring".

6.7.2 Schur rings

For the following definition, we identify any subset S of G with the formal sum of its elements in $\mathbb{C}[G]$.

Definition 6.7.3. Let G be a finite abelian group and let $C_0, C_1, \ldots, C_s$ denote finite subsets with the following properties.

- $C_0 = \{1\}$ is the singleton containing the identity.

- We have a disjoint union

$$G = C_0 \cup C_1 \cup \cdots \cup C_s,$$

with $C_i \cap C_j = \emptyset$ for all $i \neq j$.

- For each i there is a j such that $C_i^{-1} = C_j$ (and if $C_i^{-1} = C_i$ for all i, then we say the Schur ring is *symmetric*).

- For all i, j, we have

$$C_i \cdot C_j = \sum_{k=0}^{s} \rho_{ij}^{k} C_k,$$

for some integers ρ_{ij}^{k}.

The subalgebra of $\mathbb{C}[G]$ generated by $C_0, C_1, \ldots, C_s$ is called a *Schur ring* over G.

In the cases we are dealing with, the Schur ring is commutative, so $\rho_{ij}^{k} = \rho_{ji}^{k}$, for all i, j, k.

If $(G, C_0, \ldots, C_s)$ is a Schur ring, then

$$R_i = \{(g, h) \in G \times G \mid gh^{-1} \in C_i\},$$

for $0 \leq i, j \leq s$, gives rise to its corresponding association scheme.

Lemma 6.7.4. *Let G be a finite abelian multiplicative group, and let D be a subset of G such that $1 \notin D$. Suppose that D is a disjoint union $D = D_1 \cup \cdots \cup D_r$, with $D_i^{-1} = D_i$ for all i. Let $D_0 = \{1\}$ and let $D_{r+1} = G \setminus (D \cup \{1\})$. Suppose that D_{r+1} is not empty. For each i with $0 \leq i \leq r+1$, let*

$$R_i = \{(g, h) \in G \times G \mid gh^{-1} \in D_i\}.$$

Then (G, D) is a symmetric weighted PDS (defined in Definition 6.9.1) if and only if $(G, R_0, R_1, \ldots, R_{r+1})$ is a symmetric association scheme of class $s = r + 1$.

Proof. Suppose that (G, D) is a symmetric weighted PDS. To show that it determines a Schur ring, we must show that structure constants exist, i.e., we must show that there are nonnegative integers ρ_{ij}^{ℓ} such that

$$D_i \cdot D_j = \sum_{\ell=0}^{r+1} \rho_{ij}^{\ell} D_\ell, \tag{6.10}$$

for $0 \leq i, j \leq r + 1$. Symmetry of the Schur ring follows from symmetry of the weighted PDS.

Let $k_i = |D_i|$ for $0 \leq i \leq r+1$. Note that (G, D) is a symmetric weighted PDS if and only if $D_i^{-1} = D_i$ for $1 \leq i \leq r$ and the following identity holds in $\mathbb{C}[G]$, for $1 \leq i, j \leq r$:

$$D_i \cdot D_j = \delta_{ij} k_i \cdot D_0 + \sum_{\ell=1}^{r} \lambda_{i,j,\ell} D_\ell + \mu_{i,j} D_{r+1} \tag{6.11}$$

(where $\delta_{ij} = 1$ if $i = j$ and $\delta_{ij} = 0$ otherwise). Note that identity (6.11) implies identity (6.10), provided that, for $0 \leq i, j \leq r+1$, we put $\rho_{ij}^0 = \delta_{ij} k_i$, for $1 \leq i, j, \ell \leq r$, we put $\rho_{ij}^\ell = \lambda_{i,j,\ell}$ and $\rho_{ij}^{r+1} = \mu_{i,j}$. Furthermore, since $D_0 \cdot D_j = D_j \cdot D_0 = D_j$, for all j, identity (6.10) holds if we put $\rho_{0j}^\ell = \rho_{j0}^\ell = \delta_{j\ell}$ for $0 \leq j, \ell \leq r+1$.

By expanding out expressions for $D_i \cdot G$ and $D_{r+1} \cdot G$, it can be shown that

$$\rho_{i,r+1}^\ell = k_i - \delta_{i\ell} - \sum_{j=1}^{r} \lambda_{i,j,\ell}, \qquad \text{for } 1 \leq i, \ell \leq r,$$

$$\rho_{i,r+1}^{r+1} = k_i - \sum_{j=1}^{r} \mu_{i,j}, \qquad \text{for } 1 \leq i \leq r, \text{ and}$$

$$\rho_{r+1,r+1}^{r+1} = k_{r+1} - 1 - \sum_{i=1}^{r} k_i + \sum_{i=1}^{r} \sum_{j=1}^{r} \mu_{i,j}.$$

Also, $\rho_{ij}^\ell = \rho_{ji}^\ell$ for all i and j, because G is abelian.

In the converse direction, if $(G, R_0, R_1, \ldots, R_{r+1})$ is a symmetric association scheme, it follows immediately that (G, D) is a symmetric weighted PDS whose parameters are related to the intersection numbers of the association scheme by the same relations as above.   □

Remark 6.7.5. For a more general version of this, see Theorem 6.10.4 below.

Example 6.7.6. For an example of a Schur ring, we return to a partial difference set (G, D), such that $D = D^{-1}$ and D does not contain the identity element 1. Let

$$D' = G \setminus (D \cup \{1\}).$$

We have the well-known intersection

$$\begin{aligned} D \cdot D &= (k - \mu) \cdot 1 + (\lambda - \mu) \cdot D + \mu \cdot G \\ &= k \cdot 1 + \lambda \cdot D + \mu \cdot D', \end{aligned} \tag{6.12}$$

and

$$\begin{aligned} D \cdot D' &= (-k + \mu) \cdot 1 + (-1 - \lambda + \mu) \cdot D + (k - \mu) \cdot G \\ &= 0 \cdot 1 + (k - 1 - \lambda) \cdot D + (k - \mu) \cdot D'. \end{aligned} \tag{6.13}$$

Provided $k \geq \max(\mu, \lambda + 1)$, $|G| \geq \max(k + 1, 2k - \mu + 2)$, with these, one can verify that a partial difference set naturally yields an associated Schur ring, generated by D, D', and $D_0 = \{1\}$ in $\mathbb{C}[G]$, and a 2-class association scheme.

Using (6.13), one can verify that D' is (v, k', λ', μ')-partial difference set with $(D')^{-1} = D'$ and $1 \notin D'$, where

$$
\begin{aligned}
k' &= v - k - 1 \\
\lambda' &= v - 2k - 2 + \mu, \text{and} \\
\mu' &= v - 2k + \lambda.
\end{aligned}
\tag{6.14}
$$

We include a proof here for convenience.

Proof. We will show that D' is a (v, k', λ', μ')-partial difference set. The first of these three equations is immediate, from the definition of D'. The fact that $D' = (D')^{-1}$ also follows immediately from the hypotheses.

By the definition of D, and because $D^{-1} = D$, we have

$$
D \cdot D = k1 + \lambda D + \mu D'.
\tag{6.15}
$$

To find $D \cdot D'$, we note that

$$
\begin{aligned}
kG &= D \cdot G \\
&= D \cdot (\{1\} + D + D') \\
&= D + D \cdot D + D \cdot D'
\end{aligned}
$$

so that

$$
D \cdot D' = (k - \lambda - 1)D + (k - \mu)D'.
\tag{6.16}
$$

Similarly, we note that

$$
\begin{aligned}
k'G &= G \cdot D' \\
&= (\{1\} + D + D') \cdot D' \\
&= D' + D \cdot D' + D' \cdot D'
\end{aligned}
$$

so that

$$
\begin{aligned}
D' \cdot D' &= k'\{1\} + (k' - k + \lambda + 1)D + (k' - k - 1 - +\mu)D' \\
&= k'\{1\} + (v - 2k + \lambda)D + (v - 2k - 2 + \mu)D'.
\end{aligned}
\tag{6.17}
$$

Equation (6.17) shows that D' is a (v, k', λ', μ')-partial difference set, with λ' and μ' as in Equation (6.14).   □

It can be shown that $\mu' = k'\left(1 - \frac{\mu}{k}\right)$.

With the identities in the above example, one can verify that a partial difference set naturally yields an associated Schur ring and a 2-class association scheme.

6.8 The matrix-walk theorem

Suppose that $\Gamma = (V, E)$ is any edge-weighted graph (without loops or multiple edges) whose edge weights are positive integers. We fix a labeling of the set of vertices $V(\Gamma)$, which we often identify with the set $\{0, 1, \ldots, N-1\}$, where $N = |V(\Gamma)|$. If u and v are vertices of Γ, then a *walk P from u to v with weight sequence* $(w_0, w_1, \ldots, w_{k-1})$ is a sequence of edges $e_0 = (v_0, v_1) \in E$, $e_1 = (v_1, v_2) \in E$, $\ldots$, $e_{k-1} = (v_{k-1}, v_k) \in E$, where $v_0 = u$ and $v_k = v$, connecting u to v, where edge e_i has weight w_i. Let $A = (a_{ij})$ denote the $N \times N$ weighted adjacency matrix of Γ, where $i, j \in \{0, 1, \ldots, N-1\}$ and where

$$a_{ij} = \begin{cases} w, & \text{if } (i, j) \text{ is an edge of weight } w, \\ 0, & \text{if } (i, j) \text{ is not an edge of } \Gamma. \end{cases} \qquad (6.18)$$

From this adjacency matrix A, we can derive weight-specific adjacency matrices as follows. For each weight w of Γ, let $A(w) = (a(w)_{ij})$ denote the $N \times N$ $(1, 0)$-matrix defined by

$$a(w)_{ij} = \begin{cases} 1, & \text{if } (i, j) \text{ is an edge of weight } w, \\ 0, & \text{if } (i, j) \text{ is not an edge of weight } w. \end{cases} \qquad (6.19)$$

When Γ is the Cayley graph of a $GF(p)$-valued function, we identify the edge-weights with the integers $\{1, \ldots, p-1\}$. We extend the weight set by imposing the following conventions:

(a) If u and v are distinct vertices of Γ but (u, v) is not an edge of Γ, then we say the *weight* of (u, v) is $w = p$.

(b) If $u = v$ is a vertex of Γ (so (u, v) is not an edge, since Γ has no loops), then we say the *weight* of (u, v) is $w = 0$.

This allows us to define the weight-specific adjacency matrices A_p and A_0 as well, and we can (and do) extend the weight set of Γ by appending p and 0. Clearly, these weight-specific adjacency matrices have disjoint supports: if $a(w)_{ij} \neq 0$, then $a(w')_{ij} = 0$ for all weights $w' \neq w$.

Note that an alternative convention, which can be used for more general edge-weighted graphs, is to extend the weight set by appending 0 (for case (a) above) and -1 (for case (b) above).

In Corollary 6.10.5, we will give conditions under which these weight-specific adjacency matrices form a Bose–Mesner algebra.

For the reader's convenience, the well-known matrix-walk theorem is formulated as in the result below (see, e.g., [GR01] for a proof in the unweighted case).

Proposition 6.8.1. *For any vertices u, v of Γ and any sequence of nonzero edge weights $w_1, w_2, \ldots, w_k$, the (u, v)-th entry of $A(w_1)A(w_2)\ldots A(w_k)$ is* *equal to the number of walks of weight sequence $(w_1, w_2, \ldots, w_k)$ from u to v. Moreover, $\mathrm{tr}\,(A(w_1)A(w_2)\ldots A(w_k))$ is equal to the total number of closed walks of Γ of weight sequence $(w_1, w_2, \ldots, w_k)$.*

6.9 Weighted partial difference sets

Let G be a finite abelian multiplicative group of order v, and let D be a subset of G. Decompose D into a union of disjoint subsets

$$D = D_1 \cup \cdots \cup D_s, \tag{6.20}$$

and assume $1 \notin D$. Let $k_i = |D_i|$.

Definition 6.9.1. We say (G, D), or the collection $(G, D_1, D_2, \ldots, D_s)$, is a *weighted (v, k, λ, μ)-partial difference set* if the following properties hold:

- The list of "differences"

$$D_i D_j^{-1} = \{d_1 d_2^{-1} \mid d_1 \in D_i, d_2 \in D_j\},$$

  represents every non-identity element of D_ℓ exactly $\lambda_{i,j,\ell}$ times and every non-identity element of $G \setminus D$ exactly $\mu_{i,j}$ times $(1 \leq i, j, \ell \leq s)$.

- For each i there is a j such that $D_i^{-1} = D_j$ (and if $D_i^{-1} = D_i$ for all i, then we say the weighted partial difference set is *symmetric*).

Remark 6.9.2. If $D = D_1 \cup \cdots \cup D_r$ is a symmetric weighted partial difference set, then $\mu_{i,j} = \mu_{j,i}$ and $\lambda_{i,j,\ell} = \lambda_{j,i,\ell}$.

This notion can be characterized algebraically in terms of the group ring $\mathbb{C}[G]$.

Lemma 6.9.3. *Let G be a finite abelian multiplicative group, and let D be a subset of G such that $1 \notin D$. Suppose that D is a disjoint union $D = D_1 \cup \cdots \cup D_r$, with $D_i^{-1} = D_i$ for all i. Let $D_0 = \{1\}$ and let $D_{r+1} = G \setminus (D \cup \{1\})$. Suppose that D_{r+1} is not empty. Let $k_i = |D_i|$ for $0 \leq i \leq r+1$. Note that (G, D) is a symmetric weighted PDS if and only if $D_i^{-1} = D_i$ for $1 \leq i \leq r$ and the identity (6.11) holds in $\mathbb{C}[G]$, for $1 \leq i, j \leq r$.*

The straightforward proof is omitted.

We will now state a more general proposition concerning weighted partial difference sets.

Proposition 6.9.4. *Let G be a finite abelian group. Let $D_1, \ldots, D_r \subseteq G$ such that $D_i \cap D_j = \emptyset$ if $i \neq j$, and let $D_0 = \{1\}$ and let $D_{r+1} = G \setminus (D \cup \{1\})$, where D is as above. Suppose*

- G is the disjoint union $D_0 \cup \cdots \cup D_{r+1}$,

- for each i there is a j such that $D_i^{-1} = D_j$, and

- condition (6.10) holds for the structure constants ρ_{ij}^k of the D_is.

Then, the matrices $P_k = (\rho_{ij}^k)_{0 \le i,j \le l}$ satisfy the following properties:

- P_0 is a diagonal matrix with entries $|D_0|, \ldots, |D_{r+1}|$.

- For each k, the j-th column of P_k has sum $|D_j|$ $(j = 0, \ldots, r+1)$. Likewise, the i-th row of P_k has sum $|D_i|$ $(i = 0, \ldots, r+1)$.

Proof. We begin by taking the sum

$$D_i \cdot D_j = \sum_{k=0}^{r+1} \rho_{ij}^k D_k$$

over all $i, 0 \le i \le r+1$:

$$G \cdot D_j = \sum_{k=0}^{r+1} \left(\sum_{i=0}^{r+1} \rho_{ij}^k \right) D_k .$$

We know that $G \cdot D_j = |D_j| \cdot G$, and all the D_k are disjoint. Consider the above-displayed equation as an identity in the Schur ring. Each element of G must occur $|D_j|$ times on each side of this equation. Therefore,

$$|D_j| = \sum_{i=0}^{r+1} \rho_{ij}^k .$$

So the sum of the elements in the j-th row of P_k is $|D_j|$ for each j and k. The analogous claim for the row sums is proven similarly.   □

These constants ρ_{ij}^k are called the *intersection numbers* or *structure constants* of the Schur ring.

Example 6.9.5. (S. Walsh [W14]) If $G = GF(3)^2$ and

$$D_0 = \{(0,0)\},$$
$$D_1 = \{(1,0), (2,0), (0,1), (0,2)\}, \text{ and}$$
$$D_2 = \{(1,1), (2,1), (1,2), (2,2)\},$$

then (G, D_0, D_1, D_2) is a weighted partial difference set. Moreover, the intersection numbers ρ_{ij}^k are given as follows:

ρ_{ij}^0	0	1	2
0	1	0	0
1	0	4	0
2	0	0	4

ρ_{ij}^1	0	1	2
0	0	1	0
1	1	1	2
2	0	2	2

ρ_{ij}^2	0	1	2
0	0	0	1
1	0	2	2
2	1	2	1

no ρ_{ij}^3

How does the above notion of a weighted partial difference set relate to the usual notion of a partial difference set?

Lemma 6.9.6. *Let (G, D), where $D = D_1 \cup \cdots \cup D_r$ (disjoint union) is as in (6.20), be a symmetric weighted partial difference set, with parameters $(v, (k_i), (\lambda_{i,j,\ell}), (\mu_{i,j}))$. If*

$$\sum_{i,j} \lambda_{i,j,\ell}$$

does not depend on ℓ, $1 \leq \ell \leq r$, then D is also an unweighted partial difference set with parameters (v, k, λ, μ) where

$$k = \sum_i k_i, \quad \lambda = \sum_{i,j} \lambda_{i,j,\ell}, \quad \mu = \sum_{i,j} \mu_{i,j}.$$

Proof. The claim is that (G, D) is a partial difference set with parameters (v, k, λ, μ). Since $v = |G|$ and

$$k = |D| = |D_1| + \cdots + |D_r| = k_1 + \cdots + k_r,$$

we need only verify the claim regarding λ and μ.

Does each element d of D occur the same number of times in the list DD^{-1}? Suppose $d \in D_\ell$, where $1 \leq \ell \leq r$. By hypothesis, d occurs in $D_i - D_j$ exactly $\lambda_{i,j,\ell}$ times. Since DD^{-1} is the concatenation of the $D_i D_j^{-1}$, for $1 \leq i, j \leq r$, d occurs in $D \cdot D^{-1}$ exactly

$$\sum_{i,j} \lambda_{i,j,\ell}$$

times. By hypothesis, this does not depend on ℓ, so the claim regarding λ has been verified.

Does each non-identity element d of $G \setminus D$ occur the same number of times in the list $D \cdot D^{-1}$? By hypothesis, d occurs in $D_i D_j^{-1}$ exactly $\mu_{i,j}$ times. Since $D \cdot D^{-1}$ is the concatenation of the $D_i D_j^{-1}$, for $1 \leq i, j \leq r$, d occurs in $D \cdot D^{-1}$ exactly

$$\sum_{i,j} \mu_{i,j}$$

times. This verifies the claim regarding μ and completes the proof of the lemma.   □

Example 6.9.7. Consider the finite field

$$GF(9) = GF(3)[x]/(x^2 + 1) = \{0, 1, 2, x, x + 1, x + 2, 2x, 2x + 1, 2x + 2\},$$

written additively. The set of nonzero quadratic residues is given by

$$D = \{1, 2, x, 2x\}.$$

Let $D_1 = \{1, 2\}$ and $D_2 = \{x, 2x\}$. We will use square brackets [ ] when we wish to denote multi-sets, in which repetition is allowed. Translating the multiplicative notation to the additive notation, we find that as multi-sets,

$$D_1 D_1^{-1} = [d_1 - d_2 \mid d_1 \in D_1, d_2 \in D_1] = [0, 0, 1, 2],$$

$$D_1 D_2^{-1} = [d_1 - d_2 \mid d_1 \in D_1, d_2 \in D_2] = [x + 1, x + 2, 2x + 1, 2x + 2],$$

$$D_2 D_1^{-1} = [d_1 - d_2 \mid d_1 \in D_2, d_2 \in D_1] = [x + 1, x + 2, 2x + 1, 2x + 2],$$

$$D_2 D_2^{-1} = [d_1 - d_2 \mid d_1 \in D_2, d_2 \in D_2] = [0, 0, x, 2x].$$

Therefore, this describes a weighted partial difference set with parameters

$$k_1 = 2, \ k_2 = 2,$$

$$\lambda_{1,1,1} = 1, \ \lambda_{1,1,2} = 0, \ \lambda_{1,2,1} = 0, \ \lambda_{1,2,2} = 0,$$

$$\lambda_{2,1,1} = 0, \ \lambda_{2,1,2} = 0, \ \lambda_{2,2,1} = 0, \ \lambda_{2,2,2} = 1,$$

and

$$\mu_{1,1} = 0, \ \mu_{1,2} = 1, \ \mu_{2,1} = 1, \ \mu_{2,2} = 0.$$

Suppose $(G, D_1, D_2, \ldots, D_r)$ is a weighted (v, k, λ, μ)-partial difference set. The *adjacency matrix of the weighted partial difference set*, say $B = (B_{ij})$, is defined by $B_{ij} = k$ if $j - i \in D_k$ (for $i, j \in G$ and $k \in \{1, \ldots, r\}$).

6.10 Weighted Cayley graphs

Definition 6.10.1. Let G be a finite abelian multiplicative group, and let D be a subset of G such that $1 \notin D$ and such that D has a disjoint decomposition $D = D_1 \cup D_2 \cup \cdots \cup D_r$. The *edge-weighted Cayley graph* $\Gamma = \Gamma(G, D)$ *associated with* (G, D) is the edge-weighted graph constructed as follows. Let the vertices of the graph be the elements of the group G. Two vertices g_1 and g_2 are connected by an edge of weight i if $g_2 = dg_1$ for some $d \in D_i$. If $D_i^{-1} = D_i$ for all i, the graph Γ is undirected.

6.10.1 Edge-weighted strongly regular graphs

The concept of a strongly regular graph generalizes to that of an edge-weighted strongly regular graph. Let W be a set of edge weights on a graph $\Gamma = (V, E)$, i.e., suppose that to each edge e in Γ we associate a weight $w \in W$. In our examples, W will usually be $GF(p) \setminus \{0\}$.

Definition 6.10.2. For each vertex $u \in V$ and edge weight $a \in W$ we define the a-neighborhood $N(u, a)$ of u to be the set of all neighbors v of u in Γ for which the edge (u, v) has weight a.

When $W = GF(p) \setminus \{0\}$, we also define $N(u, 0)$ to be the set of all non-neighbors of u, i.e., the set of all $v \in V$ such that (u, v) is not an edge in Γ. In particular, $u \in N(u, 0)$.

Definition 6.10.3. Let Γ be a connected edge-weighted graph which is regular as a simple (unweighted) graph. The graph Γ is called an *edge-weighted strongly regular graph* (SRG) with parameters v, $k = (k_a)_{a \in W}$, $\lambda = (\lambda_a)_{a \in W^3}$, and $\mu = (\mu_a)_{a \in W^2}$, denoted $SRG_W(v, k, \lambda, \mu)$, if Γ has v vertices, and there are constants k_a, λ_{a_1, a_2, a_3}, and μ_{a_1, a_2}, for $a, a_1, a_2, a_3 \in W$, such that

$$|N(u, a)| = k_a \text{ for all } u \in V$$

and for vertices $u_1 \neq u_2$ we have

$$|N(u_1, a_1) \cap N(u_2, a_2)| = \begin{cases} \lambda_{a_1, a_2, a_3}, & \text{if } u_1 \in N(u_2, a_3), \\ \mu_{a_1, a_2}, & \text{if } u_1 \notin N(u_2). \end{cases} \tag{6.21}$$

6.10.2 Weighted partial difference sets

As we will see, there a weighted analog of the correspondence between partial difference sets and strongly regular graphs.

We have the following generalization of Theorem 6.6.5.

Theorem 6.10.4. *Let G be a finite abelian multiplicative group, and let D be a subset of G such that $1 \notin D$ and such that D has a disjoint decomposition $D = D_1 \cup D_2 \cup \cdots \cup D_r$. Let $D_0 = \{1\}$, let $D_{r+1} = G \setminus (D \cup D_0)$, and let*

$$R_i = \{(g, h) \in G \times G \mid gh^{-1} \in D_i\}, \qquad 0 \le i \le r+1.$$

The following statements are equivalent:

(a) *The set D is a symmetric weighted partial difference set.*

(b) *The graph $\Gamma(G, D)$ is an edge-weighted strongly regular graph with edge weights $\{1, 2, \ldots, r\}$.*

(c) *The tuple $(G, R_0, R_1, \ldots, R_{r+1})$ is a symmetric association scheme of class $r + 1$.*

Proof. The equivalence of (a) and (c) is just Lemma 6.7.4.

$((a) \implies (b))$ Suppose (G, D) is a weighted partial difference set satisfying $D_i^{-1} = D_i$ for all i, and having parameters (v, k, λ, μ), where $v = |G|$, $k = \{k_i\}$ with $k_i = |D_i|$, $\lambda = \{\lambda_{i,j,\ell}\}$, and $\mu = \{\mu_{i,j}\}$. The graph $\Gamma = \Gamma(G, D)$ has $v = |G|$ vertices, by definition. Each vertex g of Γ has k_i neighbors of weight i, namely, dg where $d \in D_i$. Let g_1 and g_2 be distinct vertices in Γ. Let x be a vertex which is a neighbor of each: $x \in N(g_1, i) \cap N(g_2, j)$. By definition, $x = d_1 g_1 = d_2 g_2$, for some $d_1 \in D_i$, $d_2 \in D_j$. Therefore, $d_1^{-1} d_2 = g_1 g_2^{-1}$. If $g_1 g_2^{-1} \in D_\ell$, for some $\ell \ne 0, r + 1$, then there are $\lambda_{i,j,\ell}$ solutions, by definition of a weighted PDS. If $g_1 g_2^{-1} \in D_{r+1}$, then there are $\mu_{i,j}$ solutions, by definition of a weighted PDS.

$((b) \implies (a))$ For the remainder of the proof, note that the reasoning above is reversible. Details are left to the reader. $\qquad \square$

We sometimes extend the weight set of $\Gamma = \Gamma(G, D)$ by imposing the following conventions:

(a) If u and v are distinct vertices of Γ but (u, v) is not an edge of Γ, then we say the *weight* of (u, v) is $w = r + 1$.

(b) If $u = v$ is a vertex of Γ (so (u, v) is not an edge, since Γ has no loops), then we say the *weight* of (u, v) is $w = 0$.

This allows us to extend the set of weight-specific adjacency matrices given by (6.19) to the set $A_0, A_1, \ldots, A_r, A_{r+1}$.

Corollary 6.10.5. *The graph $\Gamma(G, D)$ is an edge-weighted strongly regular graph if and only if the (extended) set of weight-specific adjacency matrices given by (6.19) form a Bose-Mesner algebra with $K = \{0\}$.*

Proof. The corollary is immediate from the definition of Bose-Mesner algebra (see Definition 6.7.2), since the weight-specific adjacency matrices of $\Gamma(G, D)$ coincide with the adjacency matrices of the binary relations R_i. $\qquad \square$

6.10.3 Level curves of p-ary functions

In this section we consider the "level curves" $f^{-1}(a) \subset GF(p)^n$ ($a \in GF(p)$, $a \neq 0$) and investigate the combinatorial structure of these sets, especially when f is bent.

Let f be a $GF(p)$-valued function on $GF(p)^n$. Recall, from (2.11), that the Cayley graph of f is defined to be the edge-weighted directed graph (digraph) $\Gamma_f = (V, E_f)$, whose vertex set is $V = V(\Gamma_f) = GF(p)^n$ and whose set of edges is

$$E_f = \{(u, v) \in GF(p)^n \times GF(p)^n \mid f(u - v) \neq 0\},$$

where the edge $(u, v) \in E_f$ has weight $f(u - v)$.

However, if a function $f \colon GF(p)^n \to GF(p)$ is even, then

- we can (and do) regard Γ_f as a weighted (undirected) graph,

- its level curves $D_i = f^{-1}(i)$ satisfy $D_i^{-1} = D_i$ (as sets).

Theorem 6.10.6. *Let $f \colon GF(p)^n \to GF(p)$ be an even function such that $f(0) = 0$. Let $G = GF(p)^n$, and let $D_i = f^{-1}(i)$, for $i = 1, 2, \ldots, p - 1$ If $(G, D_1, D_2, \ldots, D_{p-1})$ is a symmetric weighted partial difference set, then the associated edge-weighted strongly regular graph is the edge-weighted Cayley graph of f.*

Remark 6.10.7. Roughly speaking, this theorem says that "if the level curves of f form a weighted partial difference set, then the (edge-weighted) Cayley graph corresponding to f agrees with the (edge-weighted) strongly regular graph associated to the weighted partial difference set."

Proof. The adjacency matrix A for the Cayley graph of f is defined by $A_{ij} = f(j - i)$ for $i, j \in GF(p)^n$. So the top row of A is defined by $A_{0j} = f(j)$. The adjacency matrix for any Cayley graph can be determined from its top row, since it is a circulant matrix. Therefore, it is enough to show that the adjacency matrix of the weighted partial difference set has the same top row as A. Let $B = (B_{ij})$ be the adjacency matrix of the weighted partial difference set, so $B_{ij} = k$ if $j - i \in D_k$ (for all $i, j \in GF(p)^n$ and $k \in GF(p)$). The top row of B is defined by $B_{0j} = k$ if $j \in D_k$. But if $j \in D_k$, then $f(j) = k$, so $B_{0j} = f(j)$. The top rows of A and B are equivalent, so $A = B$. Therefore, the strongly regular graph associated with the weighted partial difference set (G, D) is the Cayley graph of f. $\qquad\square$

6.10.4 Intersection numbers

This section is devoted to stating some results on the ρ_{ij}^k's.

Theorem 6.10.8. *Let $f : GF(p)^n \to GF(p)$ be a function and let Γ be its Cayley graph. Assume Γ is a weighted strongly regular graph. Let $A = (a_{k,l})$ be the adjacency matrix of Γ. Let $A_i = (a_{k,l}^i)$ be the $(0,1)$-matrix where*

$$
a_{k,l}^i = \begin{cases} 1, & if\, a_{k,l} = i, \\ 0, & otherwise, \end{cases}
$$

for each $i = 1, 2, \ldots, p-1$. Let A_0 be the $p^n \times p^n$ identity matrix. Let A_p be the $(0,1)$-matrix such that $A_0 + A_1 + \cdots + A_{p-1} + A_p = J$, the $p^n \times p^n$ matrix with all entries 1. Let R denote the matrix ring generated by $\{A_0, A_1, \ldots, A_p\}$. The intersection numbers p_{ij}^k defined by

$$
A_i A_j = \sum_{k=0}^{p} p_{ij}^k A_k \tag{6.22}
$$

satisfy the formula

$$
p_{ij}^k = \left(\frac{1}{p^n |D_k|} \right) Tr(A_i A_j A_k),
$$

for all $i, j, k = 1, 2, \ldots, p$.

This is (17.13) in Cameron and van Lint [CvL91]. We provide a different proof for the reader's convenience.

Proof. By the matrix-walk theorem, $A_i A_j$ can be considered as counting walks along the Cayley graph of specific edge weights. Supposed (u, v) is an edge of Γ with weight k. If $k = 0$, then $u = v$ and the edge is a loop. If $k = p$, then (u, v) is technically not an edge in Γ, but we will label it as an edge of weight p.

The (u, v)-th entry of $A_i A_j$ is the number of walks of length 2 from u to v, where the first edge has weight i and the second edge has weight j; the entry is 0 if no such walk exists. If we consider the (u, v)-th entry on each side of the equation defining the structure constants, Equation (6.22) with $s = p$, we can deduce that p_{ij}^k is the number of walks of length 2 from u to v, where the first edge has weight i and the second edge has weight j (it equals 0 if no such walk exists) for any edge (u, v) with weight k in Γ.

Similarly, the matrix-walk theorem implies that $tr(A_i A_j A_k)$ is the total number of closed walks of length 3 having edge weights i, j, k. We claim that if $\triangle$ is any triangle with edge weights i, j, k, then, by subtracting an element $v \in GF(p)^n$, we will obtain a triangle in Γ containing the zero vector as a vertex with the same edge weights. Suppose $\triangle = (u_1, u_2, u_3)$, where (u_1, u_2) has edge weight i, (u_2, u_3) has edge weight j, and (u_3, u_1) has edge weight k. Let $\triangle' = (0, u_2 - u_1, u_3 - u_1)$. We compute the edge weights of $\triangle'$:

edge weight of $(0, u_2 - u_1) = f(0 - (u_2 - u_1)) = f(u_1 - u_2) = i$;
edge weight of $(u_2 - u_1, u_3 - u_1) = f((u_2 - u_1) - (u_3 - u_1)) = f(u_2 - u_3) = j$;
edge weight of $(u_3 - u_1, 0) = f((u_3 - u_1) - 0) = f(u_3 - u_1) = k$.

Thus the claim is proven.

Therefore,

$$\left(\frac{1}{|GF(p)^n|}\right) tr(A_i A_j A_k) = \left(\frac{1}{p^n}\right) tr(A_i A_j A_k)$$

is the number of closed walks of length 3 having edge weights i, j, k and containing the zero vector as a vertex, incident to the edge of weight i and the edge of weight k.

There are $|D_k|$ edges of weight k incident to the zero vector, so

$$\left(\frac{1}{p^n}\right)\left(\frac{1}{|D_k|}\right) tr(A_i A_j A_k)$$

is the number of (i, j)-weighted walks (of length 2) from the zero vector to any k-neighbor of it. This is equivalent to the definition of the number p_{ij}^k in the matrix-walk theorem. $\qquad\square$

The following corollary is well-known (see Cameron and van Lint [CvL91] p. 202).

Corollary 6.10.9. *Let* $G = GF(p)^n$. *Let* $D_0, \ldots, D_r \subseteq G$ *such that* $D_i \cap D_j = \emptyset$ *if* $i \neq j$, *and*

- *G is the disjoint union of $D_0 \cup \cdots \cup D_r$,*

- *for each i there is a j such that $D_i^{-1} = D_j$, and*

- *$D_i \cdot D_j = \sum_{k=0}^{r} \rho_{ij}^k D_k$ for some positive integers ρ_{ij}^k.*

Then, for all i, j, k, $|D_k|\rho_{ij}^k = |D_i|\rho_{kj}^i$.

Proof. For all i, j, k, we have the following identity of adjacency matrices:

$$\text{Tr}(A_i A_j A_k) = p^n |D_k|\rho_{ij}^k,$$

where p^n is the order of G and ρ_{ij}^k is an intersection number. Since $\text{Tr}(AB) = \text{Tr}(BA)$ for all matrices A and B, $\text{Tr}(A_i A_j A_k) = \text{Tr}(A_k A_j A_i)$, and the proposition follows. $\qquad\square$

These concepts can be reformulated in terms of association schemes.

If G is a set and $D = D_1 \cup D_2 \cup \cdots \cup D_r$ (all D_i distinct) is a weighted partial difference set of G, then we can construct an association scheme as follows:

- Define $R_0 = \Delta_G = \{(x, x) \in G \times G \mid x \in G\}$.

- For $1 \le i \le r$, define $R_i = \{(x, y) \in G \times G \mid xy^{-1} \in D_i, x \ne y\}$.

- Define $R_{r+1} = \{(x, y) \in G \times G \mid xy^{-1} \notin D, x \ne y\}$.

According to Lemma 6.7.4, this association scheme structure determines the corresponding weighted partial difference set.

6.10.5 Cayley graphs of p-ary functions

Let us return to describing the Cayley graph in (2.11) above.

As the following lemma illustrates, it is very easy to characterize the Cayley graph of an even p-ary function in terms of its adjacency matrix.

Lemma 6.10.10. *Let Γ be an undirected edge-weighted graph with weights in $GF(p)$ and with vertex set $V = GF(p)^n$ (and vertices labeled using the bijection given by the p-ary representation map of Equation 6.5). Let $A = (a_{ij})$ be the (symmetric) weighted adjacency matrix of Γ, where $i, j \in \{0, 1, \ldots, p^n - 1\}$. Let f be an even $GF(p)$-valued function on V with $f(0) = 0$. Then Γ is the Cayley graph of f if and only if Γ is regular and the following conditions hold:*

(a) *For each $i \in \{0, 1, \ldots, p^n - 1\}$, $a_{i,0} = f(\eta(i))$.*

(b) *For each $i, j \in \{0, 1, \ldots, p^n - 1\}$, $a_{i,j} = a_{k,0}$, where $\eta(k) = \eta(i) - \eta(j)$.*

Proof. Let $w \in GF(p)$. We know that $A_{i,j} = w$ if and only if there is an edge of weight w from $\eta(i)$ to $\eta(j)$ if and only if $f(\eta(i) - \eta(j)) = w$. The result follows. □

We assume, unless stated otherwise, that f is even. For each $u \in V$, define

- $N(u) = N_{\Gamma_f}(u)$ to be the set of all neighbors of u in Γ_f,

- $N(u, a) = N_{\Gamma_f}(u, a)$ to be the set of all neighbors v of u in Γ_f for which the edge $(u, v) \in E_f$ has weight a (for each $a \in GF(p)^\times = GF(p) - \{0\}$),

- $N(u, 0) = N_{\Gamma_f}(u, 0)$ to be the set of all non-neighbors v of u in Γ_f (i.e., we have $(u, v) \notin E_f$),

- $\mathrm{supp}(f) = \{v \in V \mid f(v) \ne 0\}$ to be the *support* of f.

It is clear that $\mathrm{supp}(f) = N(0)$ is the set of neighbors of the zero vector. More generally, for any $u \in V$,

$$N(u) = u + \mathrm{supp}(f), \qquad (6.23)$$

where the last set is the collection of all vectors $u + v$, for some $v \in \mathrm{supp}(f)$.

Recall that we call a map $g : GF(p)^n \to GF(p)$ *balanced* if the cardinalities $|g^{-1}(x)|$ $(x \in GF(p))$ do not depend on x. We call the *signature* of $f :$ $GF(p)^n \to GF(p)$ the list

$$|S_0|, \ |S_1|, \ |S_2|, \dots, |S_{p-1}|,$$

where, for each i in $GF(p)$,

$$S_i = \{x \mid f(x) = i\}. \tag{6.24}$$

We can extend Equation (6.23) to the more precise statement

$$N(u, a) = u + S_a, \tag{6.25}$$

for all $a \in GF(p)$. We call $N(u, a)$ the *a-neighborhood* of u.

If $f \colon V \to GF(p)$, then we let $f_{\mathbb{C}} \colon V \to \mathbb{C}$ be the function whose values are those of f but regarded as integers (i.e., we select the congruence class residue representative in the interval $\{0, 1, \dots, p-1\}$). We abuse notation and often write f in place of $f_{\mathbb{C}}$.

The weighted adjacency matrix A of the Cayley graph of f is the matrix whose entries are

$$A_{i,j} = f_{\mathbb{C}}(\eta(i) - \eta(j)), \tag{6.26}$$

where $\eta(k)$ is the p-ary representation as in (6.5). If f is even and we regard Γ_f as an unweighted graph, then Γ_f is a regular of degree

$$\omega = \omega_f = wt(f),$$

where ω denotes the cardinality of $\operatorname{supp}(f) = \{v \in V \mid f(v) \neq 0\}$. If we let

$$\sigma_f = \sum_{v \in V} f_{\mathbb{C}}(v),$$

then $\hat{f}(0) = \sigma_f \geq |\operatorname{supp}(f)|$. If f is even and we count edges according to their weights, then Γ_f is an σ_f-regular (edge-weighted) graph. If we ignore weights, then it is an ω_f-weighted graph.

If A is the adjacency matrix of an edge-weighted strongly regular graph having parameters $(v, k_a, \lambda_{a_1, a_2, a_3}, \mu_{a_1, a_2})$ and positive weights $W \subseteq \mathbb{Z}$ one can compute $(A^2)_{ij}$ explicitly, again by looking at the three separate cases (a) $i = j$, (b) $i \neq j$ and i, j adjacent, (c) $i \neq j$ and i, j non-adjacent. We obtain

$$(A^2)_{i,j} = \begin{cases} \sum_{a \in W} a^2 k_a, & i = j, \\ \sum_{(a,b) \in W^2} ab\lambda_{(a,b,c)}, & i \neq j, i \in N(j,c), \\ \sum_{(a,b) \in W^2} ab\mu_{(a,b)}, & i \neq j, i \notin N(j). \end{cases} \tag{6.27}$$

Note that

$$W_f(0) = |S_0| + |S_1|\zeta + \cdots + |S_{p-1}|\zeta^{p-1},$$

which we can regard as an identity in the $(p-1)$-dimensional $\mathbb{Q}$-vector space $\mathbb{Q}(\zeta)$. The relation

$$1 + \zeta + \zeta^2 + \cdots + \zeta^{p-1} = 0,$$

gives

$$W_f(0) - |S_0| + |S_1|$$
$$= (|S_2| - |S_1|)\zeta^2 + \cdots + (|S_{p-1}| - |S_1|)\zeta^{p-1}.$$

We have proven the following result.

Lemma 6.10.11. *If $f : GF(p)^n \to GF(p)$ has the property that $W_f(0)$ is a rational number then*

$$|S_1| = |S_2| = \cdots = |S_{p-1}|,$$

and

$$W_f(0) = |S_0| - |S_1|.$$

In particular,

$$|\mathrm{supp}(f)| \;=\; |S_1| + |S_2| + \cdots + |S_{p-1}|$$
$$= (p-1)|S_1| = (p-1)(|S_0| - W_f(0)).$$

Remark 6.10.12. It is also known that if n is even and f is bent then

$$|S_1| = |S_2| = \cdots = |S_{p-1}|.$$

6.10.6 Group actions on bent functions

We note here some useful facts about the action of nondegenerate linear transforms on p-ary functions. Suppose that $f : V = GF(p)^n \to GF(p)$ and $\phi : V \to V$ is a nondegenerate linear transformation (isomorphism of V), and $g(x) = f(\phi(x))$. The functions f and g both have the same signature, $(|f^{-1}(i)| \mid i = 1, \ldots, p-1)$.

It is straightforward to calculate that

$$W_f g(u) = W_f((\phi^{-1})^t u),$$

where t denotes transpose.

It follows that if f is bent, so is $g = f \circ \phi$, and if f is bent and regular, so is g. If f is bent and weakly regular, with μ-regular dual f^*, then g is bent and weakly regular, with μ-regular dual g^*, where $g^*(u) = f^*((\phi^{-1})^t u)$.

Next, we examine the effect of the group action on bent functions and the corresponding weighted partial difference sets (in the case that the level curves determine a weighted PDS).

Proposition 6.10.13. *Let $f : GF(p)^n \to GF(p)$ be an even, bent function such that $f(0) = 0$ and define $D_i = f^{-1}(i)$ for $i \in GF(p) - \{0\}$. Suppose $\phi : GF(p)^n \to GF(p)^n$ is a linear map that is invertible (i.e., $\det \phi \neq 0$ (mod p)). Define the function $g = f \circ \phi$; g is the composition of a bent function and an affine function, so it is also bent. If the collection of sets $\{D_1, D_2, \ldots, D_{p-1}\}$ forms a weighted partial difference set for $GF(p)^n$ then so does its image under the function ϕ.*

Proof. We can explore this question by utilizing the Schur ring generated by the sets D_i. Define $D_0 = \{0\}$, where 0 denotes the zero vector in $GF(p)^n$, and define $D_p = GF(p)^n - \cup_{0 \leq i \leq p-1} D_i$.

The collection $(D_0, D_1, D_2, \ldots, D_{p-1}, D_p)$ forms a weighted partial difference set for $GF(p)^n$ if and only if $(C_0, C_1, C_2, \ldots, C_p)$ forms a Schur ring in $\mathbb{C}[GF(p)^n]$, where

$$C_0 = \{0\} \text{ (where 0 denotes the zero element of } GF(p)^n),$$

$$C_1 = D_1, \ldots, C_{p-1} = D_{p-1}$$

$$C_p = GF(p)^n - (C_0 \cup \cdots \cup C_{p-1})$$

$$C_i \cdot C_j = \sum_{k=0}^{p} \rho_{ij}^k C_k,$$

for some intersection numbers $\rho_{ij}^k \in \mathbb{Z}$. Note that f is even, so $C_i = C_i^{-1}$ for all i, where $C_i^{-1} = \{-x \mid x \in C_i\}$. Define $S_i = g^{-1}(i) = \{v \in GF(p)^n : g(v) = i\}$. $D_i = f^{-1}(i) = (g \circ \phi^{-1})^{-1}(i) = (\phi \cdot g^{-1})(i) = \phi(S_i)$. So the map ϕ sends S_i to D_i. ϕ can be extended to a map from $\mathbb{C}[GF(p)^n] \to \mathbb{C}[GF(p)^n]$ such that $\phi(g_1 + g_2) = \phi(g_1) + \phi(g_2)$ and $\phi(S_i) = D_i$. So ϕ is a homomorphism from the Schur ring of g to the Schur ring of f. Therefore, the level curves of g give rise to a Schur ring, and the weighted partial difference set generated by f is sent to a weighted partial difference set generated by g under the map

ϕ^{-1}. We conclude that the Schur ring of g corresponds to a weighted partial difference set for $GF(p)^n$, which is the image of that for f. $\square$

Remark 6.10.14. It is known that for homogeneous[4] weakly regular bent functions, the level curves give rise to a weighted partial difference set. This weighted partial difference set corresponds to an association scheme, and the dual association scheme corresponds to the dual bent function (see Pott, Tan, Feng, and Ling [PTFL11], Corollary 3). We know that any bent function equivalent to such a bent function also has this property, thanks to the proposition above.

6.11 Fourier transforms and graph spectra

In the Boolean ($p = 2$) case, there is a nice simple relationship between the Fourier transform and the Walsh-Hadamard transform. The spectrum of Γ_f is determined by the set of values of the Walsh-Hadamard transform of f when regarded as a vector of (integer) $0, 1$-values (of length 2^n). We ask whether this result has an analog for $p > 2$. In Equation (6.29) below, we shall try to connect these two transforms, (6.1) and (6.2), in the $GF(p)$ case as well. In this context, it is worth noting that it is possible (see Proposition 6.3.9) to characterize a bent function in terms of the Fourier transform of its derivative.

Suppose we want to write the function $\zeta^{f(x)}$ as a linear combination of translates of the function f:

$$\zeta^{f(x)} = \sum_{a \in V} c_a f(x - a), \qquad (6.28)$$

for some $c_a \in \mathbb{C}$. This may be regarded as the convolution of $f_{\mathbb{C}}$ with a function, c. One way to solve for the c_a's is to write this as a matrix equation,

$$\zeta^{\vec{f}} = A \cdot \vec{c},$$

where A is the adjacency matrix of Γ_f, $\vec{c} = \vec{c}_f = (c_a \mid a \in V)$ and $\zeta^{\vec{f}} = (\zeta^{f(x)} \mid x \in V)$. If A is invertible, that is if the Fourier transform of f is always nonzero, then

$$\vec{c} = A^{-1} \zeta^{\vec{f}}.$$

If (6.28) holds then we can write the Walsh transform f,

---

[4]When regarded as a function $f \colon GF(p^n) \to GF(p)$.

$$W_f(u) = \sum_{x \in GF(p)^n} \zeta^{f(x) - \langle u, x \rangle},$$

as a linear combination of values of the Fourier transform,

$$\hat{f}(y) = \sum_{x \in V} f(x) \zeta^{-\langle x, y \rangle}.$$

In other words,

$$
\begin{aligned}
W_f(u) &= \sum_{a \in V} c_a \sum_{x \in GF(p)^n} \zeta^{-\langle u, x \rangle} f(x - a) \\
&= \sum_{a \in V} c_a \sum_{x \in GF(p)^n} \zeta^{-\langle u, x + a \rangle} f(x) \\
&= \sum_{a \in V} c_a \zeta^{-\langle u, a \rangle} \sum_{x \in GF(p)^n} \zeta^{-\langle u, x \rangle} f(x) \\
&= \hat{f}(u) \sum_{a \in V} c_a \zeta^{-\langle u, a \rangle}.
\end{aligned}
\tag{6.29}
$$

This may be regarded as the product of Fourier transforms (that of the function $f_{\mathbb{C}}$ and that of the function c, which depends on f). In other words, there is a relationship between the Fourier transform of a $GF(p)$-valued function and its Walsh-Hadamard transform. However, it is not explicit unless one knows the function c (which depends on f in a complicated way).

6.11.1 Connected components of Cayley graphs

Recall

$$\omega = \omega_f = wt(f)$$

denotes the cardinality of $\mathrm{supp}(f) = \{v \in V \mid f(v) \neq 0\}$ and let

$$\sigma_f = \sum_{v \in V} f_{\mathbb{C}}(v).$$

Note that $\hat{f}(0) = \sigma_f \geq |\mathrm{supp}(f)|$. If f is even then Γ_f is an σ_f-regular (edge-weighted) graph. If we ignore weights, then it is an ω_f-weighted graph.

Recall that, given a graph Γ and its adjacency matrix A, the spectrum

$$\sigma(\Gamma) = \{\lambda_1, \lambda_2, \ldots, \lambda_N\},$$

where $N = p^n$, is the multi-set of eigenvalues of A. Following a standard convention, we index the elements $\lambda_i = \lambda_i(A)$ of the spectrum in such a way that they are monotonically increasing. Because Γ_f is regular, the row sums of A are all σ_f, whence the all 1's vector is an eigenvector of A with eigenvalue σ_f.

Clearly, the set of vertices in Γ_f connected to $0 \in V$ is in natural bijection with $\mathrm{supp}(f)$. Let W_j denote the subset of V consisting of those vectors which can be written as the sum of j elements in $\mathrm{supp}(f)$ but not $j - 1$. Clearly,

$$W_1 = \mathrm{supp}(f) \subset W_2 \subset \cdots \subset Span(\mathrm{supp}(f)).$$

For each $v_0 \in W_1 = \mathrm{supp}(f)$, the vertices connected to v_0 are the vectors in

$$\mathrm{supp}(f_{v_0}) = \{v \in V \mid f(v - v_0) \neq 0\},$$

where $f_{v_0}(v) = f(v - v_0)$ denotes the translation of f by $-v_0$. Therefore,

$$\mathrm{supp}(f_{v_0}) = v_0 + \mathrm{supp}(f).$$

In particular, all the vectors in W_2 are connected to $0 \in V$. For each $v_0 \in W_2$, the vertices connected to v_0 are the vectors in $\mathrm{supp}(f_{v_0}) = v_0 + \mathrm{supp}(f)$, so all the vectors in W_3 are connected to $0 \in V$. Inductively, we see that $Span(\mathrm{supp}(f))$ is the connected component of 0 in Γ_f. Pick any $u \in V$ representing a nontrivial coset in $V/Span(\mathrm{supp}(f))$. Clearly, 0 is not connected with u in Γ_f. However, the above reasoning implies u is connected to v if and only if they represent the same coset in $V/Span(\mathrm{supp}(f))$. This proves the following result.

Lemma 6.11.1. *The connected components of Γ_f are in one-to-one correspondence with the elements of the quotient space $V/Span(\mathrm{supp}(f))$.*

6.12 Algebraic normal form

A p-ary function f on $GF(p)^n$ has a unique representation as a polynomial in n variables, $x_0, x_1, \ldots, x_{n-1}$ over $GF(p)$ such that each variable occurs with power at most $p - 1$. This representation is called the *algebraic normal form* of f. The highest degree of its terms is called the *degree* of f.

Carlet [C10] shows how to find the algebraic normal form of any Boolean function. Similarly, we show how to find the algebraic normal form of any p-ary function.

Definition 6.12.1. An *atomic p-ary function* is a function $GF(p)^n \to GF(p)$ supported at a single point. For $v \in GF(p)^n$, we define an atomic function $f_v : GF(p)^n \to GF(p)$ by setting

$$f_v(w) = \begin{cases} 1, & \text{if } w = v, \\ 0, & \text{if } w \neq v. \end{cases} \tag{6.30}$$

We first show how to find the algebraic normal form of the atomic function f_v, for $v = (v_0, v_1, \ldots, v_{n-1}) \in GF(p)^n$.

Theorem 6.12.2. *(Celerier) Let f_v be the atomic p-ary function defined by Equation (6.30). Then the algebraic normal form of f_v is given by*

$$f_v(x) = \prod_{i=0}^{n-1} \left(\frac{1}{(p-1)!} \prod_{j \in GF(p) \setminus \{0\}} (j + v_i - x_i) \right). \tag{6.31}$$

Proof. First, we show that $f_v(v) = 1$. Substituting $x = v$ into Equation (6.31) gives

$$\begin{aligned} f_v(v) &= \prod_{i=0}^{n-1} \left(\frac{1}{(p-1)!} \prod_{j \in GF(p) \setminus \{0\}} (j + v_i - v_i) \right) \\ &= \prod_{i=0}^{n-1} \left(\frac{1}{(p-1)!} \prod_{j \in GF(p) \setminus \{0\}} j \right) \\ &= \prod_{i=0}^{n-1} \left(\frac{(p-1)!}{(p-1)!} \right) \\ &= 1. \end{aligned}$$

Next, we show that $f_v(w) = 0$ for every $w \neq v$. Suppose that $w \neq v$. Pick k such that $w_k \neq v_k$. Then there exists a $j \in GF(p) \setminus \{0\}$ such that $j + v_k - w_k = 0$ in $GF(p)$. Thus, the inside product of (6.31) is 0 for $i = k$, and so $f_v(w) = 0$. $\qquad\square$

It follows that every p-ary function can be written in algebraic normal form.

Corollary 6.12.3. *Let $g : GF(p)^n \to GF(p)$. Then*

$$g(x) = \sum_{v \in GF(p)^n} g(v) f_v(x), \tag{6.32}$$

where f_v is the atomic function defined by Equation (6.30).

Example 6.12.4. Sage can easily list all the atomic functions over $GF(3)$ having 2 variables:

―――――――――――――――――――― Sage ――――――――――――――――――――

```
sage: V = GF(3)^2
sage: x0,x1 = var("x0,x1")
sage: xx = [x0,x1]
sage: [expand(prod([2*prod([GF(3)(j)+v[i]-xx[i] for j in range(1,3)])
       for i in range(2)])) for v in V]
[x0^2*x1^2 + 2*x0^2 + 2*x1^2 + 1,
 x0^2*x1^2 + x0*x1^2 + 2*x0^2 + 2*x0,
 x0^2*x1^2 + 2*x0*x1^2 + 2*x0^2 + x0,
 x0^2*x1^2 + x0^2*x1 + 2*x1^2 + 2*x1,
 x0^2*x1^2 + x0^2*x1 + x0*x1^2 + x0*x1,
 x0^2*x1^2 + x0^2*x1 + 2*x0*x1^2 + 2*x0*x1,
 x0^2*x1^2 + 2*x0^2*x1 + 2*x1^2 + x1,
 x0^2*x1^2 + 2*x0^2*x1 + x0*x1^2 + 2*x0*x1,
 x0^2*x1^2 + 2*x0^2*x1 + 2*x0*x1^2 + x0*x1]
sage: f = x0^2*x1^2 + x0^2*x1 + x0*x1^2 + x0*x1
sage: [f(x0=v[0],x1=v[1]) for v in V]
[0, 0, 0, 0, 1, 0, 0, 0, 0]
```

Proposition 6.12.5. *(Hou) The degree of any bent function* $f : GF(p)^n \to GF(p)$, *when represented in algebraic normal form, satisfies*

$$\deg(f) \leq \frac{n(p-1)}{2} + 1.$$

The degree of any weakly regular bent function $f : GF(p)^n \to GF(p)$, *when represented in algebraic normal form, satisfies*

$$\deg(f) \leq \frac{n(p-1)}{2},$$

provided that $(p-1)n \geq 4$.

For a proof of these results, see Hou [H04] (and see also Coulter and Matthews [CoM77] for further details).

6.13 Examples of bent functions

In this section we classify even bent functions $f : GF(p)^n \to GF(p)$ with $f(0) = 0$ for $p = 3$ and $n = 2$ and give some examples for $p = 3$ and $n = 3$, and for $p = 5$ and $n = 2$. **Sage** was used for many of the calculations.

6.13.1 Bent functions $GF(3)^2 \to GF(3)$

We focus on examples of even functions $GF(3)^2 \to GF(3)$ sending 0 to 0. There are exactly $3^4 = 81$ such functions, 18 of which are bent.

The set of such functions has some amusing combinatorial properties we shall discuss below.

Sage was used to verify the following fact (originally discovered by S. Walsh).

Proposition 6.13.1. *There are* 18 *even bent functions* $f : GF(3)^2 \to GF(3)$ *such that* $f(0) = 0$. *The functions are given here in table form and in algebraic normal form.*

$GF(3)^2$	(0, 0)	(1, 0)	(2, 0)	(0, 1)	(1, 1)	(2, 1)	(0, 2)	(1, 2)	(2, 2)
b_1	0	1	1	1	2	2	1	2	2
b_2	0	2	2	1	0	0	1	0	0
b_3	0	1	1	2	0	0	2	0	0
b_4	0	2	2	0	1	0	0	0	1
b_5	0	0	0	2	1	0	2	0	1
b_6	0	1	1	0	2	0	0	0	2
b_7	0	0	0	1	2	0	1	0	2
b_8	0	2	2	0	0	1	0	1	0
b_9	0	0	0	2	0	1	2	1	0
b_{10}	0	2	2	2	1	1	2	1	1
b_{11}	0	0	0	0	2	1	0	1	2
b_{12}	0	2	2	1	2	1	1	1	2
b_{13}	0	1	1	2	2	1	2	1	2
b_{14}	0	1	1	0	0	2	0	2	0
b_{15}	0	0	0	1	0	2	1	2	0
b_{16}	0	0	0	0	1	2	0	2	1
b_{17}	0	2	2	1	1	2	1	2	1
b_{18}	0	1	1	2	1	2	2	2	1

The algebraic normal forms of these functions are:

$$b_1 = -b_{10} = x_0^2 + x_1^2,$$
$$b_2 = -b_3 \ \ = -x_0^2 + x_1^2,$$
$$b_4 = -b_6 \ \ = -x_0^2 - x_0 x_1,$$
$$b_5 = -b_7 \ \ = -x_0 x_1 - x_1^2,$$
$$b_8 = -b_{14} = -x_0^2 + x_0 x_1,$$
$$b_9 = -b_{15} = x_0 x_1 - x_1^2,$$
$$b_{11} = -b_{16} = -x_0 x_1,$$
$$b_{12} = -b_{18} = -x_0^2 - x_0 x_1 + x_1^2,$$
$$b_{13} = -b_{17} = x_0^2 - x_0 x_1 - x_1^2.$$

The group $G = GL(2, GF(3))$ *acts on the set* $\mathbb{B}$ *of all such bent functions. There are two orbits in* $\mathbb{B}/G$:

$$B_1 = \{b_2, b_3, b_4, b_5, b_6, b_7, b_8, b_9, b_{11}, b_{14}, b_{15}, b_{16}\},$$
$$B_2 = \{b_1, b_{10}, b_{12}, b_{13}, b_{17}, b_{18}\}.$$

The functions in the orbit B_1 are all regular. The functions in the orbit B_2 are all weakly regular (but not regular).

Each of these bent functions gives rise to a weighted partial difference set and hence an edge-weighted strongly regular Cayley graph.

Exercise 6.11. Verify that b_4 is bent.

Proposition 6.13.2. *(S. Walsh [W14]) Let $f : GF(3)^2 \rightarrow GF(3)$ be an even bent function with $f(0) = 0$. Then the level curves of f,*

$$D_i = \{v \in GF(3)^2 \mid f(v) = i\},$$

yield a symmetric weighted partial difference set with intersection numbers ρ_{ij}^k given as follows.

1. *If $f \in B_1$, then $|D_1| = |D_2| = 2$, and the intersection numbers ρ_{ij}^k are given in the following tables:*

ρ_{ij}^0	0	1	2	3
0	1	0	0	0
1	0	2	0	0
2	0	0	2	0
3	0	0	0	4

ρ_{ij}^1	0	1	2	3
0	0	1	0	0
1	1	1	0	0
2	0	0	0	2
3	0	0	2	2

ρ_{ij}^2	0	1	2	3
0	0	0	1	0
1	0	0	0	2
2	1	0	1	0
3	0	2	0	2

ρ_{ij}^3	0	1	2	3
0	0	0	0	1
1	0	0	1	1
2	0	1	0	1
3	1	1	1	1

2. *If $f \in B_2$, then $|D_1| = |D_2| = 4$, $D_3 = \emptyset$, and the intersection numbers ρ_{ij}^k are given in the following tables:*

ρ_{ij}^0	0	1	2
0	1	0	0
1	0	4	0
2	0	0	4

ρ_{ij}^1	0	1	2
0	0	1	0
1	1	1	2
2	0	2	2

ρ_{ij}^2	0	1	2
0	0	0	1
1	0	2	2
2	1	2	1

no $\rho_s{ij}^3$

The proposition is verified using a case-by-case analysis, which we omit. However, see Example 6.9.5 above for more details on Part 2. Also, see [CJMPW15] for further properties of the bent functions in the orbit B_1.

Lemma 6.13.3. *The* 12 *bent functions in the orbit* B_1 *can all be obtained from* $b_6(x_0, x_1) = x_0^2 + x_0 x_1$ *by linear transformations of the coordinates, i.e.,* $(x_0, x_1) \mapsto (ax_0 + bx_1, cx_0 + dx_1)$ *where* $ad - bc \neq 0$. *Each such isomorphism of* $GF(3)^2$ *induces an isomorphism*[5] *of the associated edge-weighted Cayley graphs. Thus, the Cayley graphs of the* 12 *bent functions in the orbit* B_1 *are all isomorphic as edge-weighted strongly regular graphs. The unweighted Cayley graphs of the* 12 *bent functions in the orbit* B_1 *are strongly regular graphs with parameters* $(9, 4, 1, 2)$. *Up to isomorphism, there is only one unweighted strongly regular graph having parameters* $(9, 4, 1, 2)$ *(see Brouwer [Br] and Spence [Sp]). This unweighted graph is not complete.*

Similarly, the 6 *bent functions in the orbit* B_2 *can all be obtained from the function* $b_1(x_0, x_1) = x_0^2 + x_1^2$ *by linear transformations of the coordinates. Thus, the associated edge-weighted Cayley graphs are all isomorphic. The unweighted Cayley graphs of the* 6 *bent functions in the orbit* B_2 *are strongly regular graphs with parameters* $(9, 8, 7, 0)$. *These unweighted graphs are all isomorphic and complete.*

Example 6.13.4. We omit the Sage computations which verify the following facts (see [CJMPW15] for details).

- In the orbit B_1, the regular functions in the following pairs are dual to one another; (b_2, b_3), (b_4, b_9), (b_5, b_8), (b_6, b_{15}), (b_7, b_{14}), and (b_{11}, b_{16}).

- In the orbit B_2, the weakly regular functions b_1 and b_{10} are -1-dual to one another, while the weakly regular functions b_{12}, b_{13}, b_{17} and b_{18} are all -1-self-dual.

Exercise 6.12. Verify that b_4 and b_9 are regular and dual to each other.

Exercise 6.13. Verify the following relationships:

$$b_1 = b_6 + b_{15} = b_7 + b_{14},$$
$$b_{10} = b_4 + b_9 = b_5 + b_8,$$
$$b_{12} = b_2 + b_{11} = b_7 + b_8,$$
$$b_{13} = b_3 + b_{11} = b_6 + b_9,$$
$$b_{17} = b_2 + b_{16} = b_4 + b_{15},$$
$$b_{18} = b_3 + b_{16} = b_5 + b_{14}.$$

We now examine the supports of the 18 bent functions.

---

[5] We say edge-weighted graphs are *isomorphic* if there is a bijection of the vertices which preserves the weight of each edge.

$$\mathrm{supp}(b_2) = \mathrm{supp}(b_3) = \{(0,1),(0,2),(1,0),(2,0)\},$$
$$\mathrm{supp}(b_4) = \mathrm{supp}(b_6) = \{(1,0),(2,0),(1,1),(2,2)\},$$
$$\mathrm{supp}(b_8) = \mathrm{supp}(b_{14}) = \{(1,0),(2,0),(1,2),(2,1)\},$$
$$\mathrm{supp}(b_9) = \mathrm{supp}(b_{15}) = \{(0,1),(0,2),(1,2),(2,1)\},$$
$$\mathrm{supp}(b_5) = \mathrm{supp}(b_7) = \{(0,1),(0,2),(1,1),(2,2)\},$$
$$\mathrm{supp}(b_{11}) = \mathrm{supp}(b_{16}) = \{(1,1),(2,2),(1,2),(2,1)\},$$

and

$$\mathrm{supp}(b_1) = \mathrm{supp}(b_{10}) = \mathrm{supp}(b_{12}) = \mathrm{supp}(b_{13}) = \mathrm{supp}(b_{17}) = \mathrm{supp}(b_{18})$$

are all equal to $GF(3)^2 \setminus \{(0,0)\}$. Note that

- All 18 functions are weight 4 or weight 8.
- If S_1 and S_2 are any two weight 4 support sets, then either

$$S_1 \cap S_2 = \emptyset \quad \text{or} \quad |S_1 \cap S_2| = 2.$$

In fact, if you consider the set

$$S = \{\emptyset\} \cup \{\mathrm{supp}(f) \mid f \in B_1 \cup B_2\},$$

then S with the symmetric difference operator Δ forms a group which is isomorphic to $GF(2)^3$.

Question 6.5. *To what extent is it true that if f_1, f_2 are bent functions on $GF(p)^n$ with $f_1(0) = f_2(0) = 0$, then*

$$\mathrm{supp}(f_1) \, \Delta \, \mathrm{supp}(f_3) = \mathrm{supp}(f_3)$$

for some bent function f_3 satisfying $f_3(0) = 0$?

Question 6.6. *Over $GF(p)$, for $p \neq 2$, does the set*

$$S = \{\emptyset\} \cup \{\mathrm{supp}(f) \mid f \text{ is bent}, f(0) = 0, f \text{ is even}\}$$

form a group (under Δ)?

This does not seem to hold in the binary case[6].

Example 6.13.5. This example is intended to illustrate the bent function b_8 listed in the table above and to provide more detail on Example 6.9.7.

---

[6]What is true in the binary case is an oddly similar result: the vectors in the support of a bent function form a Hadamard difference set in the additive group $GF(2)^n$.

Consider the finite field

$$GF(9) = GF(3)[x]/(x^2 + 1) = \{0, 1, 2, x, x + 1, x + 2, 2x, 2x + 1, 2x + 2\}.$$

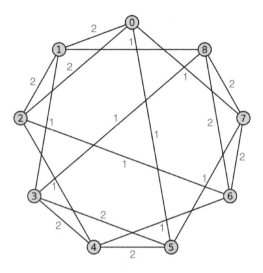

Figure 6.3:   The Cayley graph for b_8.

The set of nonzero quadratic residues is given by

$$D = \{1, 2, x, 2x\}.$$

Let Γ be the graph whose vertices are $\hat{G}F(9)$ and whose edges $e = (a, b)$ are those pairs for which $a - b \in D$.

The graph looks like the Cayley graph for b_8 in Figure 6.3, except

$$8 \to 0, 0 \to 2x + 2, 1 \to 2x + 1, 2 \to 2x,$$

$$3 \to x + 2, 4 \to x + 1, 5 \to x, 6 \to 2, 7 \to 1, 8 \to 0.$$

This is a strongly regular graph with parameters $(9, 4, 1, 2)$.

v	0	1	2	3	4	5	6	7	8
$N(v,0)$	3,4,6,8	4,5,6,7	3,5,7,8	0,2,6,7	0,1,7,8	1,2,6,8	0,1,3,5	1,2,3,4	0,2,4,5
$N(v,1)$	5,7	3,8	4,6	1,8	2,6	0,7	2,4	0,5	1,3
$N(v,2)$	1,2	0,2	0,1	4,5	3,5	3,4	7,8	6,8	6,7

The axioms of an edge-weighted strongly regular graph can be directly verified using this table.

6.13.2 Bent functions $GF(3)^3 \to GF(3)$

Sage and Mathematica were used to find and classify the even bent functions $f: GF(3)^3 \to GF(3)$ such that $f(0) = 0$. There are 2340 such functions. There are four orbits B_1, B_2, $B_3 = -B_1$, and $B_4 = -B_2$ under the action of the group $G = GL(3, GF(3))$, with sizes $|B_1| = |B_3| = 234$ and $|B_2| = |B_4| = 936$. Note that by Hou's theorem (Proposition 6.12.5) the degree d of a bent function $f: GF(3)^3 \to GF(3)$, when represented in algebraic normal form, satisfies $d \leq 4$, and if f is weakly regular then $d \leq 3$.

A representative of B_1 is

$$f_1(x_0, x_1, x_2) = x_0^2 + x_1^2 + x_2^2.$$

A representative of B_2 is

$$f_2(x_0, x_1, x_2) = x_0 x_2 + 2x_1^2 + 2x_0^2 x_1^2.$$

The bent functions in orbits B_1 and B_3 give rise to symmetric weighted partial difference sets, while those in orbits B_2 and B_4 do not. The functions in orbits B_1 and B_3 are weakly regular, but not regular. The functions in orbits B_2 and B_4 are not weakly regular.

Example 6.13.6. Let

$$
\begin{aligned}
f(x_0, x_1, x_2) = &- x_0^2 x_1^2 + x_0^2 x_1 x_2 - x_0 x_1^2 x_2 - x_0^2 x_2^2 \\
&+ x_0 x_1 x_2^2 - x_1^2 x_2^2 + x_0^2 - x_1^2 - x_0 x_2 + x_2^2.
\end{aligned}
$$

It can be shown that $W_f(0)/3^{3/2}$ is a 6-th root of unity, but not a cube root of unity. It therefore follows from Lemma 6.4.5 that f is not weakly regular.

6.13.3 Bent functions $GF(5)^2 \to GF(5)$

Sage and Mathematica were used to find and classify the even bent functions $f: GF(5)^2 \to GF(5)$ such that $f(0) = 0$. There are 1420 such functions. Note that by Hou's theorem (Proposition 6.12.5) the degree d of a bent function $f: GF(5)^2 \to GF(5)$, when represented in algebraic normal form, satisfies $d \leq 5$, and if f is weakly regular then $d \leq 4$.

There are 11 orbits B_i under the action of the group $G = GL(2, GF(5))$, with representatives given in Table 6.1. The sizes of the orbits are $|B_1| = 40$, $|B_2| = 60$, $|B_3| = \cdots = |B_9| = 120$, and $|B_{10}| - |D_{11}| = 240$

The bent functions which give rise to symmetric weighted partial difference sets are those in the orbits of f_1, f_2, f_5, f_6, and f_9 in Table 6.1. The bent functions in the orbits of the other f_i's do not.

The function f_1 is weakly regular, and the functions $f_2, \ldots, f_{11}$ are regular.

Example 6.13.7. Let $G = GF(25) = GF(5)[x]/(x^2 + 2)$,

B_1	$f_1(x_0, x_1) = -x_0^2 + 2x_1^2$
B_2	$f_2(x_0, x_1) = -x_0 x_1 + x_1^2$
B_3	$f_3(x_0, x_1) = -2x_0^4 + 2x_0^2 + 2x_0 x_1$
B_4	$f_4(x_0, x_1) = -x_1^4 + x_0 x_1 - 2x_1^2$
B_5	$f_5(x_0, x_1) = x_0^3 x_1 + 2x_1^4$
B_6	$f_6(x_0, x_1) = -x_0 x_1^3 + x_1^4$
B_7	$f_7(x_0, x_1) = x_1^4 - x_0 x_1$
B_8	$f_8(x_0, x_1) = 2x_1^4 - 2x_0 x_1 + 2x_1^2$
B_9	$f_9(x_0, x_1) = -x_0^3 x_1 + x_1^4$
B_{10}	$f_{10}(x_0, x_1) = 2x_0 x_1^3 + x_1^4 - x_1^2$
B_{11}	$f_{11}(x_0, x_1) = x_0 x_1^3 - x_1^4 - 2x_1^2$

Table 6.1: Representatives of orbits in $\mathbb{B}/G$ for $GF(5)^2 \to GF(5)$.

$$D_1 = \{1, 4, x + 2, 4x + 3\}, D_2 = \{x + 1, x + 3, 4x + 2, 4x + 4\},$$

$$D_3 = \{2x + 1, 2x + 2, 3x + 3, 3x + 4\}, D_4 = \{2, 3, 2x + 4, 3x + 1\},$$

and $D = D_1 \cup D_2 \cup D_3 \cup D_4$. If $f(x_0, x_1) = x_0^2 + x_0 x_1$ then each subset D_i ($i = 1, 2, 3, 4$) is the image of the level curve $f^{-1}(i)$ under the $GF(5)$-vector space isomorphism

$$\phi : GF(5)^2 \quad \to \quad GF(25),$$
$$(a, b) \quad \longmapsto \quad bx + a\,,$$

$D_i = \phi(f^{-1}(i))$, $i = 1, 2, 3, 4$. The associated Cayley graph Γ_f is edge-weighted strongly regular. The parameters of Γ_f, as an unweighted strongly regular graph, are $(25, 16, 9, 12)$. It is possible to compute the $k_{i,j}$'s, $\mu_{i,j}$'s, and $\lambda_{i,j}^k$'s (omitted here but see the longer arxiv version of [CJMPW15] for more details). This f is homogeneous, bent and regular (hence also weakly regular).

Exercise 6.14. Consider the bent function on $GF(5)^2$,

$$f_4(x_0, x_1) = -x_1^4 - 2x_1^2 + x_0 x_1.$$

This function represents a $GL(2, GF(5))$ orbit of size 120.

Show that the level curves of this function do not give rise to a weighted partial difference set.

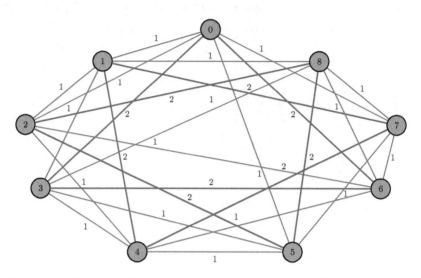

Figure 6.4:   The weighted Cayley graph of a non-bent even $GF(3)$-valued function.

6.14 Examples of Cayley graphs

Let $V = GF(p)^n$ and let $f : V \to GF(p)$. If we fix an ordering on $GF(p)^n$, then the $p^n \times p^n$ matrix $A = (A_{ij})$, where A_{ij} is as in (6.26), is a $\mathbb{Z}$-valued matrix.

Example 6.14.1. Consider the even function $f : GF(3)^2 \to GF(3)$ with the following values:

$GF(3)^2$	(0, 0)	(1, 0)	(2, 0)	(0, 1)	(1, 1)	(2, 1)	(0, 2)	(1, 2)	(2, 2)
f	0	1	1	2	0	1	2	1	0

The weighted Cayley graph Γ of f is given in Figure 6.4.

This function f is not bent. The algebraic normal form of f is

$$2x_0^2 x_1^2 + x_0^2 + x_0 x_1 + 2x_1^2.$$

In particular, f is nonhomogeneous. The weighted adjacency matrix of the Cayley graph of f is

$$A = \begin{pmatrix} 0 & 1 & 1 & 2 & 0 & 1 & 2 & 1 & 0 \\ 1 & 0 & 1 & 1 & 2 & 0 & 0 & 2 & 1 \\ 1 & 1 & 0 & 0 & 1 & 2 & 1 & 0 & 2 \\ 2 & 1 & 0 & 0 & 1 & 1 & 2 & 0 & 1 \\ 0 & 2 & 1 & 1 & 0 & 1 & 1 & 2 & 0 \\ 1 & 0 & 2 & 1 & 1 & 0 & 0 & 1 & 2 \\ 2 & 0 & 1 & 2 & 1 & 0 & 0 & 1 & 1 \\ 1 & 2 & 0 & 0 & 2 & 1 & 1 & 0 & 1 \\ 0 & 1 & 2 & 1 & 0 & 2 & 1 & 1 & 0 \end{pmatrix}.$$

The matrix N_{22} whose u, v-entry is $|N(u, 2) \cap N(v, 2)|$, (where u and v are vertices of Γ), is:

$$N_{22} = \begin{pmatrix} 2 & 0 & 0 & 1 & 0 & 0 & 1 & 0 & 0 \\ 0 & 2 & 0 & 0 & 1 & 0 & 0 & 1 & 0 \\ 0 & 0 & 2 & 0 & 0 & 1 & 0 & 0 & 1 \\ 1 & 0 & 0 & 2 & 0 & 0 & 1 & 0 & 0 \\ 0 & 1 & 0 & 0 & 2 & 0 & 0 & 1 & 0 \\ 0 & 0 & 1 & 0 & 0 & 2 & 0 & 0 & 1 \\ 1 & 0 & 0 & 1 & 0 & 0 & 2 & 0 & 0 \\ 0 & 1 & 0 & 0 & 1 & 0 & 0 & 2 & 0 \\ 0 & 0 & 1 & 0 & 0 & 1 & 0 & 0 & 2 \end{pmatrix}.$$

Exercise 6.15. Show that all the values of $N_{22}[u, v]$ are the same if u, v are distinct vertices which are not neighbors. Show $\mu_{(2,2)} = 0$.

Exercise 6.16. Show that all the values of $N_{22}[v, v]$ are the same. Show $k_2 = 2$.

Exercise 6.17. Show that all the values of $N_{22}[u, v]$ are the same if u, v are neighbors with edge-weight 2. Show $\lambda_{(2,2,2)} = 1$.

Exercise 6.18. Show that all the values of $N_{22}[u, v]$ are the same if u, v are neighbors with edge-weight 1. Show $\lambda_{(2,2,1)} = 0$.

Using these sorts of combinatorial considerations, one can show

$$\mu_{(1,1)} = 2, \ k_1 = 4, \ \lambda_{(1,1,1)} = 1, \ \lambda_{(1,1,2)} = 2,$$

$$\mu_{(1,2)} = 2, \ \lambda_{(1,2,1)} = 1, \ \lambda_{(1,2,2)} = 0,$$

$$\mu_{(2,2)} = 0, \ k_2 = 2, \ \lambda_{(2,2,1)} = 0, \ \lambda_{(2,2,2)} = 1.$$

Using these parameters, one can verify the statements in the conclusion of Analog 6.16.4 for this function. In other words, the associated edge-weighted Cayley graph is strongly regular. (However, f is not bent.)

Example 6.14.2. It can be shown that Example 6.9.7 (or an isomorphic copy) arises via the bent function b_8 (see also Example 6.13.5). For this example of b_8, we compute the adjacency matrix associated to the members R_1 and R_2 of the association scheme (G, R_0, R_1, R_2, R_3), where $G = GF(3)^2$,

$$R_i = \{(g, h) \in G \times G \mid gh^{-1} \in D_i\}, \qquad i = 1, 2,$$

and $D_i = f^{-1}(i)$.

Consider the following **Sage** computation.

```
sage: FF = GF(3)
sage: V = FF^2
sage: Vlist = V.list()
sage: flist = [0,2,2,0,0,1,0,1,0]
sage: f = lambda x: GF(3)(flist[Vlist.index(x)])
sage: F = matrix(ZZ, [[f(x-y) for x in V] for y in V])
sage: eval1 = lambda x: int((x==1))
sage: eval2 = lambda x: int((x==2))
sage: F1 = matrix(ZZ, [[eval1(f(x-y)) for x in V] for y in V])
sage: F2 = matrix(ZZ, [[eval2(f(x-y)) for x in V] for y in V])
sage: F1*F2-F2*F1 == 0
True
sage: delta = lambda x: int((x[0]==x[1]))
sage: F3 = matrix(ZZ, [[(eval0(f(x-y))+delta([x,y])) %2 for x in V] for y in V])
sage: F3*F2-F2*F3==0
True
sage: F3*F1-F1*F3==0
True
sage: F0 = matrix(ZZ, [[delta([x,y]) for x in V] for y in V])
sage: F1*F3 == 2*F2 + F3
True
```

The **Sage** computation above tells us that the adjacency matrix of R_1 is

$$A_1 = \begin{pmatrix} 0&0&0&0&0&1&0&1&0 \\ 0&0&0&1&0&0&0&0&1 \\ 0&0&0&0&1&0&1&0&0 \\ 0&1&0&0&0&0&0&0&1 \\ 0&0&1&0&0&0&1&0&0 \\ 1&0&0&0&0&0&0&1&0 \\ 0&0&1&0&1&0&0&0&0 \\ 1&0&0&0&0&1&0&0&0 \\ 0&1&0&1&0&0&0&0&0 \end{pmatrix},$$

the adjacency matrix of R_2 is

$$A_2 = \begin{pmatrix} 0 & 1 & 1 & 0 & 0 & 0 & 0 & 0 & 0 \\ 1 & 0 & 1 & 0 & 0 & 0 & 0 & 0 & 0 \\ 1 & 1 & 0 & 0 & 0 & 0 & 0 & 0 & 0 \\ 0 & 0 & 0 & 0 & 1 & 1 & 0 & 0 & 0 \\ 0 & 0 & 0 & 1 & 0 & 1 & 0 & 0 & 0 \\ 0 & 0 & 0 & 1 & 1 & 0 & 0 & 0 & 0 \\ 0 & 0 & 0 & 0 & 0 & 0 & 0 & 1 & 1 \\ 0 & 0 & 0 & 0 & 0 & 0 & 1 & 0 & 1 \\ 0 & 0 & 0 & 0 & 0 & 0 & 1 & 1 & 0 \end{pmatrix},$$

and the adjacency matrix of R_3 is

$$A_3 = \begin{pmatrix} 0 & 0 & 0 & 1 & 1 & 0 & 1 & 0 & 1 \\ 0 & 0 & 0 & 0 & 1 & 1 & 1 & 1 & 0 \\ 0 & 0 & 0 & 1 & 0 & 1 & 0 & 1 & 1 \\ 1 & 0 & 1 & 0 & 0 & 0 & 1 & 1 & 0 \\ 1 & 1 & 0 & 0 & 0 & 0 & 0 & 1 & 1 \\ 0 & 1 & 1 & 0 & 0 & 0 & 1 & 0 & 1 \\ 1 & 1 & 0 & 1 & 0 & 1 & 0 & 0 & 0 \\ 0 & 1 & 1 & 1 & 1 & 0 & 0 & 0 & 0 \\ 1 & 0 & 1 & 0 & 1 & 1 & 0 & 0 & 0 \end{pmatrix}.$$

Of course, the adjacency matrix of R_0 is the identity matrix. In the above computation, **Sage** has also verified that they commute and satisfy

$$A_1 A_3 = 2A_2 + A_3$$

in the Schur ring.

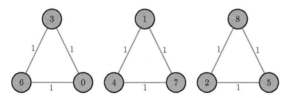

Figure 6.5: The undirected Cayley graph of an even $GF(3)$-valued function of two variables from Example 6.14.3. (The vertices are ordered as in the example.)

Example 6.14.3. We take $V = GF(3)^2$ and consider an even function whose Cayley graph has three connected components.

$GF(3)^2$	$(0, 0)$	$(1, 0)$	$(2, 0)$	$(0, 1)$	$(1, 1)$	$(2, 1)$	$(0, 2)$	$(1, 2)$	$(2, 2)$
f	0	0	0	1	0	0	1	0	0

───────────── Sage ─────────────

```
sage: V = GF(3)^2
sage: flist = [0,0,0,1,0,0,1,0,0]
sage: f = lambda x: GF(3)(flist[Vlist.index(x)])
sage: A = matrix([[f(x-y) for x in V] for y in V])
sage: Gamma = Graph(A)
sage: Gamma.connected_components_number()
3
```

The graph Γ constructed in the Sage example is shown in Figure 6.5.

Recall from Corollary 2.7.6 that the spectrum of the graph Γ_f is precisely the set of values of the Fourier transform F_f of $f_{\mathbb{C}}$. This is illustrated in the example below.

Example 6.14.4. Consider the ternary function

$$f(x_0, x_1) = -x_0^2 x_1^2 + x_0^2 + x_0 x_1 - x_1^2.$$

$GF(3)^2$	$(0, 0)$	$(1, 0)$	$(2, 0)$	$(0, 1)$	$(1, 1)$	$(2, 1)$	$(0, 2)$	$(1, 2)$	$(2, 2)$
f	0	0	0	1	0	0	1	0	0
F_f	8	2	2	-1	-1	-1	-1	-4	-4

───────────── Sage ─────────────

```
sage: V = GF(3)^2
sage: Vlist = V.list()
sage: Vlist
[(0, 0), (1, 0), (2, 0), (0, 1), (1, 1), (2, 1), (0, 2), (1, 2), (2, 2)]
sage: flist = [0,1,1,2,0,1,2,1,0]
sage: f = lambda x: GF(3)(flist[Vlist.index(x)])
sage: A = matrix([[f(x-y) for x in V] for y in V])
sage: Gamma = Graph(A)
sage: Gamma.spectrum()
[8, 2, 2, -1, -1, -1, -1, -4, -4]
```

This shows that, in this case, the spectrum of the Cayley graph of f agrees with the values of the Fourier transform of $f_{\mathbb{C}}$. See Example 2.7.7 for more details.

6.15 Cayley graphs of linear codes

We give a very brief construction of a Cayley graph attached to a binary error-correcting code.

Let C be a binary linear $[n, k, d]$-code with generator matrix G. Let B denote the columns of G and assume no column is the zero vector. Let $V = GF(2)^k$ and let $\Gamma = (V, E)$, where

$$E = \{(v, w) \in V \times V \mid v + w \in B\}.$$

This is an n-regular graph.

Lemma 6.15.1. *The Laplacian spectrum of Γ is the set of numbers $2 \cdot wt(c)$, for $c \in C$ where wt denotes the Hamming weight.*

For a proof, see Roth [Ro06], Corollary 13.14.

Example 6.15.2. Let $C \subset GF(2)^7$ denote the (Hamming) linear $[7, 4, 3]$-code with generator matrix

$$G = \begin{pmatrix} 1 & 0 & 0 & 0 & 0 & 1 & 1 \\ 0 & 1 & 0 & 0 & 1 & 0 & 1 \\ 0 & 0 & 1 & 0 & 1 & 1 & 0 \\ 0 & 0 & 0 & 1 & 1 & 1 & 1 \end{pmatrix}.$$

The weights of the code words in C are

$$0, 3, 3, 3, 3, 3, 3, 3, 4, 4, 4, 4, 4, 4, 4, 7.$$

The vertices of the associated Cayley graph $\Gamma(V, E)$ will be $V = GF(2)^4$. For example, the vertices $v_1 = (1, 1, 1, 1)$ and $(0, 1, 1, 1)$ are connected by an edge since $v_1 + v_2 = (1, 0, 0, 0)$ is a column of G. Indeed, Γ is a 7-regular graph with 56 edges. The eigenvalues of the Laplacian matrix of Γ are

$$0, 6, 6, 6, 6, 6, 6, 6, 8, 8, 8, 8, 8, 8, 8, 14,$$

as predicted by Lemma 6.15.1.

Moreover, Γ is an edge-transitive, vertex-transitive, distance-regular, bipartite graph. However, it is not strongly regular and not planar. Its automorphism group $Aut(\Gamma)$ clearly contains S_7, the symmetric group on 7 letters, since one can permute the columns of G and get the same C. However, there are extra automorphisms of Γ since $Aut(\Gamma)$ has order 80640 and a center Z of order 2.

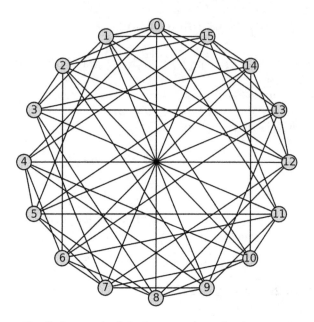

Figure 6.6:   The Cayley graph of the Hamming $[7, 4, 3]$ code, created using Sage.

See Roth [Ro06], Chapter 13 for further details.

There is an extensive and fascinating literature on Boolean functions and linear codes, a topic that would fill its own book. For example, the Reed–Muller codes, quadratic reciprocity codes, and Kerdock codes can be described in this way. Unfortunately, it would take us too far afield to discuss them in detail here. For further reading in this topic, we refer to the excellent texts Huffman and Pless [HP10], MacWilliams and Sloane [MS77], and Tokareva [T15].

6.16 Analogs and questions

Regarding the Dillon correspondence, we have the following combinatorial questions.

Analog 6.16.1. *Using the notation of Theorem 6.10.6, under what conditions on an even bent function f will the tuple*

$$(GF(p)^n, D_0, D_1, D_2, \ldots, D_{p-1}, D_p)$$

define a weighted partial difference set?

We reformulate this in an essentially equivalent way using the language of association schemes.

Analog 6.16.2. *Let f be as above and let $R_0, R_1, \ldots, R_p$ denote binary relations on $GF(p)^n$ given by*

$$R_i = \{(x,y) \in GF(p)^n \times GF(p)^n \mid f(x,y) = i\}, \quad 0 \le i \le p.$$

Under what conditions on an even bent function f will $(GF(p)^n, R_0, R_1, \ldots, R_p)$ be a p-class association scheme?

Remark 6.16.3. In Chee, Tan, and Zhang [CTZ10], it is shown that if n is even then the unweighted Cayley graph of certain[7] weakly regular even bent functions $f : GF(p)^n \to GF(p)$, with $f(0) = 0$, is strongly regular.

The (naive) analog of the Bernasconi–Codenotti–VanderKam correspondence for $p > 2$ is formalized below in Analog 6.16.4

Analog 6.16.4. *Assume n is even and $f : GF(p)^n \to GF(p)$ is even bent. Under what additional conditions on f will the following conditions hold true? For each $a \in GF(p)^\times$, we have*

- *if $u_1, u_2 \in V$ are a_3-neighbors in the Cayley graph of f then $|N(u_1, a_1) \cap N(u_2, a_2)|$ does not depend on u_1, u_2 (with a given edge-weight), for each $a_1, a_2, a_3 \in GF(p)^\times$;*

- *if $u_1, u_2 \in V$ are not distinct and neighbors in the Cayley graph of f then $|N(u_1, a_1) \cap N(u_2, a_2)|$ does not depend on u_1, u_2, for each $a_1, a_2 \in GF(p)^\times$.*

(In other words, the associated Cayley graph is strongly regular as in Definition 6.10.3.)

Unfortunately, it is not true in general.

Remark 6.16.5. 1. This analog is false when $p = 5$.
2. This analog remains false if you replace "$f : GF(p)^n \to GF(p)$ is even bent" in the hypothesis by "$f : GF(p)^n \to GF(p)$ is even bent and regular." However, when $p = 3$ and $n = 2$, see Lemma 6.13.3.
3. In general, this analog remains false if you replace "$f : GF(p)^n \to GF(p)$ is even bent" in the hypothesis by "$f : GF(p)^n \to GF(p)$ is even bent and weakly regular." However, when $p = 3$ and $n = 2$, see Lemma 6.13.3.
4. This analog is false if n is odd.
5. The converse of this analog, as stated, is false if $p > 2$.

Related conjectures can be found in [CJMPW15].

---

[7]By "certain" we mean that f, regarded as a function $GF(p^n) \to GF(p)$, is homogeneous of some degree.

6.17 Further reading

For further reading on p-ary bent functions and related topics from this chapter, see Bernasconi [B98]; Bernasconi and Codenotti [BC99]; Bernasconi, Codenotti and VanderKam [BCV01]; Carlet [C10]; Carlet and Ding [CD04]; Celerier, Joyner, Melles, and Phillips [CJMP12]; Celerier, Joyner, Melles, Phillips, and Walsh [CJMPW15]; Cesmelioglu and Meidl [CM12]; Chee, Tan, and Zhang [CTZ10]; Coulter and Matthews [CoM77]; Cusick and Stanica [CS09]; Delsarte [De73]; Fen, Wen, Xiang, and Yin [FWXY10]; Helleseth and Kholosha [HK10], [HK10], and [HK13]; Hou [H04]; Mesnager [Mes16] Pott, Tan, Feng, and Ling [PTFL11]; Stanica [St07]; and Tokareva [T10] and [T15].

Appendix A
Selected Answers

Selected answers and hints for the exercises are provided below.

Chapter 1

Exercise 1.1 The sensitivity is 4.
Exercise 1.2 (a) The adjacency matrix is

$$
\begin{pmatrix}
0 & 1 & 0 & 0 & 1 & 1 & 0 & 0 \\
1 & 0 & 1 & 0 & 1 & 0 & 1 & 0 \\
0 & 1 & 0 & 1 & 0 & 0 & 1 & 1 \\
0 & 0 & 1 & 0 & 1 & 0 & 0 & 0 \\
1 & 1 & 0 & 1 & 0 & 0 & 0 & 0 \\
1 & 0 & 0 & 0 & 0 & 0 & 1 & 1 \\
0 & 1 & 1 & 0 & 0 & 1 & 0 & 0 \\
0 & 0 & 1 & 0 & 0 & 1 & 0 & 0
\end{pmatrix}.
$$

(b) The incidence matrix is

$$
\begin{pmatrix}
1 & 1 & 1 & 0 & 0 & 0 & 0 & 0 & 0 & 0 & 0 & 0 \\
1 & 0 & 0 & 1 & 1 & 1 & 0 & 0 & 0 & 0 & 0 & 0 \\
0 & 0 & 0 & 1 & 0 & 0 & 1 & 1 & 1 & 0 & 0 & 0 \\
0 & 0 & 0 & 0 & 0 & 0 & 1 & 0 & 0 & 1 & 0 & 0 \\
0 & 1 & 0 & 0 & 1 & 0 & 0 & 0 & 0 & 1 & 0 & 0 \\
0 & 0 & 1 & 0 & 0 & 0 & 0 & 0 & 0 & 0 & 1 & 1 \\
0 & 0 & 0 & 0 & 0 & 1 & 0 & 1 & 0 & 0 & 1 & 0 \\
0 & 0 & 0 & 0 & 0 & 0 & 0 & 0 & 1 & 0 & 0 & 1
\end{pmatrix}.
$$

(c) It is Hamiltonian: $(0, 4, 3, 2, 7, 5, 6, 1]$.
(d) It's not Eulerian.
(e) The girth is 3.
(f) The diameter is 3.

© Springer International Publishing AG 2017

W.D. Joyner and C.G. Melles, *Adventures in Graph Theory*,
Applied and Numerical Harmonic Analysis,
https://doi.org/10.1007/978-3-319-68383-6

Exercise 1.4 The cycle space has basis

$$\{(1,0,-1,0,1,0),(0,1,-1,0,0,1),(0,0,0,1,-1,1)\}.$$

The cocycle space has basis

$$\{(1,0,0,-1,-1,0),(0,1,0,1,0,-1),(0,0,1,0,1,1)\}.$$

Exercise 1.5 Yes, it is connected. Its order is 11 and size is 22.
Exercise 1.6 The cycle space has basis

$$\{(1,0,-1,0,0,1,0,0,0,0,-1,0),(0,1,-1,0,0,0,-1,0,1,-1,0,-1),$$

$$(0,0,0,1,0,-1,0,0,1,0,1,-1),(0,0,0,0,1,-1,-1,0,1,-1,1,-1),$$

$$(0,0,0,0,0,0,0,1,-1,0,-1,1)\}.$$

Exercise 1.7 The cocycle space has basis

$$\{(1,0,0,0,-1,0,0,0,0,-1,1,1),(0,1,0,0,1,0,0,0,0,1,0,0),$$

$$(0,0,1,0,0,0,0,0,0,0,-1,-1),(0,0,0,1,0,0,0,-1,0,-1,0,1),$$

$$(0,0,0,0,0,1,0,1,0,0,1,0),(0,0,0,0,0,0,1,0,0,-1,0,0),$$

$$(0,0,0,0,0,0,0,0,1,0,0,1)\}.$$

Chapter 4
Exercise 4.5 $x_W^t Q x_W = 3.$
Exercise 4.10

configuration	energy		configuration	energy
$\{0,0,0,0\}$	0		$\{-2,0,2,0\}$	2
$\{-1,0,0,1\}$	$\frac{1}{2}$		$\{-3,0,2,1\}$	$\frac{7}{2}$
$\{-2,0,0,2\}$	2		$\{-2,1,0,1\}$	$\frac{3}{2}$
$\{-1,0,1,0\}$	$\frac{1}{2}$		$\{-3,1,0,2\}$	$\frac{7}{2}$
$\{-2,0,1,1\}$	$\frac{3}{2}$		$\{-3,2,0,1\}$	$\frac{7}{2}$
$\{-3,0,1,2\}$	$\frac{7}{2}$		$\{-3,2,1,0\}$	$\frac{7}{2}$

Exercise 4.16 Recall that a divisor on a graph is non-special if its degree is one less than the genus of the graph and its linear system is empty. Thus, a divisor on a cycle graph is non-special if it has degree 0 and is not linearly equivalent to any effective divisor. The only effective divisor of degree 0 is the zero divisor. Since the pullback map $\phi^* : \mathrm{Jac}(C_m) \to \mathrm{Jac}(C_n)$ is injective, it follows that if ν is a degree 0 divisor on C_m which is not linearly equivalent to the zero divisor on C_m, then $\phi^*\nu$ is a degree 0 divisor on C_n which is not linearly equivalent to the zero divisor on C_n.

Exercise 4.17 Fire the sets $\{v_2, v_3, v_4\}$, $\{v_4\}$, and $\{v_2, v_3, v_4\}$ in that order, to obtain the reduced configuration $(-2, 1, 0, 1)$.

Exercise 4.18 $(-2, 0, 1, 1)$

Exercise 4.19

D	T_D
$-3v_1 + 2v_2 + v_3$	$\{e_1, e_2, e_3\}$
$-3v_1 + v_2 + 2v_4$	$\{e_1, e_2, e_5\}$
$-3v_1 + 2v_2 + v_4$	$\{e_1, e_2, e_6\}$
$-3v_1 + v_2 + 2v_3$	$\{e_1, e_3, e_4\}$
$-3v_1 + 2v_3 + v_4$	$\{e_1, e_4, e_5\}$
$-3v_1 + v_3 + 2v_4$	$\{e_1, e_4, e_6\}$
$-2v_1 + v_3 + v_4$	$\{e_1, e_5, e_6\}$
$-2v_1 + v_2 + v_3$	$\{e_2, e_3, e_4\}$

D	T_D
$-2v_1 + 2v_3$	$\{e_2, e_3, e_5\}$
$-2v_1 + v_2 + v_4$	$\{e_2, e_4, e_6\}$
$-2v_1 + 2v_4$	$\{e_2, e_4, e_5\}$
$-v_1 + v_4$	$\{e_2, e_5, e_6\}$
$-v_1 + v_3$	$\{e_3, e_4, e_5\}$
$-v_1 + v_2$	$\{e_3, e_4, e_6\}$
0	$\{e_3, e_5, e_6\}$

Chapter 6
Exercise 6.1

$$\hat{g}^\vee(y) = \frac{1}{p^n} \sum_{z \subset V} \sum_{x \in V} g(x) \zeta^{\langle z, y - x \rangle}$$

$$= \frac{1}{p^n} \sum_{x \in V} g(x) \sum_{z \in V} \zeta^{\langle z, y - x \rangle}.$$

Let $a = y - x = (a_1, a_2, \ldots, a_n)$ and $z = (z_1, z_2, \ldots, z_n)$. Note that

$$\sum_{z \in V} \zeta^{\langle z, a \rangle} = \sum_{z \in V} \zeta^{z_1 a_1} \zeta^{z_2 a_2} \cdots \zeta^{z_n a_n}$$

$$= \prod_{i=1}^{n} \sum_{z_i=0}^{p-1} (\zeta^{a_i})^{z_i}.$$

The sum

$$\sum_{z_i=0}^{p-1} (\zeta^{a_i})^{z_i}$$

is zero if $a_i \neq 0$. Therefore, $\sum_{z \in V} \zeta^{\langle z, a \rangle} = 0$ if $a \neq 0$, i.e., if $y \neq x$. Thus

$$\hat{g}^{\vee}(y) = \frac{1}{p^n} g(y) \sum_{z \in V} \zeta^0 = g(y).$$

Exercise 6.2 Suppose that $\hat{g}(z) = c$ for all $z \in GF(p)^n$. Then

$$g(y) = \hat{g}^{\vee}(y) = \frac{1}{p^n} c \sum_{z \in V} \zeta^{\langle z, y \rangle} = \begin{cases} 0 & \text{if} \quad y \neq 0, \\ c & \text{if} \quad y = 0. \end{cases}$$

Conversely, if g is supported at 0, then

$$\hat{g}(y) = \sum_{x \in V} g(x) \zeta^{-\langle x, y \rangle} = g(0)$$

so $\hat{g}$ is constant.

Exercise 6.3 Let f be a Boolean function on $GF(2)^2$. The value of $f(0)$ must be 0, and there are 2 possible values for f at each of the other 3 elements of $GF(2)^2$, so that there are 2^3 possible functions. The bent functions must have supports of size 1 or 3, by Lemma 6.3.11. There are 4 such functions. These functions may be expressed in terms of variables x_0 and x_1 over $GF(2)$ as $x_0 x_1$, $x_0 x_1 + x_0$, $x_0 x_1 + x_1$, and $x_0 x_1 + x_0 + x_1$.

Exercise 6.4 The values of the 4 bent functions at the points of $GF(2)^2$ are given by the following table. Each function differs from each other function at exactly two points, so the Hamming distance between any two functions is exactly 2.

$GF(2)^2$	$(0,0)$	$(0,1)$	$(1,0)$	$(1,1)$
$x_0 x_1$	0	0	0	1
$x_0 x_1 + x_0$	0	1	0	0
$x_0 x_1 + x_1$	0	0	1	0
$x_0 x_1 + x_0 + x_1$	0	1	1	1

Exercise 6.5 Yes. $(7, 3, 1)$.

Exercise 6.6 Yes. $(11, 5, 2)$.

Exercise 6.7 No. There are 20 nonidentity elements in $D \cdot D^{-1}$ and only 15 in G, so the nonidentity elements in G cannot all be represented the same number of time in $D \cdot D^{-1}$.

Exercise 6.9 The incidence matrix is

$$
\begin{pmatrix}
1 & 1 & 1 & 1 & 0 & 0 & 0 & 0 & 0 & 0 & 0 & 0 & 0 & 0 & 0 & 0 & 0 & 0 \\
0 & 0 & 0 & 0 & 1 & 1 & 1 & 1 & 0 & 0 & 0 & 0 & 0 & 0 & 0 & 0 & 0 & 0 \\
0 & 0 & 0 & 0 & 0 & 0 & 0 & 0 & 1 & 1 & 1 & 1 & 0 & 0 & 0 & 0 & 0 & 0 \\
1 & 0 & 0 & 0 & 1 & 0 & 0 & 0 & 0 & 0 & 0 & 0 & 1 & 1 & 0 & 0 & 0 & 0 \\
0 & 0 & 0 & 0 & 0 & 1 & 0 & 0 & 1 & 0 & 0 & 0 & 0 & 0 & 1 & 1 & 0 & 0 \\
0 & 1 & 0 & 0 & 0 & 0 & 0 & 0 & 0 & 1 & 0 & 0 & 0 & 0 & 0 & 0 & 1 & 1 \\
0 & 0 & 1 & 0 & 0 & 0 & 0 & 0 & 0 & 0 & 1 & 0 & 1 & 0 & 1 & 0 & 0 & 0 \\
0 & 0 & 0 & 1 & 0 & 0 & 1 & 0 & 0 & 0 & 0 & 0 & 0 & 0 & 0 & 1 & 1 & 0 \\
0 & 0 & 0 & 0 & 0 & 0 & 0 & 1 & 0 & 0 & 0 & 1 & 0 & 1 & 0 & 0 & 0 & 1 \\
\end{pmatrix}
$$

and the girth is 3.

Exercise 6.10 The adjacency matrix is

$$
\begin{pmatrix}
0 & 0 & 0 & 1 & 0 & 1 & 1 & 1 & 0 \\
0 & 0 & 0 & 1 & 1 & 0 & 0 & 1 & 1 \\
0 & 0 & 0 & 0 & 1 & 1 & 1 & 0 & 1 \\
1 & 1 & 0 & 0 & 0 & 0 & 1 & 0 & 1 \\
0 & 1 & 1 & 0 & 0 & 0 & 1 & 1 & 0 \\
1 & 0 & 1 & 0 & 0 & 0 & 0 & 1 & 1 \\
1 & 0 & 1 & 1 & 1 & 0 & 0 & 0 & 0 \\
1 & 1 & 0 & 0 & 1 & 1 & 0 & 0 & 0 \\
0 & 1 & 1 & 1 & 0 & 1 & 0 & 0 & 0 \\
\end{pmatrix}
$$

and the diameter is 2.

Bibliography

[BdlH97]   R. Bacher, P. de la Harpe, The lattice of integral flows and the lattice of integral cuts on a finite graph. Bull. Soc. Math. France **125**, 167–198 (1997)

[Ba14]   S. Backman, Riemann-Roch Theory for Graph Orientations, preprint, (2014). http://arxiv.org/abs/1401.3309

[BN07]   M. Baker, S. Norine, Riemann-Roch and Abel-Jacobi theory on a finite graph. Adv. Math. **215**, 766–788 (2007)

[BN09]   M. Baker, S. Norine, Harmonic morphisms and hyperelliptic graphs. Int. Math. Res. Not. IMRN. 2914–2955 (2009)

[BS13]   M. Baker, F. Shokrieh, Chip-firing games, potential theory on graphs, and spanning trees. J. Combin. Theory Ser. A **120**, 164–182 (2013)

[BCT10]   B. Benson, D. Chakrabarty, P. Tetali, G-parking functions, acyclic orientations and spanning trees. Discret. Math. **310**, 1340–1353 (2010)

[B98]   A. Bernasconi, *Mathematical Techniques for the Analysis of Boolean Functions*, Ph.D. dissertation TD-2/98 (Universit di Pisa-Udine, 1998)

[BC99]   A. Bernasconi, B. Codenotti, Spectral analysis of Boolean functions as a graph eigenvalue problem. IEEE Trans. Comput. **48:3**, 345–351 (1999). http://ilex.iit.cnr.it/codenotti/ps_files/graph_fourier.ps

[BCV01]   A. Bernasconi, B. Codenotti, J. VanderKam, A characterization of bent functions in terms of strongly regular graphs. IEEE Trans. Comput. **50**(9), 984–985 (2001)

[Bi93]   N. Biggs, *Algebraic Graph Theory*, 2nd edn. (Cambridge University Press, Cambridge, 1993)

[Bi97]   N. Biggs, Algebraic potential theory on graphs. Bull. Lond. Math. Soc. **29**, 641–682 (1997)

[Bi99]   N. Biggs, Chip-firing and the critical group of a graph. J. Algebraic Combin. **9**, 25–45 (1999)

[Bi99b]   N. Biggs, The Tutte polynomial as a growth function. J. Algebraic Combin. **10**, 115–133 (1999)

[Bi07]   N. Biggs, The critical group from a cryptographic perspective. Bull. Lond. Math. Soc. **39**, 829–836 (2007)

[Bi08]   N. Biggs, *Codes: An Introduction to Information, Communication, and Cryptography* (Springer, Berlin, 2008)

© Springer International Publishing AG 2017
W.D. Joyner and C.G. Melles, *Adventures in Graph Theory*,
Applied and Numerical Harmonic Analysis,
https://doi.org/10.1007/978-3-319-68383-6

[BLS07]   T. Biyikogu, J. Leydold, P. Stadler, *Laplacian Eigenvectors of Graphs*, Lecture Notes in Mathematics (Springer, Berlin, 2007)

[BLoS91]   A. Björner, L. Lovász, P. Shor, Chip-firing games on graphs. Eur. J. Combin **12**, 283–291 (1991). http://www.cs.elte.hu/~lovasz/morepapers/chips.pdf

[Bl08]   S. Blackburn, Cryptanalysing the critical group: efficiently solving Biggs's discrete logarithm problem. J. Math. Cryptol. **3**, 199–203 (2009)

[Bo98]   B. Bollobás, *Modern Graph Theory*, Graduate Texts in Mathematics (Springer, Berlin, 1998)

[BM07]   J. Bondy, U. Murty, *Graph Theory with Applications* (Springer, Berlin, 2007)

[BL02]   S. Bosch, D. Lorenzini, Grothendieck's pairing on component groups of Jacobians. Invent. Math. **148**, 353–396 (2002)

[Br]   A. Brouwer, Parameters of Strongly Regular Graphs. http://www.win.tue.nl/~aeb/graphs/srg/srgtab1-50.html

[BrCN89]   A. Brouwer, A. Cohen, A. Neumaier, *Distance-Regular Graphs* (Springer, Berlin, 1989)

[BH11]   A. Brouwer, W. Haemers, *Spectra of Graphs* (Springer, Berlin, 2011). http://www.win.tue.nl/~aeb/2WF02/spectra.pdf

[CvL91]   P. Cameron, J. van Lint, Designs, Graphs, Codes and their Links. London Math. Soc. Lecture Note Ser. **19** (1991)

[C10]   C. Carlet, Boolean functions for cryptography and error correcting codes, in *Boolean Models and Methods in Mathematics, Computer Science, and Engineering*, eds. by Y. Crama and P. Hammer, (Cambridge University Press, 2010), pp. 257–397. http://www1.spms.ntu.edu.sg/~kkhoongm/chap-fcts-Bool.pdf

[CD04]   C. Carlet, C. Ding, Highly non-linear mappings. J. Complex. **20**, 205–244 (2004)

[CJMP12]   C. Celerier, D. Joyner, C. Melles, D. Phillips, On the Walsh-Hadamard transform of monotone Boolean functions. Tbilisi Math. J. (Special Issue on Sage and Research) **5**, 19–35 (2012)

[CJMPW15]   C. Celerier, D. Joyner, C. Melles, D. Phillips, S. Walsh, *Edge-weighted Cayley graphs and p-ary bent functions*, INTEGERS 16 (2016) #A35. See also *Explorations of edge-weighted Cayley graphs and p-ary bent functions*. http://www.arxiv.org/abs/1406.1087

[CM12]   A. Cesmelioglu, W. Meidl, Bent functions of maximal degree. IEEE Trans. Inform. Theory **58**, 1186–1190 (2012)

[CP05]   D. Chebikin, P. Pylyavskyy, A family of bijections between G-parking functions and spanning trees. J. Combin. Theory Ser. A **110**, 31–41 (2005)

[CTZ10]   Y. Chee, Y. Tan, and X. Zhang, Strongly Regular Graphs Constructed from p-ary Bent Functions, preprint, (2010). http://arxiv.org/abs/1011.4434

[Ch92]   F. Chung, **Spectral Graph Theory**, American Mathematical Society (1992). http://www.math.ucsd.edu/~fan/research/revised.html

[Ch05]   F. Chung, Laplacians and the Cheeger inequality for directed graphs. Ann. Comb. **9**, 1–19 (2005). http://www.math.ucsd.edu/~fan/wp/dichee.pdf

[CLP15]   J. Clancy, T. Leake, S. Payne, A note on Jacobians, Tutte polynomials, and two-variable zeta functions of graphs. Exp. Math. **24**, 1–7 (2015)

[Co76]   J. Conway, *On Numbers and Games (ONAG)* (Academic Press, New York, 1976)

[Co84]   J. Conway, Hexacode and tetracode - MINIMOG and MOG, in *Computational Group Theory*, ed. by M. Atkinson, (Academic Press, 1984)

[CS86]   J. Conway, N. Sloane, Lexicographic codes: error-correcting codes from game theory. IEEE Trans. Inform. Theory **32**, 337–348 (1986)

[CS99]   J. Conway, N. Sloane, *Sphere Packings, Lattices and Groups*, 3rd edn. (Springer, Berlin, 1999)

[CLB03]   R. Cori, Y. Le Borgne, The sandpile model and Tutte polynomials. Adv. Appl. Math. **30**, 44–52 (2003)

[Co11]   S. Corry, Genus bounds for harmonic group actions on finite graphs. Int. Math. Res. Not. **19**, 4515–4533 (2011). http://arxiv.org/abs/1006.0446

[CoM77]   R. Coulter, R. Matthews, Bent polynomials over finite fields. Bull. Aust. Math. Soc. **56**, 429–437 (1977)

[Cu76]   R. Curtis, A new combinatorial approach to M_{24}. Math. Proc. Cambridge Philos. Soc. **79**, 25–42 (1976)

[Cu84]   R. Curtis, The Steiner system $S(5, 6, 12)$, the Mathieu group M_{12}, and the kitten, in *Computational Group Theory*, ed. by M. Atkinson (Academic Press, 1984)

[CS09]   T. Cusick, P. Stanica, *Cryptographic Boolean Functions and Applications* (Academic Press - Elsevier, New York, 2009)

[DKR13]   P. Dankelmann, J. Key, B. Rodrigues, Codes form incidence matrices of graphs. Des. Codes Cryptogr. **68**, 373–393 (2013). http://www.ces.clemson.edu/~keyj/Key/dakero1Rev_UD.pdf

[DSV03]   G. Davidoff, P. Sarnak, A. Valette, *Elementary Number Theory, Group Theory, and Ramanujan Graphs*, London Mathematical Society (Cambridge University Press, Cambridge, 2003)

[De73]   P. Delsarte, An algebraic approach to the association schemes of coding theory. Philips Res. Rep. Suppl. No. **10**, (1973)

[Dh90]   D. Dhar, Self-organized critical state of sandpile automaton models. Phys. Rev. Lett. **64**, 1613–1616 (1990)

[D74]   J. Dillon, *Elementary Hadamard difference sets*, Ph.D. thesis, (University of Maryland, 1974)

[DF99]   D. Dummit, R. Foote, *Abstract Algebra*, 2nd edn. (Wiley, New York, 1999)

[DL17]   A. Doerr, K. Levasseur, *Applied Discrete Structures, Version 3.3* (2017). http://faculty.uml.edu/klevasseur/ads2/ and https://applied-discrete-structures.wiki.uml.edu/

[Du09]   N. Durgin, Abelian sandpile model on symmetric graphs, thesis, Harvey Mudd College, (2009). https://www.math.hmc.edu/seniorthesis/archives/2009/ndurgin/ndurgin-2009-thesis.pdf

[Duu01]   I. Duursma, A Riemann hypothesis analogue for self-dual codes, in *Codes and Association Schemes*, eds. by A. Barg, S. Litsyn, AMS DIMACS Series 56, (2001) pp. 115–124

[Duu03a]   I. Duursma, *Results on zeta functions for codes*, Fifth Conference on Algebraic Geometry, Number Theory, Coding Theory and Cryptography, University of Tokyo, 17–19 Jan 2003

[Duu03b]   I. Duursma, Extremal weight enumerators and ultraspherical polynomials. Discret. Math. **268**, 103–127 (2003)

[Duu04]   I. Duursma, Combinatorics of the two-variable zeta function in finite fields and applications. Lecture Notes in Comput. Sci. **2948**, 109–136 (2004)

[FWXY10]   T. Fen, B. Wen, Q. Xiang, J. Yin, Partial difference sets from quadratic forms and p -ary weakly regular bent functions, in *Number Theory and Related Areas*, ALM 27 (International Press of Boston, 2010) pp. 25–40

[Fi73]   M. Fiedler, Algebraic connectivity of graphs. Czechoslovak Math. J. **23**, 298–305 (1973)

[Fr99]   M. Frazier, *An Introduction to Wavelets through Linear Algebra* (Springer, Berlin, 1999)

[GR01]   C. Godsil, G. Royle, *Algebraic Graph Theory, Graduate Texts in Mathematics*, vol. 207 (Springer, New York, 2001)

312   BIBLIOGRAPHY

[HB68]   S. Hakimi, J. Bredeson, Graph theoretic error-correcting codes. IEEE
            Trans. Inform. Theory **14**, 584–591 (1968)
[Ha77]   R. Hartshorne, *Algebraic Geometry, Graduate Texts in Mathematics*, vol.
            52 (Springer, New York, 1977)
[HK06]   T. Helleseth, A. Kholosha, Monomial and quadratic bent functions over
            the finite fields of odd characteristic. IEEE Trans. Inform. Theory **52**,
            2018–2032 (2006)
[HK10]   T. Helleseth, A. Kholosha, On generalized bent functions, Information
            theory and applications workshop (ITA), (2010)
[HK13]   T. Helleseth, A. Kholosha, Bent functions and their connections to com-
            binatorics, in Surveys in Combinatorics, London Math. Soc. Lecture Note
            Series, Vol. 409. Cambridge University Press **2013**, 91–126 (2013)
[He11]   A. Herman, Seminar notes: Algebraic Aspects of Association Schemes
            and Scheme Rings, seminar notes, (2011). http://uregina.ca/~hermana/
            ASSR-Lecture9.pdf
[HLMPPW08] A. Holroyd, L. Levine, K. Meszaros, Y. Peres, J. Propp, D. Wilson, Chip-
            firing and rotor-routing on directed graphs, preprint, (2008). http://front.
            math.ucdavis.edu/0801.3306
[H04]   X.-D. Hou, p-Ary and q-ary versions of certain results about bent func-
            tions and resilient functions. Finite Fields Appl. **10**, 566–582 (2004)
[HP10]   W. Huffman, V. Pless, *Fundamentals of Error-Correcting Codes* (Cam-
            bridge University Press, Cambridge, 2010)
[J15]   D. Joyner, The man who found God's number. College Math. J. **45**, 258–
            266 (2014)
[JK11]   D. Joyner, J.-L. Kim, *Selected Unsolved Problems in Coding Theory*
            (Birkhaüser, New York, 2011)
[JV10]   D. Jungnickel, S. Vanstone, Graphical codes revisited. IEEE Trans. In-
            form. Theory **43**, 136–146 (1997)
[KR01]   J. Kahane, A. Ryba, The hexad game in the Fraenkel Festschrift volume.
            Electron. J. of Combin. **8**, 00–00 (2001). http://www.combinatorics.org/
            ojs/index.php/eljc/article/view/v8i2r11 #R11
[KA02]   P. Kaski, *Eigenvectors and Spectra of Cayley Graphs*, preprint,
            (2002). http://www.tcs.hut.fi/Studies/T-79.300/2002S/esitelmat/kaski_
            paper_020506.pdf
[KP09]   T. Kim, C. Praeger, On generalised Paley graphs and their automorphism
            groups. Michigan Math. J. **58**, 293–308 (2009)
[KSW85]   P. Kumar, R. Scholtz, L. Welch, Generalized bent functions and their
            properties. J. Combin. Theory Ser. A **40**, 90–107 (1985)
[La70]   S. Lang, *Algebraic Number Theory* (Addison-Wesley, 1970)
[Lu11]   Y. Luo, Rank-determining sets of metric graphs. J. Combin. Theory Ser.
            A **118**, 1775–1793 (2011)
[MS77]   *The Theory of Error-Correcting Codes*, North Holland Mathematical
            Library (North Holland Publishing Company, New York, 1977)
[Ma11]   M. Manjunath, *The Rank of a Dvisor on a Finite Graph: Geometry and
            Computation*, (2011). http://arxiv.org/abs/1111.7251
[Mar08]   D. Marcus, *Graph Theory, A Problem-Oriented Approach* (Mathematical
            Association of America, 2008)
[M15]   A. Mednykh, On the Riemann-Hurwitz Formula for Graph Coverings,
            preprint, (2015). http://arxiv.org/pdf/1505.00321.pdf
[MM14]   A. Mednykh, I. Mednykh, Graph Coverings and Harmonic Maps in Ex-
            ercises, preprint, (2014). http://kam.mff.cuni.cz/~atcagc14/materialy/
            Harmonic%20maps_1.pdf
[MvOV96]   A. Menezes, P. van Oorschot, S. Vanstone, *Handbook of Applied Cryptog-
            raphy* (CRC Press, 1996). http://www.cacr.math.uwaterloo.ca/hac/ (All
            chapters are free online.)

[Me97]   C. Merino, López, Chip-firing and the Tutte polynomial. Ann. Comb. **1**, 253–259 (1997)

[Me99]   C. Merino, Matroids, the Tutte Polynomial, and the Chip Firing Game, Ph.D. thesis, Oxford University, (1999). http://calli.matem.unam.mx/~merino/e_publications.html#2 and http://www.dmtcs.org/dmtcs-ojs/index.php/proceedings/article/viewArticle/dmAA0118 (The first link is for the thesis itself in ps format; the second link is to an associated paper.)

[Me04]   C. Merino, The chip-firing game. Discret. Math. **302**, 188–210 (2005)

[Mes16]   S. Mesnager, *Bent Functions: Fundamentals and Results* (Springer, 2016)

[Mi95]   R. Miranda, *Algebraic Curves and Riemann Surfaces No. 5, AMS*, Graduate studies in mathematics series (Providence, R.I., 1995)

[Mo91a]   B. Mohar, The Laplacian spectrum of graphs, in *Graph Theory, Combinatorics, and Applications* ed. by Y. Alavi, G. Chartrand, O. Oellermann, A. Schwenk, vol 2 (Wiley, 1991) pp. 871–898. http://www.fmf.uni-lj.si/~mohar/Papers/Spec.pdf

[Mo91b]   B. Mohar, Eigenvalues, diameter, and mean distance for graphs. Graphs Combin. **7**, 53–64 (1991). http://www.sfu.ca/~mohar/Reprints/1991/BM91_GC7_Mohar_EigenvaluesDiameterMeanDistance.pdf

[NS94]   Noam Nisan, Mario Szegedy, On the degree of boolean functions as real polynomials. Comput. Complex. **4**, 301–313 (1994)

[PPW11]   D. Perkinson, J. Perlman, and J. Wilmes, Primer for the algebraic geometry of sandpiles, preprint (2011). http://arxiv.org/abs/1112.6163

[Pe09]   J. Perlman, Sandpiles, A Bridge Between Graphs and Toric Ideals, thesis, Reed College, (2009). http://people.reed.edu/~davidp/homepage/students/perlman.pdf

[Po03]   J. Pohill, A brief survey of difference sets, partial difference sets, and relative difference sets, preprint, (2003). http://www.passhema.org/proceedings/2003/PolhillDifference2003.pdf

[PS04]   A. Postnikov, B. Shapiro, Trees, parking functions, syzygies, and deformations of monomial ideals. Trans. Am. Math. Soc. **356**, 3109–3142 (2004)

[Pot16]   V. Potapov, On Minimal Distance Between q-ary Bent Functions, preprint (2016). http://arxiv.org/abs/1606.02430

[PTFL11]   A. Pott, Y. Tan, T. Feng, S. Ling, Association schemes arising from bent functions. Des. Codes Cryptogr. **59**, 319–331 (2011)

[RKHS02]   D. Rockmore, P. Kostelec, W. Hordijk, P. Stadler, Fast Fourier transform for fitness landscapes. Appl. Comput. Harmon. Anal. **12**, 57–76 (2002)

[Ro06]   R. Roth, *Introduction to Coding Theory* (Cambridge University Press, Cambridge, 2006)

[Sh09]   F. Shokrieh, Discrete logarithm problem on the Jacobian of finite graphs preprint later retitled and published as The monodromy pairing and discrete logarithm on the Jacobian of finite graphs. J. Math. Cryptol. **4**, 43–56 (2010). http://arxiv.org/pdf/0907.4764.pdf

[Sp]   E. Spence, webpage Strongly regular graphs on at most 64 vertices. http://www.maths.gla.ac.uk/~es/srgraphs.php

[Sp10]   D. Spielman, Algorithms, graph theory, and linear equations in Laplacian matrices, in *Proceedings of the International Congress of Mathematicians* (Hyderabad, India, 2010). http://www.cs.yale.edu/homes/spielman/PAPERS/icm10post.pdf

[St07]   P. Stanica, Graph eigenvalues and Walsh spectrum of Boolean functions. Integers **7**(2), 00 (2007). #A32

[St99]   R. Stanley, *Enumerative Combinatorics*, 2nd edn. (Cambridge University Press, Cambridge, 1999)

[St14]   D. Stevanovic, *Spectral Radius of Graphs* (Academic Press, 2014)

[Su86]   T. Sunada, L-functions in geometry and some applications, in *Curvature and Topology of Riemannian Manifolds*, Lecture Notes in Mathematics, vol 1201 (Springer 1986) pp. 266–284

[Te10]   A. Terras, *A Stroll Through the Garden of Graph Zeta Functions* (Cambridge University Press, 2010)

[T10]   N. Tokareva, *Generalizations of Bent Functions: A Survey* (2010). http://eprint.iacr.org/2011/111.pdf

[T15]   N. Tokareva, *Bent Functions: Results and Applications to Cryptography* (Academic Press, 2015)

[vD01]   E. van Dam, Strongly regular decompositions of the complete graph. J. Algebraic Combin. **17**, 181–201 (2003)

[W14]   S. Walsh, *Combinatorics of p-ary Bent Functions* (United States Naval Academy honors project, 2014)

[We99]   D. Welsh, *The Tutte Polynomial.* https://www.math.ucdavis.edu/~deloera/MISC/BIBLIOTECA/trunk/Welsh/welsh-tutte-polynomial.pdf

[Wi03]   S. Weintraub, *Representation Theory of Finite Groups: Algebra and Arithmetic, Graduate Studies in Mathematics*, vol. 59 (American Mathematical Society, Providence, R.I., 2003)

Index

© Springer International Publishing AG 2017
W.D. Joyner and C.G. Melles, *Adventures in Graph Theory*,
Applied and Numerical Harmonic Analysis,
https://doi.org/10.1007/978-3-319-68383-6

Applied and Numerical Harmonic Analysis
(83 volumes)

A. Saichev and W.A. Woyczyński: *Distributions in the Physical and Engineering Sciences* (ISBN 978-0-8176-3924-2)

C.E. D'Attellis and E.M. Fernandez-Berdaguer: *Wavelet Theory and Harmonic Analysis in Applied Sciences* (ISBN 978-0-8176-3953-2)

H.G. Feichtinger and T. Strohmer: *Gabor Analysis and Algorithms* (ISBN 978-0-8176-3959-4)

R. Tolimieri and M. An: *Time-Frequency Representations* (ISBN 978-0-8176-3918-1)

T.M. Peters and J.C. Williams: *The Fourier Transform in Biomedical Engineering* (ISBN 978-0-8176-3941-9)

G.T. Herman: *Geometry of Digital Spaces* (ISBN 978-0-8176-3897-9)

A. Teolis: *Computational Signal Processing with Wavelets* (ISBN 978-0-8176-3909-9)

J. Ramanathan: *Methods of Applied Fourier Analysis* (ISBN 978-0-8176-3963-1)

J.M. Cooper: *Introduction to Partial Differential Equations with MATLAB* (ISBN 978-0-8176-3967-9)

A. Procházka, N.G. Kingsbury, P.J. Payner, and J. Uhlir: *Signal Analysis and Prediction* (ISBN 978-0-8176-4042-2)

W. Bray and C. Stanojevic: *Analysis of Divergence* (ISBN 978-1-4612-7467-4)

G.T. Herman and A. Kuba: *Discrete Tomography* (ISBN 978-0-8176-4101-6)

K. Gröchenig: *Foundations of Time-Frequency Analysis* (ISBN 978-0-8176-4022-4)

© Springer International Publishing AG 2017
W.D. Joyner and C.G. Melles, *Adventures in Graph Theory*,
Applied and Numerical Harmonic Analysis,
https://doi.org/10.1007/978-3-319-68383-6

L. Debnath: *Wavelet Transforms and Time-Frequency Signal Analysis* (ISBN 978-0-8176-4104-7)

J.J. Benedetto and P.J.S.G. Ferreira: *Modern Sampling Theory* (ISBN 978-0-8176-4023-1)

D.F. Walnut: *An Introduction to Wavelet Analysis* (ISBN 978-0-8176-3962-4)

A. Abbate, C. DeCusatis, and P.K. Das: *Wavelets and Subbands* (ISBN 978-0-8176-4136-8)

O. Bratteli, P. Jorgensen, and B. Treadway: *Wavelets Through a Looking Glass* (ISBN 978-0-8176-4280-80

H.G. Feichtinger and T. Strohmer: *Advances in Gabor Analysis* (ISBN 978-0-8176-4239-6)

O. Christensen: *An Introduction to Frames and Riesz Bases* (ISBN 978-0-8176-4295-2)

L. Debnath: *Wavelets and Signal Processing* (ISBN 978-0-8176-4235-8)

G. Bi and Y. Zeng: *Transforms and Fast Algorithms for Signal Analysis and Representations* (ISBN 978-0-8176-4279-2)

J.H. Davis: *Methods of Applied Mathematics with a MATLAB Overview* (ISBN 978-0-8176-4331-7)

J.J. Benedetto and A.I. Zayed: *Modern Sampling Theory* (ISBN 978-0-8176-4023-1)

E. Prestini: *The Evolution of Applied Harmonic Analysis* (ISBN 978-0-8176-4125-2)

L. Brandolini, L. Colzani, A. Iosevich, and G. Travaglini: *Fourier Analysis and Convexity* (ISBN 978-0-8176-3263-2)

W. Freeden and V. Michel: *Multiscale Potential Theory* (ISBN 978-0-8176-4105-4)

O. Christensen and K.L. Christensen: *Approximation Theory* (ISBN 978-0-8176-3600-5)

O. Calin and D.-C. Chang: *Geometric Mechanics on Riemannian Manifolds* (ISBN 978-0-8176-4354-6)

J.A. Hogan: *Time?Frequency and Time?Scale Methods* (ISBN 978-0-8176-4276-1)

C. Heil: *Harmonic Analysis and Applications* (ISBN 978-0-8176-3778-1)

K. Borre, D.M. Akos, N. Bertelsen, P. Rinder, and S.H. Jensen: *A Software-Defined GPS and Galileo Receiver* (ISBN 978-0-8176-4390-4)

T. Qian, M.I. Vai, and Y. Xu: *Wavelet Analysis and Applications* (ISBN 978-3-7643-7777-9)

G.T. Herman and A. Kuba: *Advances in Discrete Tomography and Its Applications* (ISBN 978-0-8176-3614-2)

M.C. Fu, R.A. Jarrow, J.-Y. Yen, and R.J. Elliott: *Advances in Mathematical Finance* (ISBN 978-0-8176-4544-1)

O. Christensen: *Frames and Bases* (ISBN 978-0-8176-4677-6)

P.E.T. Jorgensen, J.D. Merrill, and J.A. Packer: *Representations, Wavelets, and Frames* (ISBN 978-0-8176-4682-0)

M. An, A.K. Brodzik, and R. Tolimieri: *Ideal Sequence Design in Time-Frequency Space* (ISBN 978-0-8176-4737-7)

S.G. Krantz: *Explorations in Harmonic Analysis* (ISBN 978-0-8176-4668-4)

B. Luong: *Fourier Analysis on Finite Abelian Groups* (ISBN 978-0-8176-4915-9)

G.S. Chirikjian: *Stochastic Models, Information Theory, and Lie Groups, Volume 1* (ISBN 978-0-8176-4802-2)

C. Cabrelli and J.L. Torrea: *Recent Developments in Real and Harmonic Analysis* (ISBN 978-0-8176-4531-1)

M.V. Wickerhauser: *Mathematics for Multimedia* (ISBN 978-0-8176-4879-4)

B. Forster, P. Massopust, O. Christensen, K. Gröchenig, D. Labate, P. Vandergheynst, G. Weiss, and Y. Wiaux: *Four Short Courses on Harmonic Analysis* (ISBN 978-0-8176-4890-9)

O. Christensen: *Functions, Spaces, and Expansions* (ISBN 978-0-8176-4979-1)

J. Barral and S. Seuret: *Recent Developments in Fractals and Related Fields* (ISBN 978-0-8176-4887-9)

O. Calin, D.-C. Chang, and K. Furutani, and C. Iwasaki: *Heat Kernels for Elliptic and Sub-elliptic Operators* (ISBN 978-0-8176-4994-4)

C. Heil: *A Basis Theory Primer* (ISBN 978-0-8176-4686-8)

J.R. Klauder: *A Modern Approach to Functional Integration* (ISBN 978-0-8176-4790-2)

J. Cohen and A.I. Zayed: *Wavelets and Multiscale Analysis* (ISBN 978-0-8176-8094-7)

D. Joyner and J.-L. Kim: *Selected Unsolved Problems in Coding Theory* (ISBN 978-0-8176-8255-2)

G.S. Chirikjian: *Stochastic Models, Information Theory, and Lie Groups, Volume 2* (ISBN 978-0-8176-4943-2)

J.A. Hogan and J.D. Lakey: *Duration and Bandwidth Limiting* (ISBN 978-0-8176-8306-1)

G. Kutyniok and D. Labate: *Shearlets* (ISBN 978-0-8176-8315-3)

P.G. Casazza and P. Kutyniok: *Finite Frames* (ISBN 978-0-8176-8372-6)

V. Michel: *Lectures on Constructive Approximation* (ISBN 978-0-8176-8402-0)

D. Mitrea, I. Mitrea, M. Mitrea, and S. Monniaux: *Groupoid Metrization Theory* (ISBN 978-0-8176-8396-2)

T.D. Andrews, R. Balan, J.J. Benedetto, W. Czaja, and K.A. Okoudjou: *Excursions in Harmonic Analysis, Volume 1* (ISBN 978-0-8176-8375-7)

T.D. Andrews, R. Balan, J.J. Benedetto, W. Czaja, and K.A. Okoudjou: *Excursions in Harmonic Analysis, Volume 2* (ISBN 978-0-8176-8378-8)

D.V. Cruz-Uribe and A. Fiorenza: *Variable Lebesgue Spaces* (ISBN 978-3-0348-0547-6)

W. Freeden and M. Gutting: *Special Functions of Mathematical (Geo-)Physics* (ISBN 978-3-0348-0562-9)

A. Saichev and W.A. Woyczyński: *Distributions in the Physical and Engineering Sciences, Volume 2: Linear and Nonlinear Dynamics of Continuous Media* (ISBN 978-0-8176-3942-6)

S. Foucart and H. Rauhut: *A Mathematical Introduction to Compressive Sensing* (ISBN 978-0-8176-4947-0)

G. Herman and J. Frank: *Computational Methods for Three-Dimensional Microscopy Reconstruction* (ISBN 978-1-4614-9520-8)

A. Paprotny and M. Thess: *Realtime Data Mining: Self-Learning Techniques for Recommendation Engines* (ISBN 978-3-319-01320-6)

A. Zayed and G. Schmeisser: *New Perspectives on Approximation and Sampling Theory: Festschrift in Honor of Paul Butzer's 85th Birthday* (978-3-319-08800-6)

R. Balan, M. Begue, J. Benedetto, W. Czaja, and K.A Okoudjou: *Excursions in Harmonic Analysis, Volume 3* (ISBN 978-3-319-13229-7)

H. Boche, R. Calderbank, G. Kutyniok, J. Vybiral: *Compressed Sensing and its Applications* (ISBN 978-3-319-16041-2)

S. Dahlke, F. De Mari, P. Grohs, and D. Labate: *Harmonic and Applied Analysis: From Groups to Signals* (ISBN 978-3-319-18862-1)

G. Pfander: *Sampling Theory, a Renaissance* (ISBN 978-3-319-19748-7)

R. Balan, M. Begue, J. Benedetto, W. Czaja, and K.A Okoudjou: *Excursions in Harmonic Analysis, Volume 4* (ISBN 978-3-319-20187-0)

O. Christensen: *An Introduction to Frames and Riesz Bases, Second Edition* (ISBN 978-3-319-25611-5)

E. Prestini: *The Evolution of Applied Harmonic Analysis: Models of the Real World, Second Edition* (ISBN 978-1-4899-7987-2)

J.H. Davis: *Methods of Applied Mathematics with a Software Overview, Second Edition* (ISBN 978-3-319-43369-1)

M. Gilman, E. M. Smith, S. M. Tsynkov: *Transionospheric Synthetic Aperture Imaging* (ISBN 978-3-319-52125-1)

S. Chanillo, B. Franchi, G. Lu, C. Perez, E.T. Sawyer: *Harmonic Analysis, Partial Differential Equations and Applications* (ISBN 978-3-319-52741-3)

R. Balan, J. Benedetto, W. Czaja, M. Dellatorre, and K.A Okoudjou: *Excursions in Harmonic Analysis, Volume 5* (ISBN 978-3-319-54710-7)

I. Pesenson, Q.T. Le Gia, A. Mayeli, H. Mhaskar, D.X. Zhou: *Frames and Other Bases in Abstract and Function Spaces: Novel Methods in Harmonic Analysis, Volume 1* (ISBN 978-3-319-55549-2)

I. Pesenson, Q.T. Le Gia, A. Mayeli, H. Mhaskar, D.X. Zhou: *Recent Applications of Harmonic Analysis to Function Spaces, Differential Equations, and Data Science: Novel Methods in Harmonic Analysis, Volume 2* (ISBN 978-3-319-55555-3)

F. Weisz: *Convergence and Summability of Fourier Transforms and Hardy Spaces* (ISBN 978 3 319-56813-3)

C. Heil: *A Short Introduction to Metric, Banach, and Hilbert Spaces: With Operator Theory* (ISBN 978-3-319-65321-1)

S. Waldron: *An Introduction to Finite Tight Frames: Theory and Applications.* (ISBN: 978-0-8176-4814-5)

D. Joyner and C.G. Melles: *Adventures in Graph Theory: A Bridge to Advanced Mathematics.* (ISBN: 978-3-319-68381-2)

For an up-to-date list of ANHA titles, please visit
http://www.springer.com/series/4968

CPSIA information can be obtained
at www.ICGtesting.com
Printed in the USA
LVHW061355090619
620638LV00005B/448/P

9 783319 885933